New Advances in *Saccharomyces*

Edited by Antonio Morata, Carmen González, Iris Loira and Carlos Escott

Published in London, United Kingdom

New Advances in *Saccharomyces*
http://dx.doi.org/10.5772/intechopen.110980
Edited by Antonio Morata, Carmen González, Iris Loira and Carlos Escott

Contributors
Sachiko Hayashi, Lawrence Sirovich, Dmitry Karpenko, Artem Grishin, José Pérez-Navarro, Adela Mena-Morales, Sergio Gómez-Alonso, Esteban García-Romero, Pedro Miguel Izquierdo-Cañas, Antonio Morata, Carlos Escott, Juan Manuel del Fresno, Buenaventura Guamis, Iris Loira, María Antonia Bañuelos, Carmen López, Felipe Palomero, Carmen González, Julia Crespo, Margarita García, Teresa Arroyo, Valeria Romero, Juan M. Cabellos, Hervé Alexandre, Maëlys Puyo, Raphaëlle Tourdot-Maréchal, Kai Chen, Gustavo Cordero-Bueso, Marina Ruiz-Muñoz, Antonio Florido-Barba, Jesús Manuel Cantoral Fernández, Duncan Hamm, Bernardo Muñoz González, Dagnew Bitew, Berhanu Andualem, Vanesa Postigo, Margarita García, Teresa Arroyo

First published in London, United Kingdom, 2024 by IntechOpen
IntechOpen is the global imprint of INTECHOPEN LIMITED, registered in England and Wales, registration number: 11086078, 5 Princes Gate Court, London, SW7 2QJ, United Kingdom

British Library Cataloguing-in-Publication Data
A catalogue record for this book is available from the British Library

Additional hard and PDF copies can be obtained from orders@intechopen.com

New Advances in *Saccharomyces*
Edited by Antonio Morata, Carmen González, Iris Loira and Carlos Escott
p. cm.
Print ISBN 978-0-85466-128-2
Online ISBN 978-0-85466-127-5
eBook (PDF) ISBN 978-0-85466-129-9

For EU product safety concerns:
IN TECH d.o.o., Prolaz Marije Krucifikse Kozulić 3, 51000 Rijeka, Croatia,
info@intechopen.com or visit our website at intechopen.com.

Meet the editors

Antonio Morata is a Professor of Food Science and Technology at the Universidad Politécnica de Madrid (UPM), Spain, specialized in wine technology. He is a coordinator of the module of enology and winemaking in the master's program in viticulture and enology and a coordinator of the master's program in food engineering at UPM. Dr. Morata is also a Professor of Enology and Wine Technology at the European Master of Viticulture and Enology, Euromaster Vinifera-Erasmus+. He is a Spanish delegate to the group of experts in wine technology and the president of wine microbiology for the International Organizations of Vine and Wine (OIV). He has authored more than eighty-five research articles, seven books, six journal special issues, and twenty-three book chapters. Dr. Morata was listed among the top 2% of scientists worldwide by Stanford University, USA, in 2022–2023.

Professor Carmen González-Chamorro has worked as a professor at the Universidad Politécnica de Madrid (UPM), Spain, since 1993. She has dedicated her teaching work to food technology and applications in the fruit and vegetable industries and fermented meat products. From 2004 to 2016 she held management positions at the university, including Ombudsman and Deputy Director of University Extension and International Relations). Her research activity has focused on oenological biotechnology and the selection of microorganisms (yeasts and BAL) that are of special interest in wine-making processes. She has extensive experience in the use of instrumental and sensory tests to assess the quality of alcoholic beverages (wine and beer) and meat products. She has participated in different educational innovation projects and coordinated three of them. These projects have made it possible to coordinate working groups for the implementation of degrees in the United States and apply new teaching methodologies that allow the acquisition of horizontal competencies by students. Dr. González-Chamorro has also evaluated research projects and national and international degrees.

Carlos Escott graduated with a degree in chemistry from the Instituto Tecnológico y de Estudios Superiores de Monterrey (ITESM), Mexico. He also obtained a master's degree in Food Engineering Applied to Health from the Universidad Politécnica de Madrid (UPM), Spain, and a Ph.D. in Agro-environmental Technology for Sustainable Agriculture from UPM. He is an R2 food technology researcher and professor of the enology and vinification module at UPM. He is the author of forty-three papers, including thirty-six research articles and seven book chapters.

Iris Loira is an Associate Professor of Food Science and Technology at the Universidad Politécnica de Madrid (UPM), Spain, where she teaches the enology and winemaking module. She is the Academic Secretary of the master's program in food engineering at UPM. She is also a Professor of Wine Analysis at the European Master of Viticulture and Enology, Euromaster Vinifera-Erasmus+. She is the author of fifty-five research articles, four books, and twenty book chapters.

Contents

Preface

New Advances in Saccharomyces includes current information about yeast genetics, physiology, metabolism, and biotechnological applications. It is structured in three parts. The first part focuses on general contents about genetics, regulation of gene expression, and activation methods. The second part is dedicated to the biotechnology of wine fermentation, including the assembly of yeast communities in the production of ice wines and the effect on the metabolomic profile, the applications of bioprotection as a tool to produce optimal fermentation by increasing quality and reducing the use of sulfites, the use of emerging non-thermal technologies to improve the implantation of selected starters, the effect of yeast metabolites on the sensory profile of minority cultivars, the biological aging and the control of the yeast film in Sherry wines, and the effect of aging on lees to improve the sensory quality of wines. The third and final part reviews applications of beer biotechnology, such as the control of diacetyl content to avoid off-flavors and the use of non-conventional yeasts to improve beer quality.

The book includes innovative review works by knowledgeable professors and researchers from the United States, Japan, New Zealand, China, Spain, France, and Ethiopia. It is a useful resource for researchers, professors, advanced students, and professionals from different fermentation industries. We hope this work will clarify and incorporate new information on specific applications of *Saccharomyces* yeasts, especially in the food industry and beverage fermentations.

We are grateful to our families for their understanding during the time required to prepare and edit this book. We would also like to acknowledge the contributing authors and the staff at IntechOpen.

Antonio Morata, Iris Loira, Carmen González and Carlos Escott
Polytechnic University of Madrid,
Madrid, Spain

Section 1

Saccharomyces Genetics and Biotechnology

Chapter 1

Multiple Layers of Gene-Expression Regulatory Mechanisms during Fermentation and Respiration

Sachiko Hayashi

Abstract

Saccharomyces cerevisiae is widely used as a model organism for eukaryotic cells and generally prefers fermentation rather than respiration even under an aerobic environment. Only when glucose is exhausted, *S. cerevisiae* switches to aerobic respiration *via* massive reprogramming of gene expression accompanying that. These gene-expression changes are not simply achieved by the transcriptional level, rather multiple post-transcriptional regulatory steps are also involved. This chapter outlines how budding yeast cells coordinate energy metabolisms based on gene expression, with a focus on the intricate interplay of multiple post-transcriptional regulatory mechanisms. Especially, it includes the roles of RNA-binding proteins as well as non-coding RNAs for post-transcriptional regulations.

Keywords: gene expression, glucose, fermentation, respiration, mitochondria, post-transcription, signaling, RNA-binding protein, non-coding RNA

1. Introduction

Saccharomyces cerevisiae, commonly known as baker's yeast (referred to hereafter as yeast), is among the earliest eukaryotic organisms to be domesticated by humans. This unicellular microorganism serves as an excellent eukaryotic model for the elucidation of fundamental cellular processes, gene expression, and plays a pivotal role in various biotechnological applications [1].

A fermentable sugar, glucose, is the preferred carbon and energy source, while concurrently serving as a signaling molecule [2, 3]. When yeasts are cultured in glucose-rich media under aerobic conditions, they predominantly employ glucose metabolism through glycolysis, yielding pyruvate as the primary product. Following glucose depletion, the fermented ethanol emerges as a carbon source, initiating a transition to respiration. This transformative metabolic process, known as a diauxic shift, unfolds in tandem with a substantial reconfiguration of gene expression, leading to dynamic upregulation including mitochondrial biogenesis and its functional capacities [4–6].

Glucose fermentation has superior catalytic efficiency compared to respiration in terms of adenosine triphosphate (ATP) production per unit of protein mass [7]. Conversely, respiration yields a tenfold increase in ATP per glucose molecule [8].

The equilibrium between respiration and fermentation is a pivotal determinant for unicellular survival [9]. Further, the oxidative fermentation or the Crabtree effect [10, 11], grants yeast an ecological advantage by allowing it to swiftly use glucose and produce ethanol, which possesses antiseptic properties [12].

This chapter overviews yeast metabolic systems, which are meticulously regulated through multiple steps at transcriptional and post-translational levels. It encompasses well-established signaling cascades and the regulation of nuclear-encoded mitochondrial gene expression, which interact with and are regulated by RNA-binding proteins and/or non-coding RNAs.

2. Metabolic arrangements: transitions between fermentation and respiration

Yeast has evolved efficient glucose utilization, employing both respiration and fermentation pathways to generate ATP from glucose. Both processes start with glycolysis, producing two molecules of pyruvate and ATP per glucose molecule. In the fermentation process, pyruvate is subsequently metabolized into ethanol. While this process does not yield additional ATP, it recycles nicotinamide adenine dinucleotide (NAD^+) consumed during glycolysis, producing oxygen-independent ATP. In contrast, respiration involves the complete oxidation of pyruvate to CO_2 through the tricarboxylic acid (TCA) cycle and oxidative phosphorylation (OXPHOS), yielding additional ATP in the presence of oxygen. Crabtree-positive yeasts, such as *S. cerevisiae*, favor fermentation over respiration when glucose levels are high, producing ethanol as a byproduct. This metabolic strategy allows them to simultaneously engage in both fermentation and respiration when oxygen and glucose levels are abundant [9, 11]. Although the ethanol accumulating in the environment can be reutilized for ATP generation once glucose is depleted [13], it yields fewer ATP molecules than the direct oxidation of pyruvate due to the energy requirement for synthesizing acetyl-CoA from ethanol, which consumes ATP. As the rate of ATP production increases, ATP-consuming processes, such as ribosome and protein synthesis, to some extent, increasingly impose constraints on growth [14, 15]. Metabolic systems and shifts in the utilization of metabolic pathways can be considered as an array of constraints that depend on the conditions within the framework of cellular economics [9]. Transitions between energy metabolic pathways are observed in various metabolic systems, including tumor cells and bacteria, as well as yeast [9, 11, 16]. Sustaining an optimal metabolic system is a crucial point for the survival or demise of many living organisms.

3. Molecular and signaling aspects of yeast energy metabolic pathways

Glucose serves as a fundamental messenger molecule, playing a dual role as both an energy source and a signal for optimal growth conditions in cellular machinery. Yeasts, in particular, utilize glucose for this purpose. When the external glucose concentration exceeds 0.8 mM, yeast undergoes a transition into a mixed respiro-fermentative metabolism, resulting in ethanol production [17]. This shift underscores that the regulation between fermentation and respiration primarily corresponds to the sugar level [4]. Therefore, it is unsurprising that glucose plays a key element in shaping growth rate, fermentation capacity, and stress resistance [2, 18].

Multiple distinct pathways participate in responses to glucose. Some involve glucose interactions with cell surface receptors, while others require glucose import into the cell. These frequently utilized cascades can be classified into the following five signaling pathways: Ras-protein kinase A (PKA), Gpr1p–Gpa2p–PKA, the target of rapamycin (TOR)–Sch9p, Snf1p–Mig1p, or Snf3p–Rgt2p [3, 5, 19]. Among these, Ras and TOR, major global nutrient-sensing signal transduction cascades, serve pivotal roles in regulating cell growth in response to nutrient availability [18, 20]. Global glucose repression depends on an intracellular surge of cyclic adenosine monophosphate (cAMP), which activates PKA. Transcription rates significantly decrease upon Ras2p activation, which can occur independently of glucose presence and relies on a cAMP-responsive protein kinase [3, 21].

3.1 Ras–PKA and Gpr1p–Gpa2p–PKA

Ras, a guanine nucleotide-binding protein with seven transmembrane domains, activates adenylyl cyclase in its GTP-bound state. The addition of glucose to cells increases the level of GTP-bound Ras, resulting in an elevation of intracellular cAMP and subsequent activation of PKA [3, 21]. The PKA catalytic subunits, encoded by *TPK1*, *TPK2*, and *TPK3*, phosphorylate various proteins involved in metabolism and transcription. Consequently, a significant number of genes exhibit a reduction in transcription rate upon Ras2p activation [21, 22]. However, glucose-repressed genes do not uniformly respond to Ras2p and PKA activation [3]. Genes involved in trehalose or glycogen metabolism, glucose utilization, or ubiquitination exhibit weak or absent responses to Ras2p activation [3].

Gpr1p, a plasma membrane protein (**Figure 1**), can sense the presence of glucose and/or sucrose and is coupled to Gα protein Gpa2p [23, 24]. This leads to the activation of adenylyl cyclase, resulting in an increase in cAMP concentration [2, 25, 26]. Subsequently, cAMP activates PKA. Therefore, it eventually controls the transcription of genes related to ribosome biogenesis and stress-responsive genes like *RIM15*, *MSN2*, and *MSN4*. It also exerts post-translational regulation on proteins involved in storage carbohydrate synthesis, the glycolytic pathway, and gluconeogenesis [18, 27–33]. PKA directly phosphorylates and activates glycolytic enzymes Pfk2p and Pyk1p, enhancing glycolysis and showing a positive correlation between PKA activity and glycolytic flux.

3.2 TOR–Sch9p

TOR is a Ser/Thr kinase that was initially identified through yeast's genetic screening [34]. The TOR proteins assemble into two structurally and functionally distinct complexes known as TOR complex 1 (TORC1) and TOR complex 2 (TORC2), of which only TORC1 is sensitive to rapamycin [35, 36]. The central components of the TOR consist of two TOR kinases paralogs, Tor1p, and Tor2p, along with a phosphate switch composed of the type 2A-related phosphatase Sit4p, TOR kinase phosphorylates Tap42p, and its inhibitor Tip41p [37, 38]. TORC1-dependent signals are mediated *via* various effector kinases [39].

The vacuolar surface primarily serves as the location for the TORC1 signaling pathway [40]. TORC1 is involved in respiration-induced mitophagy [41, 42]. This type of mitophagy is particularly important in cells that rely on OXPHOS for energy production, as disruptions in mitochondrial respiration can have significant consequences for cellular energy balance and overall cell function [43, 44].

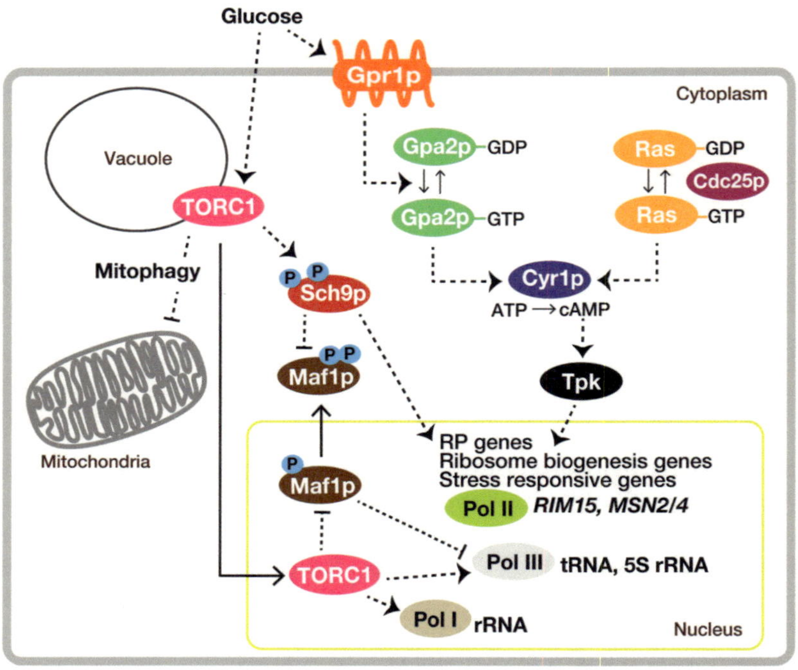

Figure 1.
Glucose signaling, facilitated by the small G-proteins Ras and Gpa2p, converges through PKA to stimulate ribosome biogenesis while concurrently suppressing the general stress response. Concurrently, within the TORC1 pathway, the kinase Sch9p plays a crucial role in strengthening the response of the PKA pathway. Dashed lines in the diagram symbolize regulatory interactions, which might not always be direct. See text for details.

An AGC family Ser/Thr kinase Sch9p is best characterized as the direct substrate of TORC1 [45]. Sch9p is a master regulator of ribosome biogenesis [45–47]. TORC1 controls all three RNA polymerase (Pol) systems *via* Sch9p (**Figure 1**). Pol I is stimulated in a Sch9p-dependent and -independent manner, partly through regulation of the transcription initiation factor Rrn3p [48–50]. Regarding Pol II, at the large cohort of the ribosome biogenesis and ribosomal protein (RP) genes, at least in part *via* Sch9p [45, 51]. Further, Sch9p regulates Pol III by phosphorylating and inactivating Maf1p, a conserved repressor of Pol III activity [50, 52–56]. Consequently, when yeasts grow logarithmically on glucose, ribosome biogenesis proceeds at maximal speed driven by the positive control of TORC1.

In contrast, gradual glucose exhaustion or abrupt withdrawal of glucose triggers a reduction in TORC1-dependent phosphorylation of five residues within the Sch9p C terminus [45, 57], leading to TORC1 inactivity. Inhibition of TORC1 provokes extensive transcriptome changes, reducing ribosomal particles by blocking the transcription of Pol I-dependent rRNA genes, Pol II-dependent RP genes, Pol III-dependent 5S rRNA, and the processing of 35S rRNA [18, 46, 47, 58]. Diminished phosphorylation of Sch9p transforms the Pol III repressor Maf1p from an inactivated to an active state [46, 59, 60]. This, in turn, leads to the suppression of Pol III transcriptome, including various small non-coding RNAs such as transfer RNA (tRNA). Consistent with TORC1's functions, the absence of Tor1p results in an increase in mitochondrial respiration during glucose-based growth, primarily due to the enhanced translation of mitochondrial DNA (mtDNA)-encoded subunits of the OXPHOS complex. This effect is not observed in cells growing on glycerol [3, 61].

Recent research has shed light on the role of Snf1p–AMPK in fine-tuning TORC1 signaling during glucose starvation. Snf1p temporarily inhibits TORC1 activity by interacting with the phosphatidylinositol-3-phosphate (PI3P) and Kog1p-binding protein Pib2p [62]. This discovery highlights the mutual interaction between TOR and Snf1p, emphasizing their significance in metabolic adaptation.

3.3 Snf1p–Mig1p

The sucrose non-fermenting (Snf1) protein kinase, the yeast ortholog of mammalian AMP-activated S/T protein kinase (AMPK), is a central component of the primary glucose repression pathway responsible for adapting to glucose limitation [63, 64]. It forms a heterotrimeric complex with Snf4p (the regulatory γ-subunit) and one of the three β-subunits, Sip1p, Sip2p, or Gal83p, alongside the catalytic α-subunit, Snf1p [65]. Snf1p is activated under glucose limitation, assisting in energy homeostasis by promoting catabolic processes and inhibiting anabolic ones related to ATP generation and consumption [66]. This Snf1p complex regulates cellular processes through different transcription factors and enzymes [65]. For instance, Mig1p, a transcriptional repressor, controls the expression of genes involved in the metabolism and transportation of alternative carbon sources (e.g., maltose, galactose, sucrose) [67, 68]. Under glucose-rich conditions, Mig1p undergoes dephosphorylation and translocases to the nucleus (**Figure 2**). Together with the Ssn6p-Tup1p corepressor, it binds to target gene promoters [67, 69, 70]. Concurrently, Snf1p also undergoes dephosphorylation to prevent its nuclear localization in glucose-rich environments [71]. This dephosphorylation process involves protein phosphatase 1 (PP1), Glc7p–Reg1/2p, and possibly Sit4p, in collaboration with the phosphatase Ptc1p [72–77].

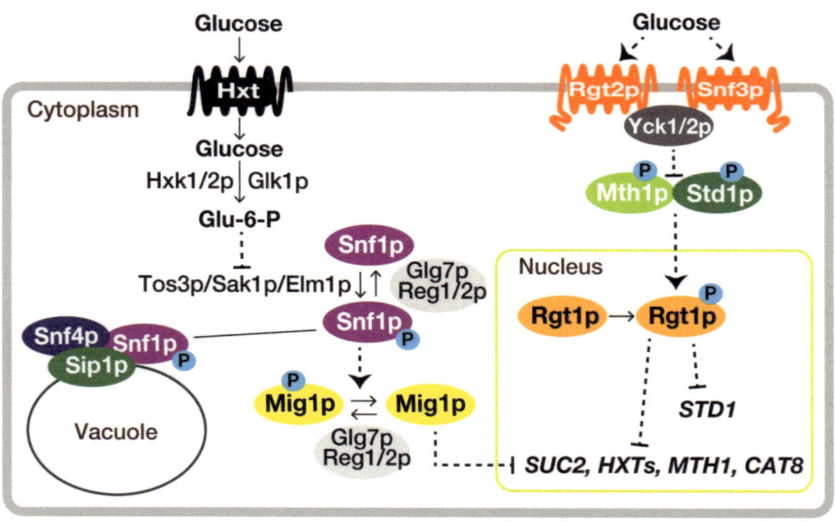

Figure 2.
The interconnected Snf and Rgt glucose signaling networks play a vital role in cellular regulation. The glucose sensor protein Snf3p, working alongside hexose transporters (Hxt), contributes to the yeast cell's ability to sense extracellular glucose levels. Rgt1p, a transcription factor, governs glucose repression by controlling the expression of genes involved in glucose sensing and metabolism. Meanwhile, the Snf1p kinase and a transcriptional repressor Mig1p also respond to glucose availability. These components form a complex interplay in the yeast glucose signaling network, revealing the sophisticated regulatory mechanisms that govern cellular responses to glucose fluctuations.

3.4 Snf3p-Rgt2p

Two plasma membrane proteins, each composed of 12 transmembrane domains and featuring putative glucose-sensing capabilities, are Snf3p and Rgt2p (**Figure 2**). They share a resemblance to the Hxt glucose transporter [78], although they lack the capacity for glucose transport [79]. Their primarily role is regulating the expression of the seven main hexose transporters (HXT) genes [78].

Snf3p and Rgt2p act as glucose sensors, modulating the activity of Rgt1p, the transcription factor, in response to low and high glucose levels, respectively [80]. In the absence of glucose, Rgt1p, along with Ssn6p, Tup1p, Mth1p, and Std1p forms a repressor complex that inhibits *HXT* gene expression by binding to their promoter regions [81–86]. In high glucose conditions, Std1p and Mth1p relocate to the plasma membrane and undergo phosphorylated by Yck1/2p [87]. These phosphorylations trigger their inactivation and subsequent degradation through the SCF-Grr1p ubiquitin ligase complex [88, 89]. Consequently, Rgt1p becomes hyperphosphorylated and dissociates from *HXT* promoters, activating *HXT* gene expression [88]. Microarray data demonstrates fluctuations in *HXT* gene expression in response to the presence of glucose and/or other carbon sources, emphasizing the efficiency and sensitivity of the pathway [79, 90].

For more in-depth information on each cascade, comprehensive reviews are available [3, 18, 27, 79, 91].

4. Mitochondria: the central organelle of energy metabolism

Mitochondria, often referred to as the "powerhouses" of the cell, are essential organelles for both fermentation and respiration. At the diauxic shift, the mitochondrial volume expands concomitantly with the upregulation of Krebs cycle enzymes and respiratory complexes replete with abundant heme and Fe–S centers. Mitochondria are not discrete or autonomous entities; instead, they form highly dynamic and interconnected networks, and their biogenesis and structure are strongly influenced by the cell's requirements [92, 93]. A classical targeting pathway for nuclear-encoded mitochondrial proteins uses mitochondrial targeting sequences (MTS) mainly located on their N-terminus [94–96], whereas approximately one-half of mRNAs for nuclear-encoded mitochondrial mRNAs are transported to the mitochondrial surface, and translated locally [97–100]. Proximity-specific ribosome profiling targeting the tagged ribosomes on the mitochondrial surface showed highly enrichment of nuclear-encoded mitochondrial mRNAs, especially those encoding proteins in the mitochondrial inner membrane [101]. Cryo-electron cryotomography (CryoET) revealed active cytosolic ribosomes attach to the mitochondrial outer membrane and interact with the TOM complex [102]. The cytosolic translation of nuclear-encoded mitochondrial mRNAs and mitochondria indeed closely interact with each other.

4.1 Gene expression strategies for mitochondrial proteins

Mitochondria originated through the permanent integration of purple non-sulfur bacteria [103]. Throughout evolution, the majority of genes originally present in ancient bacteria have been transferred to nuclear DNA. Simultaneously, the genetic code within mitochondria has diverged from the conventional genetic code, resulting in significant differences in codon utilization between these two systems [104, 105]. In yeast today, mtDNA encodes only eight proteins, primarily associated with the OXPHOS system. Over 99% of mitochondrial proteins are instead encoded by the nuclear genome.

Under fermentable conditions, it is often assumed that nuclear-encoded mitochondrial genes are completely inactive due to subdued mitochondrial biogenesis and function. However, in reality, these nuclear-encoded mitochondrial genes remain transcriptionally active but subsequently undergo translational repression and/or rapid mRNA degradation [106–108]. Conversely, when exposed to respiratory conditions, there is a substantial upregulation of nuclear-encoded mitochondrial mRNAs, followed by increased translation to support mitochondrial biogenesis and enhance oxidative catabolism of carbon substrates. This orchestrated coordination between genomic and mitochondrial gene expression, along with the accurate sorting of nuclear-encoded mitochondrial proteins, is essential for maintaining optimal mitochondrial function [95, 109–111]. Disruptions in this process, such as the abnormal buildup of mitochondrial precursors in the cytosol leading to mitochondrial precursor over-accumulation stress (mPOS), or mitochondrial dysfunction, can activate a cytosolic proteostasis system [112, 113].

4.2 Iron: a vital element for mitochondrial function

Mitochondria continuously synthesize heme and Fe/S clusters while also facilitating amino acid and lipid metabolism [114–116]. Iron availability is essential for mitochondrial function and significantly impacts cellular metabolic responses to changes in carbon availability. Interestingly, yeast can survive OXPHOS defects and even complete loss of mtDNA but not disruption of mitochondrial Fe/S assembly, which proves to be fatal [114, 116, 117]. This is because the mitochondrial iron-sulfur cluster (ISC) assembly machinery is essential for the biogenesis of all cellular Fe/S proteins, including those in the cytosol and nucleus, which are involved in DNA maintenance and protein translation [114, 118].

Because iron is essential for cellular processes, when it is scarce, yeast employs Cth2p, an RNA-binding protein induced during iron starvation, to manage iron resources efficiently. Cth2p has a dual role: it suppresses non-essential iron consumption while promoting critical iron-dependent activities, including the assembly of ribonucleotide reductase (RNR) when iron is limited. Cth2p also inhibits mRNAs with AU-rich elements (ARE), mainly those related to iron metabolism and utilization, affecting pathways such as the TCA cycle, lipid biosynthesis, amino acid synthesis, and cofactor production. Additionally, Cth2p prevents excess iron accumulation in vacuoles by degrading mRNAs responsible for iron transport, including *CCC1* [119, 120]. *CTH2* paralog, *CTH1*, functions during iron sufficiency and is regulated by Aft1/2p transcription factors [119, 121]. Aft1/2p control iron-responsive genes at the transcriptional level, and their activity is negatively regulated by Cth2p [122–124]. These mechanisms collectively ensure yeast cells adapt to changing iron conditions and maintain iron homeostasis. A partial list of genes upregulated or downregulated by Cth2p under iron depletion is presented in **Table 1**.

Comprehensive reviews with more in-depth information mitochondrial protein sorting, iron homeostasis are available [114, 119, 120, 125–127].

Gene	Function
HXK1	Hexokinase isoenzyme 1
HXT7	High-affinity glucose transporter
HXT6	High-affinity glucose transporter
SOL4	6-phosphogluconolactonase
PGM2	Phosphoglucomutase; catalyzes the conversion from glucose-1-phosphate to glucose-6-phosphate
HSP31	Methylglyoxalase that converts methylglyoxal to D-lactate; involved in diauxic shift and stationary phase survival
HPF1	Haze-protective mannoprotein
GPH1	Glycogen phosphorylase required for the mobilization of glycogen
GSY1	Glycogen synthase; expression induced by glucose limitation
IGD1	Cytoplasmic protein that inhibits Gdb1p glycogen debranching activity
ALD3	Cytoplasmic aldehyde dehydrogenase
ARG3	Ornithine carbamoyltransferase
OM45	Mitochondrial outer membrane protein
COX5B	Subunit Vb of cytochrome c oxidase
HSP12	Plasma membrane protein involved in maintaining membrane organization
COS8	Endosomal protein involved in turnover of plasma membrane proteins
REC104	Forms a complex with Rec102p and Spo11p necessary during the initiation of recombination
MSC1	Mutant is defective in directing meiotic recombination events to homologous chromatids
YHR087W	Involved in RNA metabolism
FIT1	Cell wall mannoprotein involved in siderophore-Fe uptake
FIT2	Cell wall mannoprotein involved in siderophore-Fe uptake
HMX1	Heme binding peroxidase involved in reutilization of heme Fe
CIT1	Citrate synthase
CIT3	Mitochondrial isoform of citrate synthase
ACO1	Mitochondrial aconitase, Fe-S cluster protein
KGD1	Alpha-ketoglutarate dehydrogenase
KGD2	Dihydrolipoyl transsuccinylase
SDH2	Succinate dehydrogenase (ubiquinone) Fe-S cluster subunit
SDH4	Succinate dehydrogenase membrane anchor heme-binding subunit
COX4/6/8/9	Subunit of cytochrome c oxidase
COQ6	Flavin-dependent monooxygenase, ubiquinone biosynthesis
RNR2/4	Ribonucleotide-diphosphate reductase
CCC1	Transporter that mediates vacuolar Fe storage
RLI1	RNase L inhibitor, Fe-S cluster protein
GLT1	NAD^+-dependent glutamate synthase (GOGAT)
CCP1	Mitochondrial cytochrome-c peroxidase
FET3	Ferro-O_2-oxidoreductase

Table 1.
A subset of genes upregulated or downregulated by Cth2p under iron depletion. Genes exhibiting upregulation include HXK1 to YHR087W, while those undergoing downregulation comprise FIT1 to CCP1. FET3 serves as an example of a Cth2p-independent expression gene.

5. Post-transcriptional gene expression regulation amid dynamic glucose changes

Gene expression is a multifaceted process that goes beyond mere transcriptional regulation, encompassing intricate post-transcriptional control mechanisms. It is not solely determined by transcriptional status; rather, it involves a complex interplay of factors. To optimize their growth conditions, cells undergo adaptations by adjusting their energy requirements through the modulation of vital metabolic enzymes, frequently accomplished *via* allosteric binding or post-translational modifications [128–130]. These alterations in protein abundance and interactions significantly impact cellular phenotypes, as demonstrated in proteome analyzes under various conditions [131–134]. Stringent regulation of mRNA translation and degradation is also crucial for eukaryotic cells to manage their diverse protein repertoire. In particular, this regulation becomes critical in response to glucose depletion or a diauxic shift [135, 136].

Glucose depletion triggers a swift and substantial halt in protein synthesis, which can be rapidly reversed upon glucose replenishment [137]. Additionally, this glucose depletion induces the formation of mRNA processing bodies (P-bodies), which act as central hubs where components of the 5–3′ mRNA decay pathway converge [138–140]. This compartmentalization of mRNAs in the cytosol potentially leads to translational repression and the degradation of specific mRNAs (although it has not been definitively proven). This phenomenon allows for a reduction in energy consumption while, at the same time, enabling the rapid translation of specific mRNAs. This facilitates the production of proteins necessary for adaptation [138, 141–143]. P-bodies indeed exclude translational machineries, including ribosomal components [144]. This specific response can significantly impact gene expression on a large scale. Further, during a diauxic shift, essential core components of P-bodies, Dhh1p and Pat1p, known for their roles as mRNA decapping activators and translational repressors, undergo a change in their intracellular localization. They shift from being excluded from polysomes in rapidly growing cells to co-localizing with polysomes [145].

Many aspects regarding P-bodies still remain obscure [143], but cells strategically employ adaptive mechanisms to dynamically regulate mRNA translation and degradation to manage the cellular protein repertoire. These processes are particularly important during glucose depletion and diauxic shifts. The formation of P-bodies and/or the dynamic behavior of core components such as Dhh1p and Pat1p would ensure the cell's survival and growth in response to changing environmental conditions.

6. tRNA: a dynamic player of protein synthesis and cellular adaptation

tRNA is a classical small non-coding RNA present universally in living organisms. It plays a fundamental role in translation, along with the ribosome [146]. The primary function of tRNA, transferring amino acids into ribosomes, is to guarantee the precise integration of amino acids into proteins. Alternations in nutrient availability, such as shifts in glucose levels, can impact tRNA expression and their modifications, thereby exerting a profound influence on the efficiency of protein synthesis for a diverse array of proteins.

6.1 tRNA movement

In rapidly growing yeasts, tRNAs account for approximately 15% of the total cellular RNAs [147]. The availability of tRNAs positively correlates with codon utilization, influencing the usage of corresponding codons and *vice versa* [148]. This reciprocal evolutionary process, linking tRNA abundance and codon usage, results in variations in tRNA gene copy numbers among organisms and tRNA species sharing the same anticodon identity [148, 149]. Yeast harbors 275 nucleonic tRNA genes distributed across 16 chromosomes, representing 42 tRNA species [149, 150]. All tRNA genes are typically actively transcribed, and they encompass various types of tRNAs, including modified variants [146, 151]. These tRNAs are essential for the accurate decoding of the genetic code and maintaining precise cellular proteostasis [152, 153].

Yeast's mtDNA encodes a complete set of tRNAs (24 tRNA species) required for mitochondrial translation within the organelle [105]. However, two cytosolic tRNAs, $tRNA^{Lys}{}_{CUU}$ [154] and $tRNA^{Gln}$ [155], are imported into mitochondria, potentially playing a role in stress response. In the case of $tRNA^{Lys}{}_{CUU}$, cells use a unique interaction mechanism with the mitochondrial outer membrane-attached glycolytic enzyme enolase [105]. This interaction induces a conformational change in $tRNA^{Lys}{}_{CUU}$, increasing its affinity for another protein factor, pre-mitochondrial lysyl-tRNA synthetase (preMsk1p), ultimately facilitating its co-import into the mitochondrial matrix [156]. Since the majority of mitochondrial proteins are encoded by the nuclear genome, cytosolic translation involving nuclear-encoded tRNAs significantly influences mitochondrial function. Mutants with impaired function of $tRNA^{Gln}{}_{UUG}$ or $tRNA^{Lys}{}_{UUU}$ exhibit inappropriate activation of various starvation responses during rapid growth. These responses involve the upregulation of genes related to glucose and nitrogen catabolism, along with premature and inadequate activation of autophagy. These effects can be alleviated by overexpressing $tRNA^{Gln}{}_{UUG}$ or $tRNA^{Lys}{}_{UUU}$, which lack specific modifications [157].

6.2 Balancing act: tRNA dynamics in response to cellular stress

tRNAs are very dynamic, continually shuttling between the nucleus and cytosol throughout their entire life. Approximately one-fifth of total tRNA genes (61 of 275) in yeast, contain a single intron with variable lengths ranging from 14 to 60 nucleotides [149, 158]. Intriguingly, these introns consistently occupy the canonical position, precisely one nucleotide 3′ from the anticodon. These intron-containing tRNA genes transcribe precursor tRNA (pre-tRNA) with the intron sequence forming an A-I pair with anticodon nucleotides. This A-I interaction disrupts the crucial codon-anticodon binding during translation [146, 159, 160]. Thus, tRNA splicing is a vital process to address this issue, for intron-containing pre-tRNAs. However, unlike in mammals, yeast tRNA splicing occurs at the mitochondrial periphery due to the presence of the tRNA splicing SEN complex, which is located in the mitochondrial outer membrane [146, 161]. Consequently, intron-containing pre-tRNAs are transported to the mitochondrial membrane for splicing, and some are subsequently re-imported into the nucleus post-splicing [146].

In the case of mature tRNA, they also show bidirectional movement between the cytoplasm and nucleus [146]. This movement is tightly regulated and responsive to various cellular conditions and signals (**Figure 3**). In specific instances, mature tRNAs re-enter the nucleus following their cytoplasmic function. Although the reasons for this re-import are not always fully elucidated, it is possibly linked to quality control mechanisms or other regulatory processes [162]. This bidirectional trafficking of mature tRNAs allows cells to finely tune translation processes in response to changing conditions and to maintain the precision of protein synthesis. Thus, it emphasizes

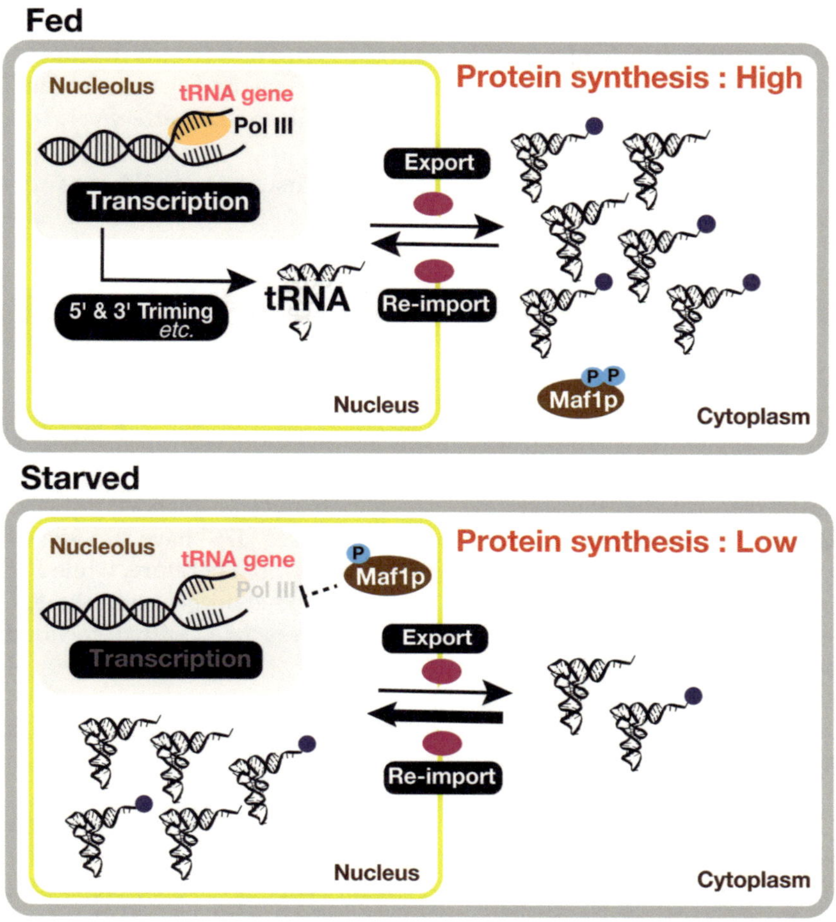

Figure 3.
Fine-tuning of tRNA dynamics in response to nutrient stress. tRNAs traverse between the nucleus and cytosol. This bidirectional movement is particularly crucial during varying cellular states, such as in nutrient-rich (fed) and nutrient-depleted (starved) conditions. Within the nucleus, tRNAs not only undergo transcription but also intricate maturation processes, including essential base modifications. Concurrently, in the cytosol, tRNAs actively engage in translation processes, fulfilling their indispensable role in protein synthesis. Under specific stress conditions, such as nutrient or glucose depletion, tRNAs are transported into the nucleus, deviating from cytosolic translation. This adaptive behavior, where tRNAs modulate both their localization and function in response to nutrient availability, underscores their critical involvement in cellular processes and protein synthesis.

the dynamic nature of tRNA trafficking, which would contribute to the accurate and adaptable synthesis of proteins in the cell [146].

Remarkably, under glucose depletion, exposure to non-fermentable carbon sources such as glycerol, or in response to certain stressors, tRNA may accumulate in the nucleus [146, 163]. This sequestration of tRNA in the nucleus serves to isolate it from the cytoplasmic protein synthesis machinery, potentially serving as a mechanism to reduce protein synthesis globally [164, 165]. However, significant nuclear accumulation of cytoplasmic tRNAs does not necessarily result in a widespread inhibition of translation [166, 167]. Instead, the decrease in cytoplasmic tRNA levels during stress may involve the regulation of nuclear-cytoplasmic tRNA shuttling or changing tRNA transcription [165]. For instance, in *maf1Δ* cells, intron-containing pre-tRNAs can accumulate in the nuclei, regardless of whether these mutant cells are cultured in glucose-rich or non-fermentative carbon sources. This accumulation appears to be due to transcriptional derepression in the absence of Maf1p, a general-negative transcriptional regulator of Pol III [146, 168]. Thus, tRNA's dynamics, trafficking across cellular compartments, would be a fundamental element in the finely tuned orchestration of protein synthesis, enabling cells to adapt effectively to shifting environmental conditions.

Comprehensive reviews with more in-depth information on tRNA are available [146, 169–172].

7. Multifaceted regulator: rNA-binding protein Puf3p

Puf3p, a Pumilio homolog RNA-binding protein, is a well-known regulator of nuclear-encoded mitochondrial mRNAs [173–175]. Global analysis showed that Puf3p physically associates with 220 transcripts at least, and more than 70% of which are nuclear-encoded mitochondrial mRNAs [176]. Multiple multi-omics studies have consistently confirmed Puf3p's binding specificity to nuclear-encoded mitochondrial mRNAs [177–179]. However, PAR-clip [178] and RIP-seq [179] have also identified numerous non-mitochondrial mRNAs as targets of Puf3p. Therefore, while Puf3p significantly influences the regulation of nuclear-encoded mitochondrial mRNAs, it also exerts a broader influence on gene expression associated with mitochondrial functions.

7.1 Molecular basis of Puf3p-RNA interaction: Structural analyzes and binding specificity determinants

Puf3p is comprised of eight Puf repeats, each composed of three α-helices, with neighboring repeats forming a crescent shape [180–182]. X-ray crystallography has revealed that three amino acid residues within each Puf repeat directly contact a single RNA base, determining binding specificity [182–188]. The Puf3p repeat domain (Puf3-RD) is sufficient to modulate mRNA metabolism and physically interacts with target mRNAs, exemplified by its binding to the 3′-UTR of *COX17* mRNA [189, 190]. The consensus Puf3p binding sequence on target mRNAs has been identified as an 8-nt UGUANAUA motif [176, 191, 192], which generally follows the pattern of PUF proteins binding to an 8-nt sequence with UGU(A/G) at the 5′ end and variable 3′ sequences specific to individual Puf proteins [176, 186, 193, 194]. The presence of cytosine at the second position from the Puf3p binding sequence is crucial for Puf3p-RNA-binding *in vitro* and biological activity *in vivo* [182]. Further, the N-terminal part of a PUF domain, responsible for recognizing the 3′ region of the Puf3p site, appears to be more flexible in accommodating target-nucleotide mutations compared to the C-terminal part [195, 196]. Indeed, some PUF proteins exhibit broader specificity by excluding certain undesirable nucleotides [188, 197]. Achieving an equilibrium between the individual binding specificity of each repeat and the total binding affinity to target mRNAs is a determinant for PUF proteins [196]. Three-hybrid analysis revealed that the N-terminal part of the PUF domain is more tolerant of combinatorial mutations in target nucleotides than the C-terminal part [196].

7.2 Multifaceted role of Puf3p in yeast physiology

Yeast Puf3p deletion mutants result in slow growth in respiratory media [176, 198], impair mitochondrial motility and biogenesis [198, 199], alter cellular oxidative stress tolerance and the glutathione redox state [200], and increase cellular oxygen consumption in a growth-dependent manner [201]. Under fermentation, Puf3p destabilizes its target mRNAs by promoting deadenylation and negatively regulates mitochondrial biogenesis [107, 108, 176, 190, 191, 201–203]. In agreement with its repressive roles in glucose-rich media, Puf3p's abundance drastically decreases during the diauxic shift [199]. However, Puf3p associates with actively translating polysomes upon glucose depletion and promotes mitochondrial biogenesis [198, 204], indicating its bidirectional functions. These dual functions of Puf3p are regulated by phosphorylation *via* casein kinase Hrr25p [198, 205]. Mutations in the *PUF3* gene, such as *PUF3*(24A), act dominantly negative and even more strongly inhibit cell growth upon the diauxic shift than the complete deletion of *PUF3* [198]. Moreover, phosphorylation of Puf3p potentially influences its activity, and Puf3p-driven regulation at the translational level could enable a more rapid response of mitochondrial biogenesis to fluctuations in glucose availability compared to transcriptional regulation [9]. Remarkably, despite not being a canonical transcription factor, Puf3p has also been implicated in the regulation of genes involved in respiration, along with the HAP2/3/4/5 DNA-binding complex [176, 206].

8. Conclusions

With its remarkable adaptability in metabolism and precision in gene regulation, yeast serves as a captivating model organism that holds significant implications for biotechnology and deepens our understanding of fundamental cellular intricacies. Its ability to expertly navigate the delicate equilibrium between glucose utilization, fermentation, and respiration underscores the core principles of cellular economics, which are vital for the survival and prosperity of all living organisms.

Furthermore, yeast's intricate mechanisms for controlling post-transcriptional gene expression, involving processes such as mRNA processing, the dynamic behavior of tRNA, and the influence of RNA-binding proteins like Puf3p, exemplify an evolved strategy that enables cells to adapt to ever-changing environmental challenges rapidly. These dynamic processes serve as the linchpin for preserving cellular proteostasis, ensuring the precise and adaptable synthesis of proteins that play a pivotal role in sustaining and nurturing the growth of life.

Acknowledgements

This work was supported by JSPS KAKENHI Grant number JP20K06491 to S. H.

Abbreviations

ATP	adenosine triphosphate
NAD^+	nicotinamide adenine dinucleotide
TCA	tricarboxylic acid
OXPHOS	oxidative phosphorylation
PKA	protein kinase A
TOR	target of rapamycin
cAMP	cyclic adenosine monophosphate
TORC1/2	TOR complex 1/2

Pol	RNA polymerase
RP	ribosomal protein
tRNA	transfer RNA
mtDNA	mitochondrial DNA
PI3P	phosphatidylinositol-3-phosphate
Snf1	sucrose non-fermenting
AMPK	AMP-activated S/T protein kinase
PP1	protein phosphatase 1
HXT	hexose transporters
MTS	mitochondrial targeting sequences
CryoET	cryo-electron cryotomography
mPOS	mitochondrial precursor over-accumulation stress
ISC	iron-sulfur cluster
RNR	ribonucleotide reductase
ARE	AU-rich elements
P-bodies	mRNA processing bodies
pre-tRNA	precursor tRNA
Puf3-RD	Puf3p repeat domain

Author details

Sachiko Hayashi
Graduate School of Science, University of Hyogo, Ako-gun, Japan

*Address all correspondence to: shayashi@sci.u-hyogo.ac.jp

References

[1] Tullio V. Yeast genomics and its applications in biotechnological processes: What is our present and near future? Journal of Fungi. 2022;**8**(7):752. DOI: 10.3390/jof8070752

[2] Rolland F, Winderickx J, Thevelein JM. Glucose-sensing and -signalling mechanisms in yeast. FEMS Yeast Research. 2002;**2**(2):183-201. DOI: 10.1073/pnas.0305901101

[3] Gancedo JM. The early steps of glucose signalling in yeast. FEMS Microbiology Reviews. 2008;**32**(4):673-704. DOI: 10.1111/j.1574-6976.2008.00117.x

[4] Otterstedt K, Larsson C, Bill RM, Ståhlberg A, Boles E, Hohmann S, et al. Switching the mode of metabolism in the yeast *Saccharomyces cerevisiae*. EMBO Reports. 2004;**5**(5):532-537. DOI: 10.1038/sj.embor.7400132

[5] Zaman S, Lippman SI, Zhao X, Broach JR. How *Saccharomyces* responds to nutrients. Annual Review of Genetics. 2008;**42**:27-81. DOI: 10.1146/annurev.genet.41.110306.130206

[6] De Virgilio C. The essence of yeast quiescence. FEMS Microbiology Reviews. 2012;**36**(2):306-339. DOI: 10.1111/j.1574-6976.2011.00287.x

[7] Nilsson A, Nielsen J. Metabolic trade-offs in yeast are caused by F1F0-ATP synthase. Scientific Reports. 2016;**6**:22264. DOI: 10.1038/srep22264

[8] van Dijken J, Bauer J, Brambilla L, Duboc P, Francois J, Gancedo C, et al. An interlaboratory comparison of physiological and genetic properties of four *Saccharomyces cerevisiae* strains. Enzyme and Microbial Technology. 2000;**26**(9-10):706-714. DOI: 10.1016/s0141-0229(00)00162-9

[9] Molenaar D, Van Berlo R, De Ridder D, Teusink B. Shifts in growth strategies reflect tradeoffs in cellular economics. Molecular Systems Biology. 2009;**5**:323. DOI: 10.1038/msb.2009.82

[10] Crabtree HG. Observations on the carbohydrate metabolism of tumours. The Biochemical Journal. 1929;**23**(3): 536-545. DOI: 10.1042/bj0230536%0A

[11] Pfeiffer T, Morley A. An evolutionary perspective on the Crabtree effect. Frontiers in Molecular Biosciences. 2014;**1**:17. DOI: 10.3389/fmolb.2014.00017

[12] Dai Z, Huang M, Chen Y, Siewers V, Nielsen J. Global rewiring of cellular metabolism renders *Saccharomyces cerevisiae* Crabtree negative. Nature Communications. 2018;**9**(1):3059. DOI: 10.1038/s41467-018-05409-9

[13] Piskur J, Rozpedowska E, Polakova S, Merico A, Compagno C. How did *Saccharomyces* evolve to become a good brewer? Trends in Genetics. 2006;**22**(4):183-186. DOI: 10.1016/j.tig.2006.02.002

[14] Scott M, Gunderson CW, Mateescu EM, Zhang Z, Hwa T. Interdependence of cell growth. Science. 2010;**330**(6007):1099-1102. DOI: 10.1126/science.1192588

[15] Kussell E. Evolution in microbes. Annual Review of Biophysics. 2013;**42**(1):493-514. DOI: 10.1146/annurev-biophys-083012-130320

[16] Warburg O. On the origin of cancer. Science. 1956;**123**:309-314. DOI: 10.1126/science.123.3191.309

[17] Verduyn C, Zomerdijk TPL, van Dijken JP, Scheffers WA. Continuous

measurement of ethanol production by aerobic yeast suspensions with an enzyme electrode. Applied Microbiology and Biotechnology. 1984;**19**(3):181-185. DOI: 10.1007/BF00256451

[18] Broach JR. Nutritional control of growth and development in yeast. Genetics. 2012;**192**(1):73-105. DOI: 10.1534/genetics.111.135731

[19] Santangelo GM. Glucose signaling in *Saccharomyces cerevisiae*. Microbiology and Molecular Biology Reviews. 2006;**70**(1):253-282. DOI: 10.1128/MMBR.70.1.253-282.2006

[20] Schmelzle T, Beck T, Martin DE, Hall MN. Activation of the RAS/cyclic AMP pathway suppresses a TOR deficiency in yeast. Molecular and Cellular Biology. 2004;**24**(1):338-351. DOI: 10.1128/MCB.24.1.338-351.2004

[21] Tamanoi F. Ras signaling in yeast. Genes & Cancer. 2011;**2**(3):210-215. DOI: 10.1177/1947601911407322

[22] Wang Y, Pierce M, Schneper L, Güldal CG, Zhang X, Tavazoie S, et al. Ras and Gpa2 mediate one branch of a redundant glucose signaling pathway in yeast. PLoS Biology. 2004;**2**(5):610-622. DOI: 10.1371/journal.pbio.0020128

[23] Yun C-W, Tamaki H, Nakayama R, Yamamoto K, Kumagai H. G-protein coupled receptor from yeast *Saccharomyces cerevisiae*. Biochemical and Biophysical Research Communications. 1997;**240**(2):287-292. DOI: 10.1006/bbrc.1997.7649

[24] Xue Y, Batlle M, Hirsch JP. GPR1 encodes a putative G protein-coupled receptor that associates with the Gpa2p Gα subunit and functions in a Ras-independent pathway. The EMBO Journal. 1998;**17**(7):1996-2007. DOI: 10.1093/emboj/17.7.1996

[25] Yun C-W, Tamaki H, Nakayama R, Yamamoto K, Kumagai H. Gpr1p, a putative G-protein coupled receptor, regulates glucose-dependent cellular cAMP level in yeast *Saccharomyces cerevisiae*. Biochemical and Biophysical Research Communications. 1998;**252**(1):29-33. DOI: 10.1006/bbrc.1998.9600

[26] Kraakman L, Lemaire K, Ma P, Teunlssen AW, Donaton MC, Van Dijck P, et al. A *Saccharomyces cerevisiae* G-protein coupled receptor, Gpr1, is specifically required for glucose activation of the cAMP pathway during the transition to growth on glucose. Molecular Microbiology. 1999;**32**(5):1002-1012. DOI: 10.1046/j.1365-2958.1999.01413.x

[27] Conrad M, Schothorst J, Kankipati HN, Van Zeebroeck G, Rubio-Texeira M, Thevelein JM. Nutrient sensing and signaling in the yeast *Saccharomyces cerevisiae*. FEMS Microbiology Reviews. 2014;**38**(2):254-299. DOI: 10.15698/mic2021.01.740

[28] Wera S, De Schrijver E, Geyskens I, Nwaka S, Thevelein JM. Opposite roles of trehalase activity in heat-shock recovery and heat-shock survival in *Saccharomyces cerevisiae*. The Biochemical Journal. 1999;**343**(3):621-626

[29] François JM, Walther T, Parrou JL. Genetics and regulation of glycogen and trehalose metabolism in *Saccharomyces cerevisiae*. In: Systems Biology. New York: SpringerLink; 2012. pp. 29-55. DOI: 10.1007/978-3-642-21467-7

[30] Wingender-Drissen R, Becker JU. Regulation of yeast phosphorylase by phosphorylase kinase and cAMP-dependent protein kinase. FEBS Letters. 1983;**163**(1):33-36. DOI: 10.1016/0014-5793(83)81156-9

[31] Dihazi H, Kessler R, Eschrich K. Glucose-induced stimulation of the

Ras-cAMP pathway in yeast leads to multiple phosphorylations and activation of 6-phosphofructo-2-kinase. Biochemistry. 2003;**42**(20):6275-6282. DOI: 10.1021/bi034167r

[32] Portela P, Moreno S, Rossi S. Characterization of yeast pyruvate kinase 1 as a protein kinase A substrate, and specificity of the phosphorylation site sequence in the whole protein. The Biochemical Journal. 2006;**396**(1): 117-126. DOI: 10.1042/BJ20051642

[33] Mazón MJ, Gancedo JM, Gancedo C. Inactivation of yeast fructose-1,6-bisphosphatase. In vivo phosphorylation of the enzyme. The Journal of Biological Chemistry. 1982;**257**(3):1128-1130

[34] Heitman J, Movva NR, Hall MN. Targets for cell cycle arrest by the immunosuppressant rapamycin in yeast. Science. 1991;**253**(5022):905-909. DOI: 10.1126/science.1715094 DOI: 10.1126/science.1715094

[35] González A, Hall MN. Nutrient sensing and TOR signaling in yeast and mammals. The EMBO Journal. 2017;**36**(4):397-408. DOI: 10.15252/embj.201696010

[36] Loewith R, Jacinto E, Wullschleger S, Lorberg A, Crespo JL, Bonenfant D, et al. Two TOR complexes, only one of which is rapamycin sensitive, have distinct roles in cell growth control. Molecular Cell. 2002;**10**(3):457-468. DOI: 10.1016/s1097-2765(02)00636-6

[37] Düvel K, Broach JR. The role of phosphatases in TOR signaling in yeast. In: Current Topics in Microbiology and Immunology. New York: SpringerLink; 2004. pp. 19-38. DOI: 10.1007/978-3-642-18930-2_2

[38] Jacinto E, Guo B, Arndt KT, Schmelzle T, Hall MN. TIP41 interacts with TAP42 and negatively regulates the TOR signaling pathway. Molecular Cell. 2001;**8**(5):1017-1026. DOI: 10.1016/s1097-2765(01)00386-0

[39] Breitkreutz A, Choi H, Sharom JR, Boucher L, Neduva V, Larsen B, et al. A global protein kinase and phosphatase interaction network in yeast. Science. 2010;**328**(5981):1043-1046. DOI: 10.1126/science.1176495

[40] Sturgill TW, Cohen A, Diefenbacher M, Trautwein M, Martin DE, Hall MN. TOR1 and TOR2 have distinct locations in live cells. Eukaryotic Cell. 2008;**7**(10):1819-1830. DOI: 10.1128/EC.00088-08

[41] Liu Y, Okamoto K. The TORC1 signaling pathway regulates respiration-induced mitophagy in yeast. Biochemical and Biophysical Research Communications. 2018;**502**(1):76-83. DOI: 10.1016/j.bbrc.2018.05.123

[42] Innokentev A, Kanki T. Mitophagy in yeast: Molecular mechanism and regulation. Cell. 2021;**10**(12):3569. DOI: 10.3390/cells10123569

[43] May AI, Prescott M, Ohsumi Y. Autophagy facilitates adaptation of budding yeast to respiratory growth by recycling serine for one-carbon metabolism. Nature Communications. 2020;**11**(1):5052. DOI: 10.1038/s41467-020-18805-x

[44] Kumar R, Reichert AS. Autophagy promotes mitochondrial respiration by providing serine for one-carbon-metabolism. Autophagy. 2021;**17**(12):4480-4483. DOI: 10.1080/15548627.2021.1909408

[45] Urban J, Soulard A, Huber A, Lippman S, Mukhopadhyay D, Deloche O, et al. Sch9 is a major target of TORC1 in *Saccharomyces cerevisiae*.

Molecular Cell. 2007;**26**(5):663-674. DOI: 10.1016/j.molcel.2007.04.020

[46] Soulard A, Cohen A, Hall MN. TOR signaling in invertebrates. Current Opinion in Cell Biology. 2009;**21**(6):825-836. DOI: 10.1016/j.ceb.2009.08.007

[47] Eltschinger S, Loewith R. TOR complexes and the maintenance of cellular homeostasis. Trends in Cell Biology. 2016;**26**(2):148-159. DOI: 10.1016/j.tcb.2015.10.003

[48] Peyroche G, Milkereit P, Bischler N, Tschochner H, Schultz P, Sentenac A, et al. The recruitment of RNA polymerase I on rDNA is mediated by the interaction of the A43 subunit with Rrn3. The EMBO Journal. 2000;**19**(20):5473-5482. DOI: 10.1093/emboj/19.20.5473

[49] Laferté A, Favry E, Sentenac A, Riva M, Carles C, Chédin S. The transcriptional activity of RNA polymerase I is a key determinant for the level of all ribosome components. Genes & Development. 2006;**20**(15):2030-2040. DOI: 10.1101/gad.386106

[50] Huber A, Bodenmiller B, Uotila A, Stahl M, Wanka S, Gerrits B, et al. Characterization of the rapamycin-sensitive phosphoproteome reveals that Sch9 is a central coordinator of protein synthesis. Genes & Development. 2009;**23**(16):1929-1943. DOI: 10.1101/gad.532109

[51] Jorgensen P, Rupeš I, Sharom JR, Schneper L, Broach JR, Tyers M. A dynamic transcriptional network communicates growth potential to ribosome synthesis and critical cell size. Genes & Development. 2004;**18**(20):2491-2505. DOI: 10.1101/gad.1228804

[52] Upadhya R, Lee JH, Willis IM. Maf1 is an essential mediator of diverse signals that repress RNA polymerase III transcription. Molecular Cell. 2002;**10**(6):1489-1494. DOI: 10.1016/s1097-2765(02)00787-6

[53] Oficjalska-Pham D, Harismendy O, Smagowicz WJ, Gonzalez de Peredo A, Boguta M, Sentenac A, et al. General repression of RNA polymerase III transcription is triggered by protein phosphatase type 2A-mediated dephosphorylation of Maf1. Molecular Cell. 2006;**22**(5):623-632. DOI: 10.1016/j.molcel.2006.04.008

[54] Roberts DN, Wilson B, Huff JT, Stewart AJ, Cairns BR. Dephosphorylation and genome-wide association of Maf1 with Pol III-transcribed genes during repression. Molecular Cell. 2006;**22**(5):633-644. DOI: 10.1016/j.molcel.2006.04.009

[55] Lee JH, Moir RD, Willis IM. Regulation of RNA polymerase III transcription involves SCH9-dependent and SCH9-independent branches of the target of rapamycin (TOR) pathway. The Journal of Biological Chemistry. 2009;**284**(19):12604-12608. DOI: 10.1074/jbc.C900020200

[56] Wei Y, Zheng XFS. Sch9 partially mediates TORC1 signaling to control ribosomal RNA synthesis. Cell Cycle. 2009;**8**(24):4085-4090. DOI: 10.4161/cc.8.24.10170

[57] Hughes Hallett JE, Luo X, Capaldi AP. State transitions in the TORC1 signaling pathway and information processing in *Saccharomyces cerevisiae*. Genetics. 2014;**198**(2):773-786. DOI: 10.1534/genetics.114.168369

[58] Powers T, Walter P. Regulation of ribosome biogenesis by the rapamycin-sensitive TOR-signaling pathway in *Saccharomyces cerevisiae*. Molecular Biology of the Cell. 1999;**10**(4):987-1000. DOI: 10.1091/mbc.10.4.987

[59] Boguta M. Maf1, a general negative regulator of RNA polymerase III in yeast. Biochimica et Biophysica Acta: Gene Regulatory Mechanisms. 2013;**1829**(3-4):376-384. DOI: 10.1016/j.bbagrm.2012.11.004

[60] Peisker K, Chiabudini M, Rospert S. The ribosome-bound Hsp70 homolog Ssb of *Saccharomyces cerevisiae*. Biochimica et Biophysica Acta: Molecular Cell Research. 2010;**1803**(6):662-672. DOI: 10.1016/j.bbamcr.2010.03.005

[61] Bonawitz ND, Chatenay-Lapointe M, Pan Y, Shadel GS. Reduced TOR signaling extends chronological life span via increased respiration and upregulation of mitochondrial gene expression. Cell Metabolism. 2007;**5**(4):265-277. DOI: 10.1016/j.cmet.2007.02.009

[62] Caligaris M, Nicastro R, Hu Z, Tripodi F, Hummel JE, Pillet B, et al. Snf1/AMPK fine-tunes TORC1 signaling in response to glucose starvation. eLife. 2023;**12**:e84319. DOI: 10.7554/eLife.84319

[63] Carlson M, Osmond BC, Botstein D. Mutants of yeast defective in sucrose utilization. Genetics. 1981;**98**(1):25-40. DOI: 10.1093/genetics/98.1.25

[64] Celenza JL, Carlson M. A yeast gene that is essential for release from glucose repression encodes a protein kinase. Science. 1986;**233**(4769):1175-1180. DOI: 10.1126/science.3526554

[65] Hedbacker K, Carlson M. SNF1/AMPK pathways in yeast. Frontiers in Bioscience. 2008;**13**(13):2408. DOI: 10.2741/2854

[66] Hardie DG, Ross FA, Hawley SA. AMPK: A nutrient and energy sensor that maintains energy homeostasis. Nature Reviews Molecular Cell Biology. 2012;**13**(4):251-262. DOI: 10.1038/nrm3311

[67] Nehlin JO, Ronne H. Yeast MIG1 repressor is related to the mammalian early growth response and Wilms' tumour finger proteins. The EMBO Journal. 1990;**9**(9):2891-2898. DOI: 10.1002/j.1460-2075.1990.tb07479.x

[68] Treitel MA, Kuchin S, Carlson M. Snf1 protein kinase regulates phosphorylation of the Mig1 repressor in *Saccharomyces cerevisiae*. Molecular and Cellular Biology. 1998;**18**(11):6273-6280. DOI: 10.1128/MCB.18.11.6273

[69] Schuller HJ, Entian KD. Extragenic suppressors of yeast glucose derepression mutants leading to constitutive synthesis of several glucose-repressible enzymes. Journal of Bacteriology. 1991;**173**(6):2045-2052. DOI: 10.1128/jb.173.6.2045-2052.1991

[70] Flick JS, Johnston M. Analysis of URS(G)-mediated glucose repression of the GAL1 promoter of *Saccharomyces cerevisiae*. Genetics. 1992;**130**(2):295-304. DOI: 10.1093/genetics/130.2.295

[71] Vincent O, Townley R, Kuchin S, Carlson M. Subcellular localization of the Snf1 kinase is regulated by specific β subunits and a novel glucose signaling mechanism. Genes & Development. 2001;**15**(9):1104-1114. DOI: 10.1101/gad.879301

[72] Tu J, Carlson M. REG1 binds to protein phosphatase type 1 and regulates glucose repression in *Saccharomyces cerevisiae*. The EMBO Journal. 1995;**14**(23):5939-5946. DOI: 10.1002/j.1460-2075.1995.tb00282.x

[73] Sanz P, Alms GR, Haystead TAJ, Carlson M. Regulatory interactions between the Reg1-Glc7 protein phosphatase and the Snf1 protein kinase. Molecular and Cellular Biology. 2000;**20**(4):1321-1328. DOI: 10.1128/MCB.20.4.1321-1328.2000

[74] Rubenstein EM, McCartney RR, Zhang C, Shokat KM, Shirra MK, Arndt KM, et al. Access denied: Snf1 activation loop phosphorylation is controlled by availability of the phosphorylated threonine 210 to the PP1 phosphatase. The Journal of Biological Chemistry. 2008;**283**(1):222-230. DOI: 10.1074/jbc.M707957200

[75] Castermans D, Somers I, Kriel J, Louwet W, Wera S, Versele M, et al. Glucose-induced posttranslational activation of protein phosphatases PP2A and PP1 in yeast. Cell Research. 2012;**22**(6):1058-1077. DOI: 10.1038/cr.2012.20

[76] Ruiz A, Liu Y, Xu X, Carlson M. Heterotrimer-independent regulation of activation-loop phosphorylation of Snf1 protein kinase involves two protein phosphatases. Proceedings of the National Academy of Sciences of the United States of America. 2012;**109**(22):8652-8657. DOI: 10.1073/pnas.1206280109

[77] Ruiz A, Xu X, Carlson M. Ptc1 protein phosphatase 2C contributes to glucose regulation of SNF1/AMP-activated protein kinase (AMPK) in *Saccharomyces cerevisiae*. The Journal of Biological Chemistry. 2013;**288**(43):31052-31058. DOI: 10.1074/jbc.M113.503763

[78] Özcan S, Dover J, Rosenwald AG, Wölfl S, Johnston M. Two glucose transporters in *Saccharomyces cerevisiae* are glucose sensors that generate a signal for induction of gene expression. Proceedings of the National Academy of Sciences of the United States of America. 1996;**93**(22):12428-12432. DOI: 10.1073/pnas.93.22.12428

[79] Shashkova S, Welkenhuysen N, Hohmann S. Molecular communication: Crosstalk between the Snf1 and other signaling pathways. FEMS Yeast Research. 2015;**15**(4):fov026. DOI: 10.1093/femsyr/fov026

[80] Özcan S, Dover J, Johnston M. Glucose sensing and signaling by two glucose receptors in the yeast *Saccharomyces cerevisiae*. The EMBO Journal. 1998;**17**(9):2566-2573. DOI: 10.1093/emboj/17.9.2566

[81] Özcan S, Johnston M. Three different regulatory mechanisms enable yeast hexose transporter (HXT) genes to be induced by different levels of glucose. Molecular and Cellular Biology. 1995;**15**(3):1564-1572. DOI: 10.1128/MCB.15.3.1564

[82] Tomás-Cobos L, Sanz P. Active Snf1 protein kinase inhibits expression of the *Saccharomyces cerevisiae* HXT1 glucose transporter gene. The Biochemical Journal. 2002;**368**(2):657-663. DOI: 10.1042/BJ20020984

[83] Kim J-H, Polish J, Johnston M. Specificity and regulation of DNA binding by the yeast glucose transporter gene repressor Rgt1. Molecular and Cellular Biology. 2003;**23**(15):5208-5216. DOI: 10.1128/MCB.23.15.5208-5216.2003

[84] Lakshmanan J, Mosley AL, Özcan S. Repression of transcription by Rgt1 in the absence of glucose requires Std1 and Mth1. Current Genetics. 2003;**44**(1):19-25. DOI: 10.1007/s00294-003-0423-2

[85] Mosley AL, Lakshmanan J, Aryal BK, Özcan S. Glucose-mediated phosphorylation converts the transcription factor Rgt1 from a repressor to an activator. The Journal of Biological Chemistry. 2003;**278**(12):10322-10327. DOI: 10.1074/jbc.M212802200

[86] Polish JA, Kim JH, Johnston M. How the Rgt1 transcription factor of *Saccharomyces cerevisiae* is regulated by

glucose. Genetics. 2005;**169**(2):583-594. DOI: 10.1534/genetics.104.034512

[87] Moriya H, Johnston M. Glucose sensing and signaling in *Saccharomyces cerevisiae* through the Rgt2 glucose sensor and casein kinase I. Proceedings of the National Academy of Sciences of the United States of America. 2004;**101**(6):1572-1577. DOI: 10.1073/pnas.0305901101

[88] Flick KM, Spielewoy N, Kalashnikova TI, Guaderrama M, Zhu Q, Chang H-C, et al. Grr1-dependent inactivation of Mth1 mediates glucose-induced dissociation of Rgt1 from HXT gene promoters. Molecular Biology of the Cell. 2003;**14**(8):3230-3241. DOI: 10.1091/mbc.e03-03-0135

[89] Kim JH, Brachet V, Moriya H, Johnston M. Integration of transcriptional and posttranslational regulation in a glucose signal transduction pathway in *Saccharomyces cerevisiae*. Eukaryotic Cell. 2006;**5**(1):167-173. DOI: 10.1128/EC.5.1.167-173.2006

[90] Zaman S, Lippman SI, Schneper L, Slonim N, Broach JR. Glucose regulates transcription in yeast through a network of signaling pathways. Molecular Systems Biology. 2009;**5**(245):1-14. DOI: 10.1038/msb.2009.2

[91] Gagiano M, Bauer FF, Pretorius IS. The sensing of nutritional status and the relationship to filamentous growth in *Saccharomyces cerevisiae*. FEMS Yeast Research. 2002;**2**(4):433-470. DOI: 10.1111/j.1567-1364.2002.tb00114.x

[92] Lackner LL. Shaping the dynamic mitochondrial network. BMC Biology. 2014;**12**(1):35. DOI: 10.1186/1741-7007-12-35

[93] Tsuboi T, Viana MP, Xu F, Yu J, Chanchani R, Arceo XG, et al. Mitochondrial volume fraction and translation duration impact mitochondrial mRNA localization and protein synthesis. eLife. 2020;**9**(529289):e57814. DOI: 10.7554/eLife.57814

[94] Chacinska A, Koehler CM, Milenkovic D, Lithgow T, Pfanner N. Importing mitochondrial proteins: Machineries and mechanisms. Cell. 2009;**138**(4):628-644. DOI: 10.1016/j.cell.2009.08.005

[95] Wiedemann N, Pfanner N. Mitochondrial machineries for protein import and assembly. Annual Review of Biochemistry. 2017;**86**(1):685-714. DOI: 10.1146/annurev-biochem-060815-014352

[96] Avendaño-Monsalve MC, Ponce-Rojas JC, Funes S. From cytosol to mitochondria: The beginning of a protein journey. Biological Chemistry. 2020;**401**(6-7):645-661. DOI: 10.1515/hsz-2020-0110

[97] Garcia M, Darzacq X, Delaveau T, Jourdren L, Singer RH, Jacq C. Mitochondria-associated yeast mRNAs and the biogenesis of molecular complexes. Fox T, editor. Molecular Biology of the Cell. 2007;**18**(2):362-368. DOI: 10.1091/mbc.e06-09-0827

[98] Marc P, Margeot A, Devaux F, Blugeon C, Corral-Debrinski M, Jacq C. Genome-wide analysis of mRNAs targeted to yeast mitochondria. EMBO Reports. 2002;**3**(2):159-164. DOI: 10.1093/embo-reports/kvf025

[99] Lesnik C, Golani-Armon A, Arava Y. Localized translation near the mitochondrial outer membrane: An update. RNA Biology. 2015;**12**(8):801-809. DOI: 10.1080/15476286.2015.1058686

[100] Sylvestre J, Vialette S, Corral Debrinski M, Jacq C. Long mRNAs

coding for yeast mitochondrial proteins of prokaryotic origin preferentially localize to the vicinity of mitochondria. Genome Biology. 2003;**4**:R44. DOI: 10.1186/gb-2003-4-7-r44

[101] Williams CC, Jan CH, Weissman JS. Targeting and plasticity of mitochondrial proteins revealed by proximity-specific ribosome profiling. Science. 2014;**346**(6210):748-751. DOI: 10.1126/science.1257522

[102] Gold VA, Chroscicki P, Bragoszewski P, Chacinska A. Visualization of cytosolic ribosomes on the surface of mitochondria by electron cryotomography. EMBO Reports. 2017;**18**(10):1786-1800. DOI: 10.15252/embr.201744261

[103] Cavalier-Smith T. Origin of mitochondria by intracellular enslavement of a photosynthetic purple bacterium. Proceedings of the Royal Society B: Biological Sciences. 2006;**273**(1596):1943-1952. DOI: 10.1098/rspb.2006.3531

[104] Bonitz SG, Berlani R, Coruzzi G, Li M, Macino G, Nobrega FG, et al. Codon recognition rules in yeast mitochondria. Proceedings of the National Academy of Sciences of the United States of America. 1980;**77**(6I):3167-3170. DOI: 10.1073/pnas.77.6.3167

[105] Salinas-Giegé T, Giegé R, Giegé P. tRNA biology in mitochondria. International Journal of Molecular Sciences. 2015;**16**(3):4518-4559. DOI: 10.3390/ijms16034518

[106] Scheffler IE, De La Cruz BJ, Prieto S. Control of mRNA turnover as a mechanism of glucose repression in *Saccharomyces cerevisiae*. The International Journal of Biochemistry & Cell Biology. 1998;**30**(11):1175-1193. DOI: 10.1016/s1357-2725(98)00086-7

[107] Miller MA, Russo J, Fischer AD, Leban FAL, Olivas WM. Carbon source-dependent alteration of Puf3p activity mediates rapid changes in the stabilities of mRNAs involved in mitochondrial function. Nucleic Acids Research. 2014;**42**(6):3954-3970. DOI: 10.1093/nar/gkt1346

[108] Foat BC, Houshmandi SS, Olivas WM, Bussemaker HJ. Profiling condition-specific, genome-wide regulationof mRNA stability in yeast. Proceedings of the National Academy of Sciences of the United States of America. 2005;**102**(49):17675-17680. DOI: 10.1073/pnas.0503803102

[109] Endo T, Yamano K, Kawano S. Structural insight into the mitochondrial protein import system. Biochimica et Biophysica Acta (BBA) - Biomembranes. 2011;**1808**(3):955-970. DOI: 10.1016/j.bbamem.2010.07.018 DOI: 10.1016/j.bbamem.2010.07.018

[110] Priesnitz C, Becker T. Pathways to balance mitochondrial translation and protein import. Genes & Development. 2018;**32**(19-20):1285-1296. DOI: 10.1101/gad.316547.118

[111] Couvillion MT, Soto IC, Shipkovenska G, Churchman LS. Synchronized mitochondrial and cytosolic translation programs. Nature. 2016;**533**(7604):499-503. DOI: 10.1038/nature18015

[112] Wang X, Chen XJ. A cytosolic network suppressing mitochondria-mediated proteostatic stress and cell death. Nature. 2015;**524**(7566):481-484. DOI: 10.1038/nature14859

[113] Andréasson C, Ott M, Büttner S. Mitochondria orchestrate proteostatic and metabolic stress responses. EMBO Reports. 2019;**20**(10):e47865. DOI: 10.15252/embr.201947865

[114] Lill R, Mühlenhoff U. Maturation of iron-sulfur proteins in eukaryotes: Mechanisms, connected processes, and diseases. Annual Review of Biochemistry. 2008;**77**(1):669-700. DOI: 10.1146/annurev.biochem.76.052705.162653

[115] Attardi G, Schatz G. Biogenesis of mitochondria. Annual Review of Cell Biology. 1988;**4**:289-333. DOI: 10.1146/annurev.cb.04.110188.001445

[116] Malina C, Larsson C, Nielsen J. Yeast mitochondria: An overview of mitochondrial biology and the potential of mitochondrial systems biology. FEMS Yeast Research. 2018;**18**(5):foy040. DOI: 10.1093/femsyr/foy040

[117] Gancedo JM. Yeast carbon catabolite repression. Microbiology and Molecular Biology Reviews. 1998;**62**(2):334-361. DOI: 10.1128/MMBR.62.2.334-361.1998

[118] Sharma AK, Pallesen LJ, Spang RJ, Walden WE. Cytosolic iron-sulfur cluster assembly (CIA) system: Factors, mechanism, and relevance to cellular iron regulation. The Journal of Biological Chemistry. 2010;**285**(35):26745-26751. DOI: 10.1074/jbc.R110.122218

[119] Ramos-Alonso L, Romero AM, Martínez-Pastor MT, Puig S. Iron regulatory mechanisms in *Saccharomyces cerevisiae*. Frontiers in Microbiology. 2020;**11**:582830. DOI: 10.3389/fmicb.2020.582830

[120] Romero AM, Martínez-Pastor MT, Puig S. Iron in translation: From the beginning to the end. Microorganisms. 2021;**9**(5):1058. DOI: 10.3390/microorganisms9051058

[121] Puig S, Vergara SV, Thiele DJ. Cooperation of two mRNA-binding proteins drives metabolic adaptation to iron deficiency. Cell Metabolism. 2008;**7**(6):555-564. DOI: 10.1016/j.cmet.2008.04.010

[122] Yamaguchi-Iwai Y, Dancis A, Klausner RD. AFT1: A mediator of iron regulated transcriptional control in *Saccharomyces cerevisiae*. The EMBO Journal. 1995;**14**(6):1231-1239. DOI: 10.1002/j.1460-2075.1995.tb07106.x

[123] Haurie V, Boucherie H, Sagliocco F. The Snf1 protein kinase controls the induction of genes of the iron uptake pathway at the diauxic shift in *Saccharomyces cerevisiae*. The Journal of Biological Chemistry. 2003;**278**(46):45391-45396. DOI: 10.1074/jbc.M307447200

[124] Philpott CC, Protchenko O. Response to iron deprivation in *Saccharomyces cerevisiae*. Eukaryotic Cell. 2008;**7**(1):20-27. DOI: 10.1128/EC.00354-07

[125] Lill R, Mühlenhoff U. Iron-sulfur protein biogenesis in eukaryotes: Components and mechanisms. Annual Review of Cell and Developmental Biology. 2006;**22**:457-486. DOI: 10.1146/annurev.cellbio.22.010305.104538

[126] Holmes-Hampton GP, Jhurry ND, McCormick SP, Lindahl PA, Jo WJ, Hyoun JH, et al. Response to iron deprivation in *Saccharomyces cerevisiae*. Biochemistry. 2008;**10**(1):105-114. DOI: 10.1128/EC.00354-07

[127] Martins TS, Costa V, Pereira C. Signaling pathways governing iron homeostasis in budding yeast. Molecular Microbiology. 2018;**109**(4):422-432. DOI: 10.1111/mmi.14009

[128] den Ridder M, Daran-Lapujade P, Pabst M. Shot-gun proteomics: Why thousands of unidentified signals matter. FEMS Yeast Research. 2020;**20**(1):foz088. DOI: 10.1093/femsyr/foz088

[129] Daran-Lapujade P, Rossell S, Van Gulik WM, Luttik MAH, De Groot MJL, Slijper M, et al. The fluxes through glycolytic enzymes in *Saccharomyces cerevisiae* are predominantly regulated at posttranscriptional levels. Proceedings of the National Academy of Sciences of the United States of America. 2007;**104**(40):15753-15758. DOI: 10.1073/pnas.0707476104

[130] Soares Rodrigues CI, den Ridder M, Pabst M, Gombert AK, Wahl SA. Comparative proteome analysis of different *Saccharomyces cerevisiae* strains during growth on sucrose and glucose. Scientific Reports. 2023;**13**(1):2126. DOI: 10.1038/s41598-023-29172-0

[131] Kolkman A, Olsthoorn MMA, Heeremans CEM, Heck AJR, Slijper M. Comparative proteome analysis of *Saccharomyces cerevisiae* grown in chemostat cultures limited for glucose or ethanol. Molecular & Cellular Proteomics. 2005;**4**(1):1-11. DOI: 10.1074/mcp.M400087-MCP200

[132] Garcia-Albornoz M, Holman SW, Antonisse T, Daran-Lapujade P, Teusink B, Beynon RJ, et al. A proteome-integrated, carbon source dependent genetic regulatory network in: *Saccharomyces cerevisiae*. Molecular Oral Microbiology. 2020;**16**(1):59-72. DOI: 10.1039/C9MO00136K

[133] Paulo JA, O'Connell JD, Everley RA, O'Brien J, Gygi MA, Gygi SP. Quantitative mass spectrometry-based multiplexing compares the abundance of 5000 S. cerevisiae proteins across 10 carbon sources. Journal of Proteomics. 2016;**148**:85-93. DOI: 10.1016/j.jprot.2016.07.005

[134] Paulo JA, O'Connell JD, Gaun A, Gygi SP. Proteome-wide quantitative multiplexed profiling of protein expression: Carbon-source dependency in *Saccharomyces cerevisiae*. Molecular Biology of the Cell. 2015;**26**(22):4063-4074. DOI: 10.1091/mbc.E15-07-0499

[135] Pichon X, Wilson LA, Stoneley M, Bastide A, King HA, Somers J, et al. RNA binding protein/RNA element interactions and the control of translation. Current Protein & Peptide Science. 2012;**13**(4):294-304. DOI: 10.2174/138920312801619475

[136] Simpson CE, Ashe MP. Adaptation to stress in yeast: To translate or not? Biochemical Society Transactions. 2012;**40**(4):794-799. DOI: 10.1042/BST20120078

[137] Ashe MP, De LSK, Sachs AB. Glucose depletion rapidly inhibits translation initiation in yeast. Molecular Biology of the Cell. 2000;**11**:833-848. DOI: 10.1091/mbc.11.3.833

[138] Buchan JR, Nissan T, Parker R. Analyzing P-bodies and stress granules in *Saccharomyces cerevisiae*. Methods in Enzymology. 2010;**470**:619-640. DOI: 10.1016/S0076-6879(10)70025-2

[139] Kedersha N, Stoecklin G, Ayodele M, Yacono P, Lykke-Andersen J, Fitzler MJ, et al. Stress granules and processing bodies are dynamically linked sites of mRNP remodeling. The Journal of Cell Biology. 2005;**169**(6):871-884. DOI: 10.1083/jcb.200502088

[140] Xing W, Muhlrad D, Parker R, Rosen MK. A quantitative inventory of yeast P body proteins reveals principles of composition and specificity. eLife. 2020;**9**:e56525. DOI: 10.7554/eLife.56525

[141] Balagopal V, Fluch L, Nissan T. Ways and means of eukaryotic mRNA decay. Biochimica et Biophysica Acta: Gene Regulatory Mechanisms.

2012;**1819**(6):593-603. DOI: 10.1016/j.bbagrm.2012.01.001

[142] Balagopal V, Parker R. Polysomes, P bodies and stress granules: States and fates of eukaryotic mRNAs. Current Opinion in Cell Biology. 2009;**21**(3):403-408. DOI: 10.1016/j.ceb.2009.03.005

[143] Luo Y, Na Z, Slavoff SA. P-bodies: Composition, properties, and functions. Biochemistry. 2018;**57**(17):2424-2431. DOI: 10.1021/acs.biochem.7b01162

[144] Decker CJ, Parker R. P-bodies and stress granules: Possible roles in the control of translation and mRNA degradation. Cold Spring Harbor Perspectives in Biology. 2012;**4**(9):a012286. DOI: 10.1101/cshperspect.a012286

[145] Drummond SP, Hildyard J, Firczuk H, Reamtong O, Li N, Kannambath S, et al. Diauxic shift-dependent relocalization of decapping activators Dhh1 and Pat1 to polysomal complexes. Nucleic Acids Research. 2011;**39**(17):7764-7774. DOI: 10.1093/nar/gkr474

[146] Phizicky EM, Hopper AK. The life and times of a tRNA. RNA. 2023;**29**(7):898-957. DOI: 10.1261/rna.079620.123

[147] Warner JR. The economics of ribosome biosynthesis in yeast. Trends in Biochemical Sciences. 1999;**24**(11):437-440. DOI: 10.1016/s0968-0004(99)01460-7

[148] Quax TEF, Claassens NJ, Söll D, van der Oost J. Codon bias as a means to fine-tune gene expression. Molecular Cell. 2015;**59**(2):149-161. DOI: 10.1016/j.molcel.2015.05.035

[149] Chan PP, Lowe TM. GtRNAdb 2.0: An expanded database of transfer RNA genes identified in complete and draft genomes. Nucleic Acids Research. 2016;**44**(D1):D184-D189. DOI: 10.1093/nar/gkv1309

[150] Hani J, Feldmann H. tRNA genes and retroelements in the yeast genome. Nucleic Acids Research. 1998;**26**(3):689-696. DOI: 10.1093/nar/26.3.689

[151] Guimarães AR, Correia I, Sousa I, Oliveira C, Moura G, Bezerra AR, et al. tRNAs as a driving force of genome evolution in yeast. Frontiers in Microbiology. 2021;**12**:634004. DOI: 10.3389/fmicb.2021.634004

[152] Tavares JF, Davis NK, Poim A, Reis A, Kellner S, Sousa I, et al. tRNA-modifying enzyme mutations induce codon-specific mistranslation and protein aggregation in yeast. RNA Biology. 2021;**18**(4):563-575. DOI: 10.1080/15476286.2020.1819671

[153] Edwards AM, Addo MA, Dos Santos PC. Extracurricular functions of tRNA modifications in microorganisms. Genes (Basel). 2020;**11**(8):907. DOI: 10.3390/genes11080907

[154] Tarassov I, Entelis N, Martin RP. Mitochondrial import of a cytoplasmic lysine-tRNA in yeast is mediated by cooperation of cytoplasmic and mitochondrial lysyl-tRNA synthetases. The EMBO Journal. 1995;**14**(14):3461-3471. DOI: 10.1002/j.1460-2075.1995.tb07352.x

[155] Rinehart J, Krett B, Rubio MAT, Alfonzo JD, Söll D. *Saccharomyces cerevisiae* imports the cytosolic pathway for Gln-tRNA synthesis into the mitochondrion. Genes & Development. 2005;**19**(5):583-592. DOI: 10.1101/gad.1269305

[156] Rubio MAT, Hopper AK. Transfer RNA travels from the cytoplasm to

organelles. Wiley Interdisciplinary Reviews RNA. 2011;**2**(6):802-817. DOI: 10.1002/wrna.93

[157] Bruch A, Laguna T, Butter F, Schaffrath R, Klassen R. Misactivation of multiple starvation responses in yeast by loss of tRNA modifications. Nucleic Acids Research. 2020;**48**(13):7307-7320. DOI: 10.1093/nar/gkaa455

[158] Hayashi S, Mori S, Suzuki T, Suzuki T, Yoshihisa T. Impact of intron removal from tRNA genes on *Saccharomyces cerevisiae*. Nucleic Acids Research. 2019;**47**(11):5936-5949. DOI: 10.1093/nar/gkz270

[159] Baldi MI, Mattoccia E, Bufardeci E, Fabbri S, Tocchini-Valentini GP. Participation of the intron in the reaction catalyzed by the xenopus tRNA splicing endonuclease. Science. 1992;**255**(5050):1404-1408. DOI: 10.1126/science.1542788

[160] Negri EDN, Fabbri S, Bufardeci E, Baldi MI, Attardi DG, Mattoccia E, et al. The eucaryal tRNA splicing endonuclease recognizes a tripartite set of RNA elements. Cell. 1997;**89**(6):859-866. DOI: 10.1016/s0092-8674(00)80271-8

[161] Yoshihisa T, Yunoki-Esaki K, Ohshima C, Tanaka N, Endo T. Possibility of cytoplasmic pre-tRNA splicing: The yeast tRNA splicing endonuclease mainly localizes on the mitochondria. Molecular Biology of the Cell. 2003;**14**(8):3266-3279. DOI: 10.1091/mbc.E02-11-0757

[162] Hopper AK, Nostramo RT. tRNA processing and subcellular trafficking proteins multitask in pathways for other RNAs. Frontiers in Genetics. 2019;**10**(FEB):1-14. DOI: 10.3389/fgene.2019.00096

[163] Huang HY, Hopper AK. Multiple layers of stress-induced regulation in tRNA biology. Life. 2016;**6**(2):13-16. DOI: 10.3390/life6020016

[164] Holcik M, Sonenberg N. Translational control in stress and apoptosis. Nature Reviews Molecular Cell Biology. 2005;**6**(4):318-327. DOI: 10.1038/nrm1618

[165] Chatterjee K, Nostramo RT, Wan Y, Hopper AK. tRNA dynamics between the nucleus, cytoplasm and mitochondrial surface: Location, location, location. Biochimica et Biophysica Acta: Gene Regulatory Mechanisms. 2018;**1861**(4):373-386. DOI: 10.1016/j.bbagrm.2017.11.007

[166] Chu H-Y, Hopper AK. Genome-wide investigation of the role of the tRNA nuclear-cytoplasmic trafficking pathway in regulation of the yeast *Saccharomyces cerevisiae* transcriptome and proteome. Molecular and Cellular Biology. 2013;**33**(21):4241-4254. DOI: 10.1128/MCB.00785-13

[167] Whitney ML, Hurto RL, Shaheen HH, Hopper AK. Rapid and reversible nuclear accumulation of cytoplasmic tRNA in response to nutrient availability. Fox T, editor. Molecular Biology of the Cell. 2007;**18**(7):2678-2686. DOI: 10.1091/mbc.e07-01-0006

[168] Karkusiewicz I, Turowski TW, Graczyk D, Towpik J, Dhungel N, Hopper AK, et al. Maf1 protein, repressor of RNA polymerase III, indirectly affects tRNA processing. The Journal of Biological Chemistry. 2011;**286**(45):39478-39488. DOI: 10.1074/jbc.M111.253310

[169] Phizicky EM, Hopper AK. tRNA biology charges to the front. Genes & Development. 2010;**24**(17):1832-1860. DOI: 10.1101/gad.1956510

[170] Yoshihisa T. Handling tRNA introns, archaeal way and eukaryotic way.

Frontiers in Genetics. 2014;**5**(JUL):1-16. DOI: 10.3389/fgene.2014.00213

[171] Fujishima K, Kanai A. tRNA gene diversity in the three domains of life. Frontiers in Genetics. 2014;**5**:142. DOI: 10.3389/fgene.2014.00142

[172] Schmidt CA, Matera AG. tRNA introns: Presence, processing, and purpose. WIREs RNA [Internet]. 2020;**11**(3):e1583. DOI: 10.1002/wrna.1583

[173] Walters R, Parker R. Is there quality control of localized mRNAs? The Journal of Cell Biology. 2014;**204**(6):863-868. DOI: 10.1083/jcb.201401059

[174] Quenault T, Lithgow T, Traven A. PUF proteins: Repression, activation and mRNA localization. Trends in Cell Biology. 2011;**21**(2):104-112. DOI: 10.1016/j.tcb.2010.09.013

[175] Crawford RA, Pavitt GD. Translational regulation in response to stress in *Saccharomyces cerevisiae*. Yeast. 2019;**36**(1):5-21. DOI: 10.1002/yea.3349

[176] Gerber AP, Herschlag D, Brown PO. Extensive association of functionally and cytotopically related mRNAs with Puf family RNA-binding proteins in yeast. PLoS Biology. 2004;**2**(3):E79. DOI: 10.1371/journal.pbio.0020079

[177] Hogan DJ, Riordan DP, Gerber AP, Herschlag D, Brown PO. Diverse RNA-binding proteins interact with functionally related sets of RNAs, suggesting an extensive regulatory system. PLoS Biology. 2008;**6**(10):2297-2313. DOI: 10.1371/journal.pbio.0060255

[178] Freeberg MA, Han T, Moresco JJ, Kong A, Yang YC, Lu ZJ, et al. Pervasive and dynamic protein binding sites of the mRNA transcriptome in *Saccharomyces cerevisiae*. Genome Biology. 2013;**14**(2):R13. DOI: 10.1186/gb-2013-14-2-r13

[179] Kershaw CJ, Costello JL, Talavera D, Rowe W, Castelli LM, Sims PFG, et al. Integrated multi-omics analyses reveal the pleiotropic nature of the control of gene expression by Puf3p. Scientific Reports. 2015;**5**:15518. DOI: 10.1038/srep15518

[180] Edwards TA, Pyle SE, Wharton RP, Aggarwal AK. Structure of Pumilio reveals similarity between RNA and peptide binding motifs. Cell. 2001;**105**(2):281-289. DOI: 10.1016/s0092-8674(01)00318-x

[181] Wang X, Zamore PD, Tanaka Hall TM. Crystal structure of a Pumilio homology domain. Molecular Cell. 2001;**7**(4):855-865. DOI: 10.1016/s1097-2765(01)00229-5

[182] Zhu D, Stumpf CR, Krahn JM, Wickens M, Tanaka Hall TM. A 5′ cytosine binding pocket in Puf3p specifies regulation of mitochondrial mRNAs. Proceedings of the National Academy of Sciences of the United States of America. 2009;**106**(48):20192-20197. DOI: 10.1073/pnas.0812079106

[183] Cheong CG, Tanaka Hall TM. Engineering RNA sequence specificity of Pumilio repeats. Proceedings of the National Academy of Sciences of the United States of America. 2006;**103**(37):13635-13639. DOI: 10.1073/pnas.0606294103

[184] Koh YY, Opperman L, Stumpf C, Mandan A, Keles S, Wickens M. A single C. elegans PUF protein binds RNA in multiple modes. RNA. 2009;**15**(6):1090-1099. DOI: 10.1261/rna.1545309

[185] Miller MT, Higgin JJ, Tanaka Hall TM. Basis of altered RNA-binding specificity by PUF proteins revealed

by crystal structures of yeast Puf4p. Nature Structural & Molecular Biology. 2008;**15**(4):397-402. DOI: 10.1038/nsmb.1390

[186] Opperman L, Hook B, DeFino M, Bernstein DS, Wickens M. A single spacer nucleotide determines the specificities of two mRNA regulatory proteins. Nature Structural & Molecular Biology. 2005;**12**(11):945-951. DOI: 10.1038/nsmb1010

[187] Varshney U, Lee CP, RajBhandary UL. Direct analysis of aminoacylation levels of tRNAs in vivo: Application to studying recognition of *Escherichia coli* initiator tRNA mutants by glutaminyl-tRNA synthetase. The Journal of Biological Chemistry. 1991;**266**(36):24712-24718

[188] Wang Y, Opperman L, Wickens M, Tanaka Hall TM. Structural basis for specific recognition of multiple mRNA targets by a PUF regulatory protein. Proceedings of the National Academy of Sciences of the United States of America. 2009;**106**(48):20186-20191. DOI: 10.1073/pnas.0812076106

[189] Houshmandi SS, Olivas WM. Yeast Puf3 mutants reveal the complexity of Puf-RNA binding and identify a loop required for regulation of mRNA decay. RNA. 2005;**11**(11):1655-1666. DOI: 10.1261/rna.2168505

[190] Jackson JS, Houshmandi SS, Leban FL, Olivas WM. Recruitment of the Puf3 protein to its mRNA target for regulation of mRNA decay in yeast. RNA. 2004;**10**(10):1625-1636. DOI: 10.1261/rna.7270204

[191] Olivas W, Parker R. The Puf3 protein is a transcript-specific regulator of mRNA degradation in yeast. The EMBO Journal. 2000;**19**(23):6602-6611. DOI: 10.1093/emboj/19.23.6602

[192] Riordan DP, Herschlag D, Brown PO. Identification of RNA recognition elements in the *Saccharomyces cerevisiae* transcriptome. Nucleic Acids Research. 2011;**39**(4):1501-1509. DOI: 10.1093/nar/gkq920

[193] Bernstein D, Hook B, Hajarnavis A, Opperman L, Wickens M. Binding specificity and mRNA targets of a C. elegans PUF protein, FBF-1. RNA. 2005;**11**(4):447-458. DOI: 10.1261/rna.7255805

[194] Wickens M, Bernstein DS, Kimble J, Parker R. A PUF family portrait: 3'UTR regulation as a way of life. Trends in Genetics. 2002;**18**(3):150-157. DOI: 10.1016/S0168-9525(01)02616-6

[195] Campbell ZT, Valley CT, Wickens M. A protein-RNA specificity code enables targeted activation of an endogenous human transcript. Nature Structural & Molecular Biology. 2014;**21**(8):732-738. DOI: 10.1038/nsmb.2847

[196] Zhou W, Melamed D, Banyai G, Meyer C, Tuschl T, Wickens M, et al. Expanding the binding specificity for RNA recognition by a PUF domain. Nature Communications. 2021;**12**:5107. DOI: 10.1038/s41467-021-25433-6

[197] Gupta YK, Nair DT, Wharton RP, Aggarwal AK. Structures of human Pumilio with noncognate RNAs reveal molecular mechanisms for binding promiscuity. Structure. 2008;**16**(4):549-557. DOI: 10.1016/j.str.2008.01.006

[198] Lee CD, Tu BP. Glucose-regulated phosphorylation of the PUF protein Puf3 regulates the translational fate of its bound mRNAs and association with RNA granules. Cell Reports. 2015;**11**(10):1638-1650. DOI: 10.1016/j.celrep.2015.05.014

[199] García-Rodríguez LJ, Gay AC, Pon LA. Puf3p, a Pumilio family

RNA binding protein, localizes to mitochondria and regulates mitochondrial biogenesis and motility in budding yeast. The Journal of Cell Biology. 2007;**176**(2):197-207. DOI: 10.1083/jcb.200606054

[200] Rowe W, Kershaw CJ, Castelli LM, Costello JL, Ashe MP, Grant CM, et al. Puf3p induces translational repression of genes linked to oxidative stress. Nucleic Acids Research. 2014;**42**(2):1026-1041. DOI: 10.1093/nar/gkt948

[201] Chatenay-Lapointe M, Shadel GS. Repression of mitochondrial translation, respiration and a metabolic cycle-regulated gene, SLF1, by the yeast Pumilio-family protein Puf3p. PLoS One. 2011;**6**(5):e20441. DOI: 10.1371/journal.pone.0020441

[202] Gupta I, Clauder-Münster S, Klaus B, Järvelin AI, Aiyar RS, Benes V, et al. Alternative polyadenylation diversifies post-transcriptional regulation by selective RNA-protein interactions. Molecular Systems Biology. 2014;**10**(2):1-11. DOI: 10.1002/msb.135068

[203] Lee D, Ohn T, Chiang Y-C, Quigley G, Yao G, Liu Y, et al. PUF3 acceleration of deadenylation in vivo can operate independently of CCR4 activity, possibly involving effects on the PAB1–mRNP structure. Journal of Molecular Biology. 2010;**399**(4):562-575. DOI: 10.1016/j.jmb.2010.04.034

[204] Wang Z, Sun X, Wee J, Guo X, Gu Z. Novel insights into global translational regulation through Pumilio family RNA-binding protein Puf3p revealed by ribosomal profiling. Current Genetics. 2019;**65**(1):201-212. DOI: 10.1007/s00294-018-0862-4

[205] Bhondeley M, Liu Z. Mitochondrial biogenesis is positively regulated by casein kinase I Hrr25 through phosphorylation of Puf3 in *Saccharomyces cerevisiae*. Genetics. 2020;**215**(2):463-482. DOI: 10.1534/genetics.120.303191

[206] Forsburg SL, Guarente L. Identification and characterization of HAP4: A third component of the CCAAT-bound HAP2/HAP3 heteromer. Genes & Development. 1989;**3**(8):1166-1178. DOI: 10.1101/gad.3.8.1166

Chapter 2

Coherent Gene Assemblies: Example, Yeast Cell Division Cycle, CDC

Lawrence Sirovich

Abstract

A novel approach to the dynamics of gene assemblies is presented. Central concepts are high-value genes; correlated activity; orderly unfolding of gene dynamics; dynamic mode decomposition; DMD unraveling dynamics. This is carried out for the Orlando et al. yeast database. It is shown that the yeast cell division cycle, CDC, only requires a six-dimensional space, formed by three complex temporal modal pairs: (1) a fast clock mother cohort; (2) a slower clock daughter cell cohort; and (3) an unrelated inherent gene expression. A derived set of sixty high-value genes serves as a model for the correlated unfolding of gene activity. Confirmation of this choice comes from an independent database and other considerations. The present analysis leads to a Fourier description, for the very sparsely sampled laboratory data. From this, resolved peak times of gene expression are obtained. This in turn leads to precise times of expression in the unfolding of the CDC genes. The activation of each gene appears as uncoupled dynamics originating in the mother and daughter cohorts, and of different durations. This leads to estimates of the composition of the original laboratory data. A theory-based yeast modeling framework is proposed, and additionally new experiments are suggested.

Keywords: cell division cycle, co-regulated genes, high-value genes, dynamic mode decomposition, low dimensional models

1. Introduction

The blueprint of a life form is contained in its DNA genome, as formed from the four bases, [A, C, G, T], as assembled in the double helix [1]. The genome contains instructions for decoding itself, constructing itself, duplicating itself, and inserting these instructions in the duplicate. This *instruction set* conforms to the Von Neumann [2] vision of an automaton, an *imitation of life*, also see [3]. The genetic code [4] codes base triplet codons for amino acids, the molecular building blocks of polypeptides. Embedded genes, under *transcription*, synthesize mRNA from DNA, followed by *translation*, the synthesis of protein from mRNA, regarded as the *central dogma* of molecular biology [5].

Budding yeast, (Saccharomyces) *S. cerevisiae*, a single cell eukaryote, has been well studied using gene arrays [6–8]. Recently, the use of some novel mathematical methods has been applied in the analysis of the yeast CDC [9]. The present paper significantly refines and extends the previous results. In particular "co-regulated" genes of [8] are considered.

To avoid misunderstanding, the present goal is the dynamic unfolding of gene dynamics in contrast to the dynamics of individual genes, as for example in [10].

2. Yeast

The fate of a budding yeast cell is to divide asymmetrically into a mother and daughter cell. The cell period is in the range of an hour or two. In their pioneering paper Spellman et al. [8] describe several different means by which to assemble a population of quasi-identical daughter cells, for purposes of tracking the dynamics of the CDC. As in [6] elutriation will be used here to assemble a population of mother and daughter cells. Laboratory data monitors 5716 genes by sampling, at 16-min intervals, 15 times, covering roughly 2 cycles of the CDC. The database contains two wild type sets, WT1 and WT2 (referred to here as G1 and G2, and two mutant sets). Each experimental dataset is thus represented by a matrix of 5716 rows and 15 columns, that describes the temporal expression of all genes, denoted by,

$$G = G\left(g_j, t\right) = G(j, t), j = 1, \dots, M, \tag{1}$$

where $M = 5716$, that specifies the expression levels of each gene, g_j, uniformly sampled,

$$t = [dt, 2dt, \dots, Ndt], \tag{2}$$

where $N = 15 \times dt = 16$ min. = 240 min. Presently, there is no generally accepted set of CDC genes, nor is there an accepted framework for describing the mechanics of the CDC. In [8] the concept of co-regulated genes is proposed as a framework by which to explore related genes of the CDC. One goal of the present study is to shed light on these concepts, solely based on data.

For our purposes, instead of (1) the mean subtracted form,

$$G_S = G - \overline{G}, \overline{G} = \langle G \rangle_t, \tag{3}$$

is an effective beginning. G_S is also composed of 15 time snapshots of the 5716 genes. (Since $\langle G_S \rangle_t = 0$, 14 is the correct figure.) The *method of snapshots* [11] demonstrates that an exact treatment of G_S requires no more than 15 dimensions, as demonstrated by the SVD [12], of G_S, (3),

$$G_S = U \Lambda V^{\dagger}, \tag{4}$$

where $U = {}_{5716}U^{15}$; $V = {}_{5716}V^{15}$, and Λ is the 15×15 diagonal matrix of singular values. Note that (4) is exact. (The 15×15 matrix $G_S^{\dagger} G_S$ generatesΛ and V, from which U follows by elementary considerations. The plot below is a log-log plot of the spectrum of variances (squares of singular values).)

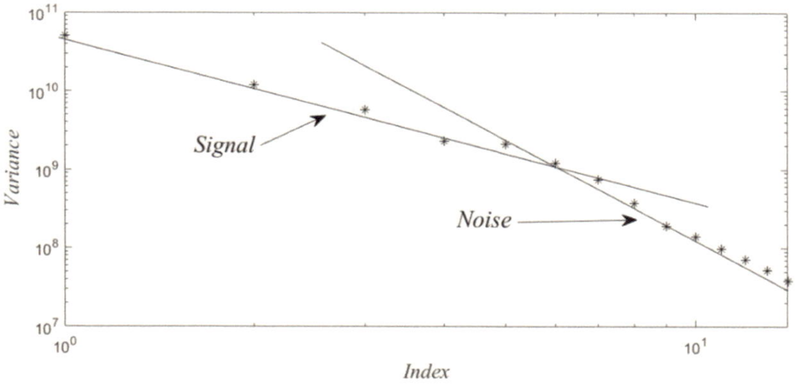

Figure 1.
Log-log plots of the variances reveal that these follow two different power laws, which by routine arguments can be associated with signal or noise.

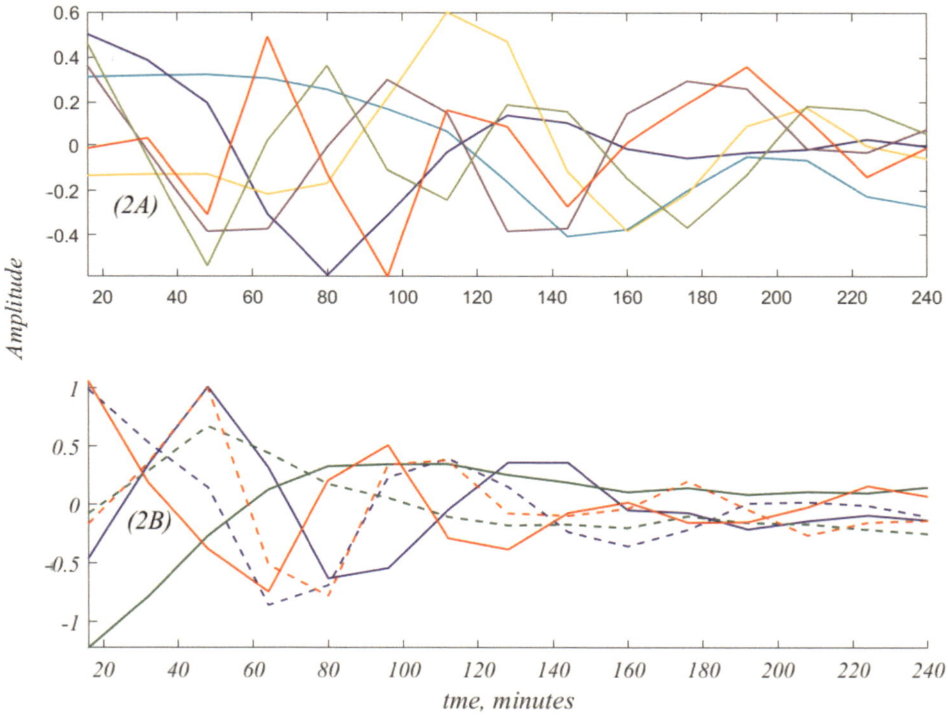

Figure 2.
The six time histories of V_t *are shown in (A). The true underlying dynamics of the data is shown in (B), as obtained by dynamic mode decomposition, DMD [13], also see [14] and Section 8. This is composed of three complex modes, the real and imaginary parts of which, are depicted by the three colored pairs in (B). Further details are given below.*

One can reasonably conclude from **Figure 1**, that the analysis can be reduced to the six modes, ahead of the *knee*, or intersection. Henceforth, we consider the induced six-dimensional representation,

$$G_S = U\Lambda V_t, \tag{5}$$

where V_t refer to the 6×15 time courses that result from SVD as plotted in **Figure 2(A)**.

The plots of **Figure 2(A)** may be regarded as artifactual consequences of SVD, that only hint at the true dynamics. The goal of DMD is to recast the data in terms of the true exponential frequencies that are latent in the data. Thus, the curves of **Figure 2(A)** are entangled versions of the true dynamics, shown in **Figure 2(B)**. In Section 8 we obtain the 6×6 matrix, E, that disentangles V_t, into the three complex modes shown in **Figure 2(B)**. For the present, note that the decomposition

$$G_S = (U\Lambda E)\left(E^{-1}V^{\dagger}\right), \tag{6}$$

provides a generalization of SVD, (4), that is appropriate for data that have a rational dynamic ordering of columns.

3. High value genes

One can reasonably anticipate two limiting forms of gene expression; steady expression, as might be the case, for the proteins that form the cell wall and membrane; and a briefer activation and later inactivation on a shorter timescale, as might be the case for formation of the cell nucleus and its components. As a criterion for distinguishing these two limits consider the coefficient of variation,

$$C_V = \sigma_G/\overline{G}, \tag{7}$$

where σ_G denotes the standard deviation of the time history of G, and $\overline{G}$, the mean value. For example, there are more than 1500 genes for which $C_V < 0.1$ and over 1100 for $C_V > 0.25$. In **Table 1**, the second line specifies six known CDC genes [6], with their coefficient of variation in the line above. Large C_V suggests that gene expression is due to transitory mechanisms, as will be explained below.

Co-regulation implies correlated activity. The fourth line of the table shows genes highly correlated to those of the second row, shown on the third line, to the genes of the second line. The implication of the Table is that the genes of the fourth line are better gene representatives. Peak times of two like genes are virtually identical.

In general, any gene can be well correlated to many genes. This is illustrated in the next figure the for genes correlated with GR108W, with gene names and coefficients of variation in the legend. YGL021W with a C_V = 0.7222 the best exemplar of this set of highly correlated genes.

C_V	0.1845	0.7216	0.3936	0.2337	0.6861	0.2903
Gene	YDL155W	YGR108W	YGR109C	YLR210W	YPR119W	YPR120C
CorrCoef	0.9843	0.9878	1	1	0.9950	0.9864
Gene	YMR144W	YGL021W	YGR109C	YLR210W	YGL021W	YJL181W
C_V	0.3475	0.7222	0.3936	0.2337	0.7222	0.3128

Table 1.
Six known CDC genes [6], line 2. Their coefficients of variation, shown on line 1. Each are highly correlated, line 3, to the genes of line 4; each of which have the higher coefficients of variation, shown on line 5.

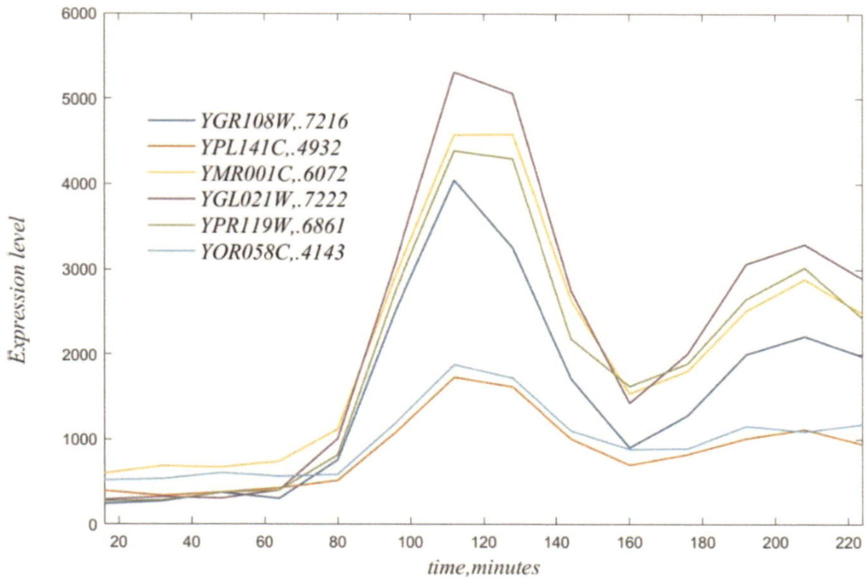

Figure 3.
Time courses of genes highly correlated to YGR108W, see legend. Note that each of these has a peak at t = 112, *based on the experimental sampling.*

As is clear from **Figure 3**, peak locations, ipso facto, must occur at sampling locations, but would be better resolved by interpolation.

3.1 Gene selection

As mentioned above there are 1192 genes for which $C_V > 0.25$. These will be regarded as a starting point for selecting high value genes. A large number of traces exhibit pure exponential decay, starting at extremely high expression values, and are regarded as artefactual. Thus, a second criterion is restriction to time traces that start with relatively small expression, as is the case in **Figure 3**. For example, a restriction to initial value of a gene at 16 min, of <450, results in a well-correlated set of 109 genes. **Figure 4** shows the correlation image based on the correlation criterion, $\rho > .85$ for the chosen set of 109 genes, in given order.

The above figure is based on $r = 6$ and the 109 high-value genes. The SVD of the derived G_S for this case has the form,

$$G^*{}_S = U^* \Lambda^* V_t{}^*, \tag{8}$$

where U^* is 109×6, $\Lambda^* = 6 \times 6$ and $V_t^* = 6 \times 15$. A straightforward calculation shows that this captures more than 99% of the variance of G_S. **Figure 2** is also based on these choices.

4. Analysis of the high value genes

The DMD analysis of G^* produces three complex modes with complex frequencies,

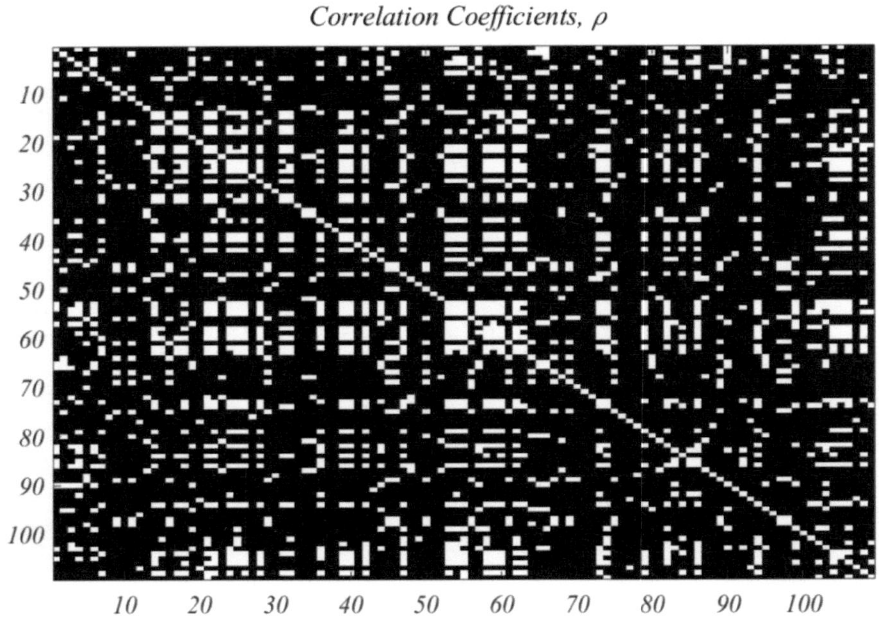

Figure 4.
Correlation coefficient of the 109 selected genes, with correlation coefficient, ρ, greater than .85 shown as white pixels.

$$[\Omega_1, \Omega_2, \Omega_3] \approx [-.018 \pm .032i; -.022 \pm 0.090\,6i; -.013 \pm .069i]. \quad (9)$$

Ω_3 is identified with daughter cells, with *period* $T_3 = 2\pi/\Omega_3 \approx 92$ min; Ω_2 is identified with mother cells, with period $T_2 = 2\pi/\Omega_2 \approx 71\,min$. (Smaller daughter cells are well known to take longer to mature.) Depending upon the complexion of the gene sets chosen, there is a range of values for these two periods, however the big picture remains the same. The first period, T_1, is roughly 200 min, close to the duration of the experiment. The time courses following from DMD, and take the form,

$$V_t \approx E \exp(\Omega t) E^{-1} V_0, \quad (10)$$

where V_o is the initial value, and E is derived from the data, and t is given by (2). The exponential matrix in (10) is given by

$$\exp(\Omega t) = \begin{bmatrix} \exp(\Omega_1 t) & 0 & 0 \\ 0 & \exp(\Omega_2 t) & 0 \\ 0 & 0 & \exp(\Omega_3 t) \end{bmatrix}. \quad (11)$$

This form speaks volumes. Since each $\exp(\Omega t)$ is a diagonal matrix, these entries are pure exponentials, and thus from (10) $E^{-1}V_t$ disentangles the modes into exponentials. Support for this assertion is shown in **Figure 2**. On this basis, for data sets of a dynamical type, SVD is generalized by

$$G - \overline{G} = G_S = (U\Lambda E) \times \left(E^{-1} V_t\right) \quad (12)$$

where $\left(E^{-1}V_t\right)$, produces un-entangled temporal modes, with amplitudes, $(U\Lambda E)$.

And enduring puzzle of SVD analysis has been the origin of the time courses of V. This was solved by Schmid [13] with his discovery of DMD, as discussed above. This observation applies, generally, to data of dynamical type, and in particular for our data with the result,

$$G_s = (U\Lambda E)\left(\exp(\Omega t)E^{-1}V_0.\right) \tag{13}$$

4.1 Unfolding the CDC

The goal is acquisition of a data-based model of the CDC, under the assumption that CDC is a temporal unfolding of genes, with defined activation times. As is clear from **Figure 3**, time resolution is limited by course sample times. The exponential representation of (13) induces a natural Fourier representation that overcomes this limitation. From the calculated frequencies, (9), we can *sample* every minute, including the first 16 min. In **Figure 5**, the smooth version of the indicated gene is shown as a dashed curve, and compared with the disentangled experimental modes, (12), in order to convey a sense of the Fourier form, as well as the improved peak locations.

Inspection of the 109 highly sampled genes reveals that 43 take on negative values, and 8 have a peak at $t = 1$. Both conditions are regarded as unacceptable and the corresponding genes will be dropped from consideration. The result is a set of 60 genes that have peak expression times, $\{T_j\}$, shown in **Figure 6**.

In **Figure 7**, the left image shows the trajectory of gene expression for peak times arranged in ascending order. This compares favorably with the phase plots that appear in [6]. At the right is the comparable plot for WT2 under the same gene ordering.

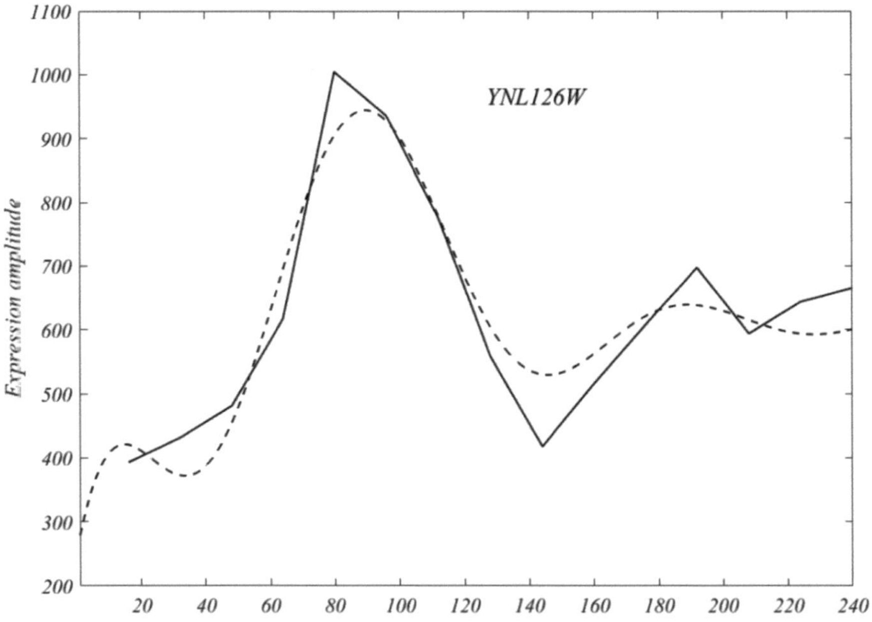

Figure 5.
The dashed curve represents the Fourier interpolated version of the original polygonal data of the gene YNL126W, and is the first of the 109 genes. Note the interpolation covers the initial 16 min, and also runs over the second attenuated cycle.

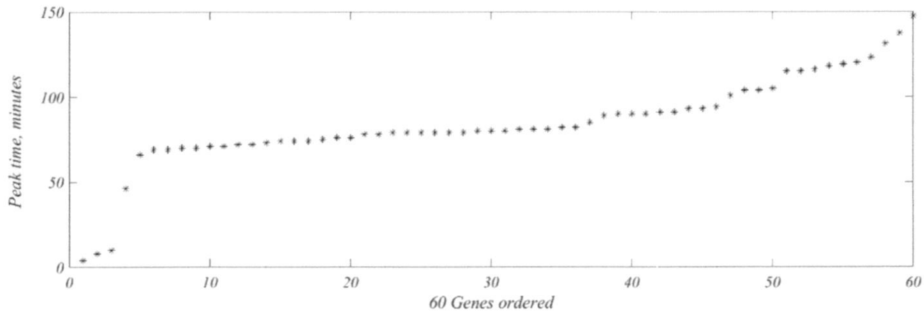

Figure 6.
Times, $\{T_j\}$, of peak appearance, y-axis, of the 60 choice genes.

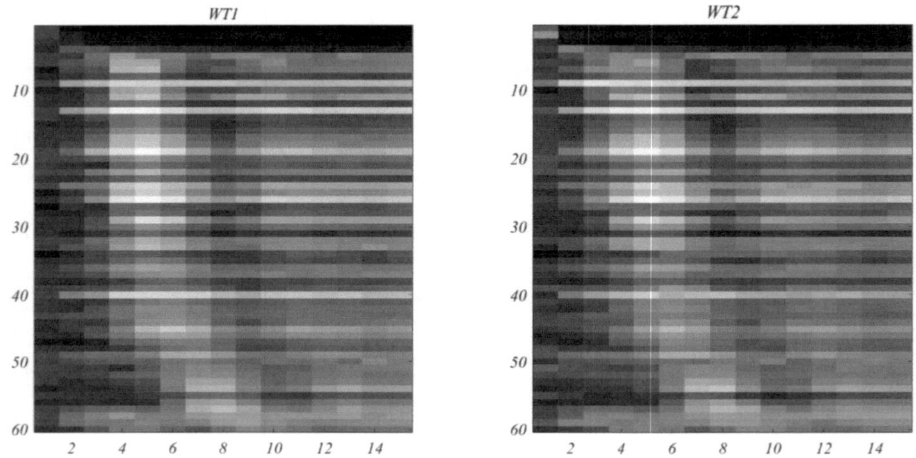

Figure 7.
Track of mRNA expression of the 60 selected genes, G^{60}_{ord}, arranged in ascending order of peak activation time. As in Orlando et al. [6], a log transformation has been applied to enhance the image.

It should be noted that the Orlando et al. plots are based on their 440 *consensus* genes. That set contains seven of the present high-value 60 genes, while the co-regulated set of 800 genes in [8], contain 42 of the 60 high-value genes. The table of 60 high-value genes, and their properties appear in the Supplement.

5. Coherent gene sets

Here, *co-regulated* implies *well-correlated*, and we next construct the temporal correlation coefficients. For this purpose, the genes will be ordered as they unfold. This arrangement will be denoted by G^{60}_{ord}. The correlation coefficients of G^{60}_{ord}, are shown at the left of **Figure 8**.

The figure on the right is the result of the same calculation, based on the selected ordering applied to WT2. This provides a compelling demonstration that the 60 choice genes are "strongly co-regulated", in general. A much larger set, 413 genes, similarly constructed, produces an intersection of 204 genes with the Spellman co-regulated set of 800.

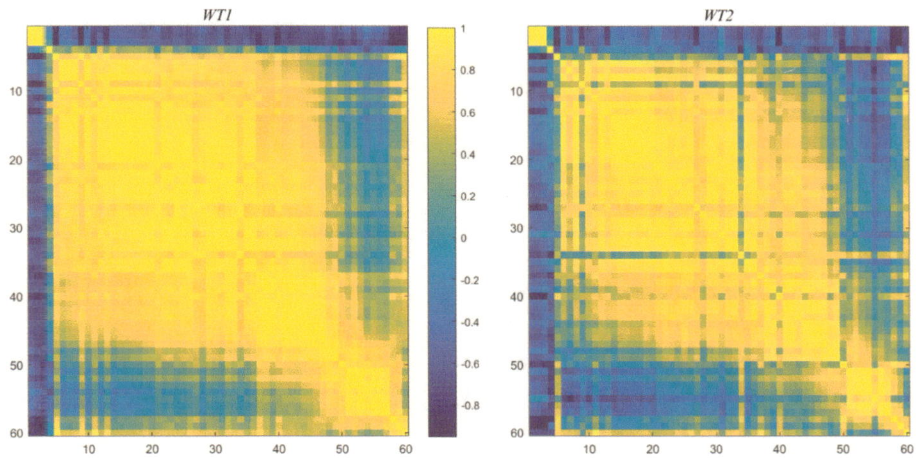

Figure 8.
Correlation coefficients, ρ, of G^{60}_{ord}, as time ordered, for WT1 and WT2.

Unfolding times is regarded as a reasonable hypothesis for gene ordering; though other possibilities may be considered. For example, ordering the genes in terms of descending correlations, ρ, or equivalently increasing temporal distance of time histories. This leads to heat plots and dendrograms, none of which were deemed productive.

6. The single yeast cell

Figure 3, exhibits a typical gene time course, and displays single gene expression duration over many tens of minutes. However, the accepted estimate for the duration of gene transcription and translation is 1–2 min [15], and for convenience this value is taken to be 1 min in the calculations performed below. To explain what might appear to be an inconsistency, we review the data acquisition procedure, and as will be seen is due to the different maturation periods of mother and daughter cells, and randomness.

In experiments, after assembly of a suitable pool of yeast cells, aliquot removals along with genetic snapshots are obtained at 16 min intervals, and repeated 15 times. The result is the report of mRNA expression for each gene at each sampling instant. According to [6] each sampling, contains more than 200 yeast cells. To obtain a sense of the process consider gene YGR174-A, the earliest activated gene, of the 60 gene set, as deconstructed in **Figure 9**.

The mean subtracted form of this genes is denoted by g_S; DMD produces the four traces related by,

$$g_S - i = m + d, \tag{14}$$

where i, m, d represent the inherent, mother, daughter time courses.

The mother cohort, m, has a peak of 90 at t = 1, of period 71 min; and daughter cohort, d, a peak of 45 at 16 min and has period 92 min. The two entries in the left-hand side of (14) are defined over the full interval, and their difference has a peak of 115 at 5 min. If we denote by N_M and N_D the unknown number of mother and

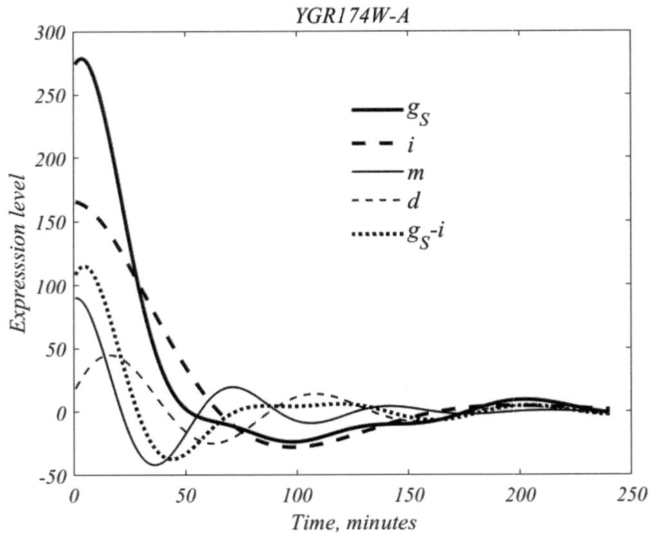

Figure 9.
The dynamical composition of gene YGR174W-A, *the earliest activated gene. See text.*

daughter cells that are participating in the signal, then straightforward back of the envelope calculation, based on peak quantities and locations, shows that $N_M/N_D \sim .285$, from which the population fraction,

$$f = N_M/(N_D + N_M) \approx .22, \tag{15}$$

follows. Thus, there are more than three times as many daughter cells as mother cells in an average aliquot.

For formal purposes the inherent signal and the mean will be divided into 2 parts, as follows,

$$\begin{aligned} Mo &= m + f \times (i + g_m) \\ Da &= d + (1 - f) \times (i + g_m), \end{aligned} \tag{16}$$

so that the total signal is equals $Mo + Da$.

where mother, daughter, inherent and mean our shown in **Figure 9**. As shown in **Figure 10**, the full signal this gene equals $Mo + Da$.

To summarize, genes are expressed both in a background manner, inherent expression, and in a scheduled manner, to peak at some specific time. It is also clear from this analysis that the activities of mother and daughter cells are not coupled.

6.1 A yeast model

Next, we consider a proposed computational model of the yeast cell. For this we focus on the gene traces displayed in **Figure 10**. While additional genes might be included in the model, the uncertainties in experimental results [15, 16] do not justify such generalizations. Our purpose in this exercise is merely to demonstrates that a practical framework can be created.

To start, it is noted that estimates of protein molecules per yeast cell are in the range of $\sim 5 \times 10^7$ [15]. Since there are roughly 6000 genes in the yeast cell under

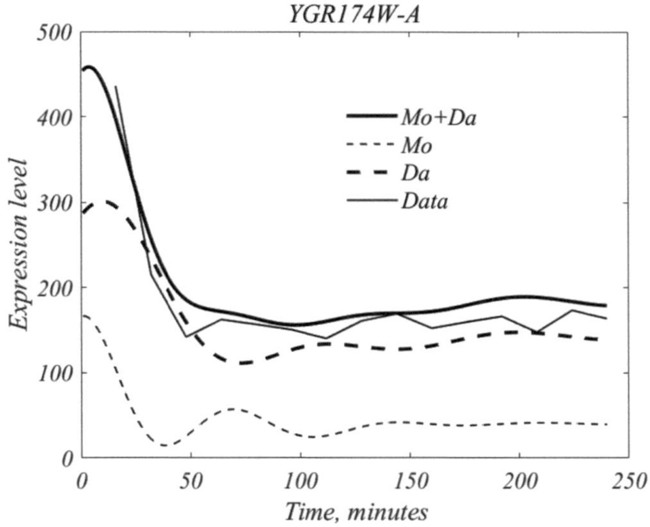

Figure 10.
Mother cell dynamics are initiated at roughly t = 0*, and daughter cell dynamics at roughly* 10 *min, with each being responsible for a steady production at large* t*. The two add to yield the overall gene activity shown in the heavy curve.*

consideration, the estimated number of such molecules per gene is $\sim 5 \times 10^7/6000 \approx 8.3 \times 10^3$. This is a *ballpark* figure, and in this spirit other considerations such as protein degradation will be ignored. Further, the duration of proteins expression will be taken as 1 min. If we denote the mother and daughter dynamics of **Figure 10** by $m(t)\&d(t)$, respectively, the number of proteins at maturation for mother and daughter cells is estimated from by,

$$
\begin{aligned}
p_M &= \sum_{i=1}^{71} Mo(t_i) \approx 4.53 \times 10^3, \\
p_D &= \sum_{i=1}^{92} Da(t_i) \approx 1.75 \times 10^4.
\end{aligned} \tag{17}
$$

The ratio p_D/p_M = .26 is remarkably close to the above mentioned $N_M/N_D \sim .285$. Since these estimates are in the range of the ballpark gene estimate of 8.3×10^3, it is surmised that the Orlando et al. [6] data was normalized by the estimated number of cells in an aliquots, >200. These considerations also suggest that cell size differences of mother and daughter cells plays little or no role in protein content at maturation.

On the basis of these deliberations one might contemplate creating an algorithmic model of the CDC. Randomization can be introduced through variations in mother and daughter CDC periods, and variations in the number of mother and daughter cells, say adding up to roughly 200. This is a future project, which can useful only with better knowledge and precision of the quantities involved.

7. Additional comments

The high degree of correlation, seen in **Figure 6** tells little beyond timing. For example, it does not imply anything certain about gene interactions, nor is there any information about the activation and deactivation times of genes. The mechanism by which budding yeast cell assembles itself, is an open question. Since no outside intervention is in play, it is noncontroversial to presume, that the cell self-assembles. Just how this self-assembly takes place is another open question, it might e.g. only be a matter of proteins falling into their proper place, based on the timing of gene expression. In this case the cell model is an *assembly line* for proteins that arrive in an orderly fashion.

In an effort to introduce some additional theory it is noted activation and inactivation of gene expression may be likened to an equilibrium disturbance, followed by restoration, (suggestive of wave phenomena.). In this connection it is noted that the mother and daughter modes that describe the CDC each follow oscillator dynamics [9]

$$\frac{d^2g}{dt^2} + 2\lambda\frac{dg}{dt} + \left(\lambda^2 + \omega^2\right)g = 0, \tag{18}$$

which has solutions in the form,

$$g \propto \exp[(-\lambda \pm i\omega)t]. \tag{19}$$

In this connection, we can define a new variable, v, by,

$$\frac{\partial g}{\partial t} = \frac{\omega}{2\pi}v, \tag{20}$$

which since $2\pi/\omega$ gives the period of mRNA, might be reasonably associated with rate of protein production, and from (18) v is governed by

$$\frac{\partial v}{\partial t} + 2\lambda v + \left(\lambda^2 + \omega^2\right)\frac{2\pi g}{\omega} = 0. \tag{21}$$

The pair of Eqs. (20) and (21) may be viewed as a coarse-grained version of the *central dogma.*mentioned the dynamics of transcription and translation occurs on a scale of ~1 min, or so. As dictated by the experiments, our deliberations are based on timescales large compared to 1 min, and thus, only genes and their proteins can figure in the description. This is consistent with "one gene-one polypeptide" view of [17]. Impressionistically, this Beadle-Tatum view is a macroscopic description.

A key result of the present investigation is the remarkable ability of DMD, to distinguish dynamic characteristics of the mother and daughter cohorts. Gene experiments typically attempt to sequester daughter cohorts, and it is natural on the basis of the methods pursued here to consider what the outcome would have been of considering complementary populations, and furthermore to monitor the base population, without any form of sequestering. Given the remarkable ability of DMD to parse dynamic activity more complicated population yeast populations might be considered. Hopefully, this ability to distinguish yeast subpopulations will lead to new ways to probe into the yeast life form. Another future goal, is that the algorithmic model touched on here can be further advanced, since a falsifiable model is always desirable.

8. Methods

8.1 Dynamic mode decomposition

Traditional signal analysis is based on the hypothesis that a *signal* is an admixture of sinusoids, and therefore that Fourier analysis can decompose the signal into his

Fourier components. DMD may be regarded as an extension of this idea. Suppose for real λ & ω, we define a complex frequency,

$$z = \lambda + i\omega, \tag{22}$$

with corresponding complex signal,

$$s \sim \alpha \exp(zt) = \alpha \exp(\lambda t) \cdot [\cos \omega t + i \sin \omega t]. \tag{23}$$

DMD can be regarded as a method for extracting such a signal, and more generally, admixtures of such signals. In this respect Fourier analysis is a subset of *DMD*.

Typically, a laboratory signal is a uniformly sampled version of the continuous case. Suppose for example the uniformly sampled times are,

$$t = [dt, 2dt, \cdots, Ndt]; \tag{24}$$

in which case (23) becomes the geometric sequence.

$$S = [s_1, \ldots, s_N] = \left[\delta^1, \ldots, \delta^N\right]\alpha \tag{25}$$

where

$$\delta = \exp(\lambda \cdot dt), \tag{26}$$

which is therefore the generator of the sampled signal. Reciprocally, if $\{s_j\}$ is (noisy) data, we can seek a generator, *A*, such that

$$s_{j+1} = As_j, j = 1, \ldots, N-1, \tag{27}$$

which can be viewed as a case of many equations for the one unknown, *A*. Least squares is the suggested approach in this case. The error in applying (27) is

$$Er = \sum_{k=1}^{N-1} \|s_{k+1} - As_k\|^2, \tag{28}$$

and the (least-squares) minimization of (28) produces the solution

$$A = \sum_{k=1}^{N-1} s_{k+1}\theta_k; \theta_k = s_k / \sum_{k=1}^{N-1} s_k^2. \tag{29}$$

A more compact form is obtained by defining the vectors,

$$S1 = [s_1, \ldots, s_{N-1}] \& S2 = [s_2, \ldots, s_N] \tag{30}$$

in which terms (29) can be written as

$$\theta = S1/\|S1\|^2.A = (S2, \theta). \tag{31}$$

for later purposes observe that if the problem is posed as,

$$A \times S1 = S2, \tag{32}$$

then it is solved by

$$A = S2 \times S1^{+}, \tag{33}$$

where $S1^{+}$ is referred to as the Moore-Penrose inverse [18]. Clearly, $S1^{+} = \theta$, and (33) is equivalent to minimizing (28). Briefly stated, the Moore Penrose inverse has the smallest norm of the many solutions that solve (32).

In the general case, of multiple complex signals, we are confronted by a matrix $V(t)$,

$$V_{ij}, i = 1, \dots, M; j = 1, \dots, N, \tag{34}$$

with M time histories, each sampled uniformly N times. For example V_t^* (8) is 6×15. Under the hypothesis that the data is composed of complex exponentials, one can seek the generator matrix, A by defining

$$\begin{aligned} V1 &= V(i, 1 \le j \le N-1) \\ V2 &= V(i, 2 \le j \le N) \end{aligned} \tag{35}$$

so that

$$AV1 = V2 \tag{36}$$

which is solved by the Moore-Penrose inverse, $V1^{+}$.

$$A = V2 \times V1^{+}. \tag{37}$$

The spectral decomposition of the generator matrix, A,

$$A = E\Sigma(t)E^{-1}, \tag{38}$$

produce complex frequencies as eigenvalues of the diagonal matrix Σ, and that the corresponding eigenvectors in E disentangle the time courses produced by SVD. Evidence supporting both assertions appear in the data analysis. The construction (38) permits generalization of SVD, (6). If we write

$$\Sigma(t) = \begin{pmatrix} \sigma_1(t) & 0 & 0 \\ 0 & \ddots & 0 \\ 0 & 0 & \sigma_n(t) \end{pmatrix}, \tag{39}$$

where n = 6 in the yeast application, then it proves to be convenient to transform to the hypothesized exponential form

$$\Omega = -\log \sigma / dt, \tag{40}$$

which is just the inverse of (26). dt is the sampling interval (2) and (39) transforms to (11).

In the interest of brevity, we forgo examples. Consideration of synthetic data generated by,

$$D(\Omega, M) = \sum_{j=1}^{m} \exp(\lambda_j t)\left(\alpha_{ij}\cos(\omega_j t) + \beta_{ij}\sin(\omega_j t)\right); i = 1, 2\ldots, M, \tag{41}$$

with each of the M trials an admixture of m complex signals and randomly chosen coefficients shows a remarkable accuracy in recovering the frequency content after relatively few trials.

9. Conclusions

This paper represents a substantial extension of an earlier preliminary analysis [9] of the high quality yeast database assembled by Orlando et al. [6].

The principal focus of this paper is the introduction of advanced mathematical methods that should be useful under a wider set of circumstances. For example, for general dynamical molecular biological data sets, and for when better resolved data becomes available for *S. cerevisiae* studies. Since the Orlando data is sampled at a large fraction of the CDC cycle, and near the Nyquist limit [19], the data is broadly speaking, and out of necessity, under-sampled. For this reason and for purposes of exposition the Orlando data has been converted to interpolated version, by Fourier methods to get a grip on what might potentially t be achieved with better sampled data, see **Figure 5**. This leads to a description of the CDC in terms of precise times of gene activation and deactivation, which are likely only a semblance of the true times. Another focus has been on the concept of co-regulation, as introduced by Spellman et al. [3], which in the present analysis is reformulated as obtaining gene collections that are highly correlated with one another. Within this very approximate framework a set of roughly 100 genes is proposed as being central to the CDC, and which again is only a semblance of what truly are the *co-regulated* genes.

The Singular Value Decomposition, SVD, is the chosen mathematical framework for dealing with the dynamical structure of the yeast data and is believed likely to play an important role in examining dynamical biological data of a general nature. A shortcoming of SVD is that the dynamics generated by SVD is severely constrained by the underlying methodology of SVD. This shortcoming was repaired by Schmidt [13], by a method that is termed dynamic mode decomposition, DMD. This is fully treated in Section 8 of this paper, in particular see (13). In brief, the result is that the CDC is well approximated by a single mode that depicts the dynamics in terms of timescales representative of mother cells, and daughter cells.

It is an opinion that future more highly sampled data lead to the same qualitative description that is more refined, and accurate.

Author details

Lawrence Sirovich
Division of Applied Mathematics, Brown University, Providence, Rhode Island, USA

*Address all correspondence to: lawrence_sirovich@brown.edu

References

[1] Watson JD, Crick FH. Molecular structure of nucleic acids: A structure for deoxyribose nucleic acid. Nature. (Wiley, New York). 1953;**171**(4356):737-738

[2] Neumann JV. The General and Logical Theory of Automata. Vol. 1951. New York: Wiley; 1951. pp. 1-41

[3] Schrödinger E. What is Life?: With Mind and Matter and Autobiographical Sketches. Cambridge University Press; 1992

[4] Nirenberg MW, Matthaei JH. The dependence of cell-free protein synthesis in *E. coli* upon naturally occurring or synthetic polyribonucleotides. Proceedings of the National Academy of Sciences. 1961;**47**(10):1588-1602

[5] Crick F. Central dogma of molecular biology. Nature. 1970;**227**(5258):561-563

[6] Orlando DA et al. Global control of cell-cycle transcription by coupled CDK and network oscillators. Nature. 2008; **453**(7197):944-947

[7] Cho RJ et al. A genome-wide transcriptional analysis of the mitotic cell cycle. Molecular Cell. 1998;**2**(1):65-73

[8] Spellman PT et al. Comprehensive identification of cell cycle–regulated genes of the yeast *Saccharomyces cerevisiae* by microarray hybridization. Molecular Biology of the Cell. 1998; **9**(12):3273-3297

[9] Sirovich L. A novel analysis of gene array data: Yeast cell cycle. Biology Methods and Protocols. 2020;**5**(1)1-10

[10] Tyson JJ. Modeling the cell division cycle: cdc2 and cyclin interactions. Proceedings of the National Academy of Sciences. 1991;**88**(16):7328-7332

[11] Sirovich L. Turbulence and the dynamics of coherent structures. I. Coherent structures. Quarterly of Applied Mathematics. 1987;**45**(3): 561-571

[12] Lax PD. Linear Algebra and its Applications. New York: Wiley; 2007. p. 2007

[13] Schmid PJ. Dynamic mode decomposition of numerical and experimental data. Journal of Fluid Mechanics. 2010;**656**:5-28

[14] Kutz JN et al. Dynamic Mode Decomposition: Data-Driven Modeling of Complex Systems. Philadelphia: SIAM; 2016

[15] Milo R, Phillips R. Cell Biology by the Numbers. New York: Garland Science; 2015

[16] Gerstein MB et al. What is a gene, post-ENCODE? History and updated definition. Genome Research. 2007; **17**(6):669-681

[17] Beadle GW, Tatum EL. Genetic control of biochemical reactions in Neurospora. Proceedings of the National Academy of Sciences of the United States of America. 1941;**27**(11):499

[18] Golub GH, Van Loan CF. Matrix Computation. 1989. Johns Hopkins. Baltimore, MD: University Press; 1989

[19] Nyquist H. Certain topics in telegraph transmission theory. Transactions of the American Institute of Electrical Engineers. 1928;**47**(2):617-644

Chapter 3

Yeast Activation Methods Used in Fermentation Industries

Dmitry Karpenko and Artem Grishin

Abstract

The reasons why it is practically impossible to maintain optimal conditions for the development of cultural yeast populations under production conditions are briefly substantiated. A simplified classification of yeast activation methods is given: chemical, physical, and combined. In each of the mentioned groups, the varieties of the proposed technological methods and the modes of their implementation are considered. Experimental data obtained in recent years on the influence of the sound in the audible range (20–20,000 Hz) and light in the visible range on the development of Saccharomyces cerevisiae yeast used in brewing are presented. An attempt made to compare the effectiveness of various ways to improve technological indicators: the increase in the total titer of cells, the percentage of nonviable cells, the accumulation of ethyl alcohol.

Keywords: *Saccharomyces cerevisiae*, fermentation industries, chemical, physical and combined activation methods, effectiveness of activation methods

1. Introduction

A wide range of productions, including food production, are based on the growth of microbial populations, such as yeast, among others. A technologist in a production of this nature should be seeking to culture a yeast population as intensively as possible. The aim is to obtain biomass or metabolism products within the shortest possible period of time with the maximum possible utilization of nutrients within the cultivation medium.

The most logical approach to address this issue would be to maintain ideal conditions for microorganisms, specifically for cultivating yeast of a particular genus and species [1]. However, achieving this objective is often unfeasible for various reasons. One clear example is the production of bottom-fermentated beer. In traditional brewing techniques, brewers cultivate a yeast population at the stage of fermentation under highly unfavorable conditions for this yeast type [2]. The process begins with a high osmotic pressure and ends with a relatively high ethanol concentration. The temperature remains too low, and the physiological condition of the inoculated yeast is suboptimal, especially in the case of pure-culture yeast or late-generations seed yeast. Furthermore, despite the technologist's effort, the composition of the growth medium can quite often be unbalanced, with low-molecular nitrogenous compounds typically being a limiting factor.

In contemporary conditions, the growth medium may also contain abiotic substances such as heavy metals, pesticides, and radionuclides that have entered the process flow from the raw materials, although their content in the latter is restricted and monitored. These factors, either individually or collectively, create suboptimal conditions for yeast population growth. Consequently, the activation of the process through additional processing methods becomes desirable or even necessary. Considerable attention has been directed toward the development of such methods. This chapter aims to provide a concise overview of these techniques (addition of activators of various natures; removal of undesirable components from a nutrient medium; physical processing of inoculum yeast, wort, or fermentation medium; or a combination of these methods) and, if possible, to compare their efficacy. It should be noted that we have intentionally excluded from the discussion any approach that relies based on enhancing the characteristics of yeast through genetic manipulation [3].

A comprehensive classification system [4, 5] for methods aimed at enhancing yeast activity has been proposed. This classification is based on various criteria, including teleological, genetic, and technological aspects. Additionally, it was suggested distinguishing activating agents based on their chemical properties, composition, origin, production stage, and intended final objectives. Nevertheless, we believe that even a complex classification fails to take into account every difference. Therefore, we find it more reasonable to categorize all yeast activation methods, whether proposed or implemented on an industrial scale, into three main groups:

- Chemical methods: These involve the addition of specific agents and substances to the food medium, or conversely, the removal of unwanted components from the food medium.

- Physical methods: These rely on the utilization of specialized equipment, which does not form a part of the traditional hardware circuit typically used for the given process.

- Combination of the above (combined methods).

The information presented in this chapter is organized in accordance with this subdivision.

2. Chemical methods of activation

Agents employed as chemical yeast activators can be categorized based on their intended functions. These categories encompass the following:

Deficient (limiting) compounds: These agents are employed to supplement the medium with essential compounds that may be lacking for yeast growth.

Biostimulators promoting yeast activity.

Extraction of unwanted components: Specific agents are used to remove undesirable components from the medium that could hinder yeast development.

Stressors.

Antibacterial agents: These agents are introduced to create an environment conductive to the growth of the main yeast culture.

Biopolymer degraders: Certain agents are employed to break down biopolymers found in raw and semifinished products.

Adjustments to the medium composition can be made at various stages - from the preparation of the nutrient medium to the stage of cultivation of the yeast population. These additions or agents may vary in terms of their chemical nature, falling into categories such as organic, nonorganic, or mixed (complex). They can be obtained through chemical synthesis, microbiological processes, or have a natural origin. The proposed classification system enables a targeted approach in selecting agents capable of altering yeast culture metabolic activity by adjusting the composition of the culture media [4]. Below, we provide examples of agents belonging to the discussed groups.

In order to fulfill the yeast's requirements for microelements and vitamins, various specialized preparations are currently in widespread use ("yeast feedings", "yeast nutrition"). These preparations came in both single-component and multicomponent forms, containing amino acids, vitamins, and minerals: zinc sulfate, "Istex," "Eastfield," "Hi Vit," "Eastfood," "Alkoten," "Rhodium Zumesit," and others [6]. The utilization of these compounds accelerates the start of wort fermentation, prevents fermentation from slowing down and halting, reduces fermentation duration, fosters thorough fermentation of wort sugars, increases yeast growth and resistance to autolysis, and, in some instances, assists in reducing diacetyl content in beer. Typically, these feedings are added to the wort before fermentation. However, it is important to note that the composition of these feedings includes inorganic substances as mineral components (such as diammonium phosphate, manganese and zinc sulfates, and potassium metabisulfite), which are foreign to the food product and considered undesirable from a hygiene standpoint.

A study of the chemical composition of sweet potatoes revealed the presence of a complex of biologically active substances. In order to increase their yield, the use of enzyme preparations became necessary. A biologically active additive, derived from sweet potatoes, was employed as a fermentation activator, exhibiting a positive impact on brewer's yeast and beer quality [7]. These dietary supplements were added to the wort during the fermentation stage, at concentration of 1% and 2%. Fermentation was carried out at temperatures ranging from 5 to 8° C. The research findings indicate that the use of dietary supplements in beer production positively influences several parameters of young beer, including the degree of fermentation (increased by 5.8%) and ethanol content (elevated by 3.6%). Furthermore, it was observed that the experimental sample, utilizing the sweet potato preparation, outperformed the control sample by 44% in terms of yeast biomass growth. The experiment results demonstrated that the introduction of the sweet potato preparation into the wort accelerates the fermentation process by 2–3 days compared to the control sample. Moreover, the use of this biologically active sweet potato additive during fermentation led to a more significant reduction of diacetyl, with the control and experimental beer samples having diacetyl content of 0.14 and 0.09 mg/dm^3, respectively.

The organs of the Far Eastern wild plant *Aralia mandshurica* were chosen as a biostimulant for *Saccharomyces cerevisiae* yeast 34/70 [8]. Alcohol-containing extracts of aralia (derived from roots, branches, or leaves) were introduced after the main fermentation, before the start of 16 days post-fermentation period. This increased the physiological activity of the yeast, as assessed by the number of physiologically active yeast cells containing glycogen. In the control sample, there was a gradual decline in the number of physiologically active cells containing glycogen (from 47 to 15%). This decline is explaining by the natural, modest reduction in the functional (fermentation) activity of yeast during the beer fermentation process. In the experimental samples with *Aralia* extracts, the initial number of glycogen-containing cells was lower compared to that in the control sample, averaging around 33%. According to the authors, this can be attributed to a certain "shock" experienced by microorganisms due to the ethanol content in *Aralia* extracts. However, in the days following fermentation, the number of viable cells in all experimental samples increased by the 8th to 10th day, surpassing the control sample by an average of 10% by the end of the experiment. The most significant impact in terms of the number of glycogen-containing microorganisms was observed in the experiment using an extract from *Aralia* leaves.

The study examined the results of using *S. cerevisiae* yeast strains 11 and 8a (M) in the production of kvass—a traditional low-alcohol fermentation beverage in the Russian Federation. The authors propose methods to enhance the accumulation of physiologically active yeast biomass using succinic acid and yeast autolysate obtained through their developed technology [9].

The impact of succinic acid at concentrations ranging from 0.01% to 0.2% on the metabolism of the pure culture of alcohol yeast S. cerevisiae (strain XII) is demonstrated. Changes were established in the consumption of dry substances, yeast biomass accumulation, active and titrated acidity values, CO_2 production, concentration of dry substances, and yeast cells titer during alcohol yeast reproduction. After 24 hours of cultivation with the introduction of 0.1% succinic acid, the number of yeast cells increased by 14%, budding cells by 71%, and cells with glycogen by 11.5% compared to the control sample [10].

The results [11] of applying an amino acid-vitamin activator (AVA), a natural stimulant obtained from residual brewer's yeast cells of the first-generation strain Saflager S-189, were also studied. AVA serves as a source of amino acids, vitamins, and other essential compounds required for cell development. The addition of the activator to a 12% wort in quantities of 0.2 and 0.5% stimulated the growth and accumulation of reserve nutrients by yeast cells. After 5 days of fermentation, the increase in cell count in the control sample was 2.46 million/cm^3, whereas in samples with 0.2% AVA, it reached 4.6 million/cm^3, and with 0.5%, it was 1.9 million/cm^3. After 7 days of fermentation, the percentage of budding cells in the aforementioned samples was 52%, 74%, and 57%, respectively. Additionally, the proportion of cells containing glycogen was 62%, 83%, and 70%, respectively. However, it was observed that an excessive amount of AVA in a dosage of 1.0% somewhat suppressed yeast growth and activity. During testing in a small-scale brewery, the activator was found to be more effective as the generation number of seed yeast increased. When fermenting 12% wort with seed yeast of the third generation, the experimental sample with 0.2% AVA achieved a visible extract of 3.8% and a visible degree of fermentation of 68.3% after 6 days. In contrast, the control sample reached similar values of 4.3% and 64.1%, respectively, but only after 7 days. Similar results were obtained when assessing the effectiveness of using the AVA activator in combination with a preparation derived from the microalgae *Spirulina platensis* [12].

A method has been developed to increase the viability of brewing yeast using exclusively preparations derived from *Spirulina platensis* algae. Yeast activated with these preparations exhibit an increased reproduction rate. They also demonstrate efficient flocculation, resulting in the formation of a dense precipitate by the conclusion of the main fermentation. The use of this preparation enables yeast to maintain their activity over an extended period. The incorporation of this preparation into the wort is recommended not only as a routine practice but also as a means of providing "first aid" to reactivate yeast that may have been weakened during technological processes or have been reused multiple times. This approach also serves to stabilize their fermentation characteristics [13].

The impact of baking yeast hydrolysate on fermentation and the physiological state of *Saccharomyces cerevisiae* yeast strains 34, 776, 8Am was investigated [14]. In order to assess the influence of hydrolysate on the extent of wort fermentation, it was added in quantities of 50, 100, and 150 mg per 100 cm^3 of wort. The fermentation process employed sterile hopped wort with varying concentrations of dry substances, specifically 8%, 10%, 11%, 12%, and 14%. Fermentation was conducted within a temperature range of 7 to 9° C for a duration of 7 days, with daily measurements of the visible extract. The experiment revealed that the hydrolysate did not significantly impact the extent of wort fermentation when the initial dry substances concentration was 8% and 10%. However, in wort with an initial dry matter concentration of 11%, 12%, and 14% and containing 100–150 mg per 100 cm^3 of hydrolysate, a more profound fermentation occurred, resulting in the degree of attenuation enhancement of 20–30%. This allowed for a reduction in the fermentation duration by 1 to 1.5 day. Beer produced with the inclusion of hydrolysate exhibited deeper attenuation, higher alcohol content, lower diacetyl levels, and superior organoleptic characteristics compared to the control sample.

Yeast extracts also exhibit effectiveness in fermentation high-gravity beer wort with dry substances concentration ranging from 16.2% to 20.3%. The addition of 1% of the extract, on average, reduced the fermentation duration by 1.5 days. Furthermore, it was found to stimulate yeast flocculation ability, fermentation activity, and the number of budding cells [15].

Yeast extracts have proven to be efficient in activating *Saccharomyces cerevisiae* yeast, commonly used in alcohol production [16, 17]. As a result, they contribute to the intensification of alcohol wort fermentation [16].

Complex yeast feeding (CYF) was developed, and its impact on the wort fermentation process and beer quality using dry yeast of the Saflager strain W-34/70 was evaluated. CYF can be added to yeast or to the wort before yeast introduction. CYF consists of a mixture of coarsely ground natural zeolite-containing tuffs and yeast. The use of CYF was found to significantly increase the activity of yeast cell enzymes compared to the control sample: α-glucosidase increased by 1.7–2.7 times, zymase by 1.6–2.4 times, and invertase by 9–30%. The positive impact of CYF on the physiological state of the yeast culture was observed even at a dosage of 0.05 g per 100 cm^3 and further improved with higher doses. The number of budding cells increases by 1.5–2.0 times compared to the control sample, and cells containing glycogen increased by 1.6–1.7 times. Additionally, there was a reduction in the concentration of dead cells, ranging from 8 to 30%. In order to study the wort fermentation process, yeast was added to the wort at a rate of 20 million cells/cm^3. The fermentation was conducted at a temperature of 12° C for 7–8 days, followed by secondary fermentation lasting 30 days at a temperature of 2–3° C. The following samples were compared: experiment sample 1—fermentation of wort with yeast pre-activated using CYF at a dose of 0.1 g/100 cm^3; experiment sample 2—fermentation of wort into which CYF was previously introduced at a rate of 0.05 g/100 cm^3, with reactivated yeast without prior treatment; control sample - wort fermented with yeast starter without prior activation and without CYF application. In the experiment samples, the processes of yeast propagation and consumption of extract and amino nitrogen were notably more vigorous. In the experiment samples of young beer, the concentration of yeast cells in the suspended state by the end of fermentation was significantly lower than in the control sample. This is expected to have a positive impact on the beer clarification process during secondary fermentation and filtration and potentially enhance the colloidal stability of the finished drink. In the samples initially enriched with the yeast feeding, several favorable outcomes were observed. These include a higher degree of attenuation and increased volume fraction of alcohol, a reduced content of polyphenolic compounds, and a lower presence of high-molecular-weight proteins. In addition, in the samples where CYF was applied, the beer exhibited a harmonious and full flavor profile, accompanied by a gentle hop bitterness [18].

Many small beer production enterprises employ active dry brewing yeast for pitching the wort, but in most cases, the viability of the yeast is reduced. Therefore, yeast activation is necessary before fermentation. The effectiveness of increasing the activity of brewing dry yeast [19] Saflager W-34/70 was established through the use of preparations of natural organic and inorganic origin. Yeast activation involved the utilization of antler-containing raw materials in the form of a dry preparation as a source of various biologically active organic substances. The antler-containing preparation composes lipids, nitrogenous compounds, calcium, phosphorus, and other components. The lipid complex includes phospholipids; mono-, di-, and triglycerides; sterols; fatty acids; and sterol esters. The medium for processing the yeast culture, as well as for fermentation, was industrial hopped beer wort with an extra activity of 12%. Biochemical parameters of the culture were assessed both in the initial yeast and during its activation, focusing on the activity of α-glucosidase (maltase), invertase (β-fructofuronosidase), and the zymase complex. The results indicate that adding the antler preparation to the yeast starter during its preparation for wort fermentation leads to a significant increase in the activity of the yeast cell enzymes under investigation when compared to the control sample: α-glucosidase (maltase) by 2.0–6.8 times, zymase by 1.5–2.5 times, and invertase by 4.5–6.2 times. The introduction of the antler-containing preparation, either separately or in combination with CYF, during both yeast preparation for fermentation and directly into the wort before yeast pitching, ensures high yeast viability and high biochemical and physiological activity of the culture. The effectiveness of these feedings is further enhanced when applied under aeration conditions during the dry yeast preparation stage for wort fermentation. Recommended doses of the antlers-containing preparation (% by volume of the medium) are $(0.10–0.75) \times 10^{-3}$, and CYF should be in the range of 0.05–0.075.

The study presents the results demonstrating the positive impact of multifunctional components of geothermal water in the yeast culture medium on ethanol biosynthesis [20]. It was observed that an increase in carbohydrate content and the fermentation duration in a nutrient medium with geothermal water resulted in a 20% increase in alcohol synthesis compared to the previously established technology. Morphophysiological parameters of the cells further validate the active state of the experimental yeast.

The positive effects of silicon on the carbohydrate and nitrogen metabolism of *Saccharomyces* yeast were revealed. Silicon dioxide was sourced from natural quartz sand found in the Sarykum dune and industrial quartz glass. This led to an increase in the biological activity of the yeast population impacting growth, vital activity, generative activity, and adaptive variability in the shape and size of cells [21].

The research also explored the potential for activating dry brewing yeast using an energy exchange regulator: a mixture of organic acids from the Krebs cycle, including succinic, malic, fumaric, citric, and oxalic acid in a 1:1:1:1:1 ratio, at concentrations ranging from 10^{-8} to 10^{-10} mol/dm^3). The influence of various concentrations of this acid mixture was evaluated, demonstrating a positive effect on yeast cell enzymes activity and the physiological state of the yeast culture due to increased cell membranes permeability. The use of yeast activated in the acids solution at a concentration of $1x10^{-10}$ mol/dm^3 positively impacted the beer wort fermentation process, as evidenced by production tests. This resulted in a reduction of the fermentation process duration by 1 day and improved physiological parameters of the yeast culture compared to the control sample without special treatment. Additionally, the quality indicators and organoleptic characteristics of the beer met the standard requirements [22].

The utilization of milk whey as an activator of brewer's yeast was investigated [23]. The efficacy of yeast treatment through exposure in a medium consisting of whey, as well as a mixture of whey and beer wort, at a temperature of 4–6° C, was also defined. During the exposure, the number of budding cells was shown to increase twofold, cells with glycogen by 1.3 times, the zymase activity of yeast by over 2-fold, and maltase activity by 21–42.4%. This led to a reduction in the main fermentation duration by 1 day. At the same time, in experimental beer samples, higher degree of attenuation and alcohol content were achieved, with reduced diacetyl formation. An additional effect is achieved through a joint rotary-pulsation process, as described in the section on combined processing methods.

Furthermore, the optimization of the nutrient medium composition for yeast cultivation can be achieved not only by introducing components with a positive impact but also by eliminating those with a negative influence [24]. This task can be addressed, in part, by employing sorbents.

A proposal is made to employed oyster mushroom mycelium (*Pleurotos ostreatus*), a food additive, for the removal of undesirable components from the wort, thereby optimizing the fermentation process. The sorbent was added to the 12% beer wort before commencing the primary fermentation stage. After 7 days of fermentation, the alcohol concentration in the experimental samples, compared to the control sample, exhibited a significant increase of 145.0% and 152.95% when using biosorbent dosage of 0.5 and 0.3 wt. %/wort volume, respectively [25].

Biosorbent "OD-2," derived from initiated autolysis of sedimentary brewer's yeast, when added to the wort before pitching at dosages ranging from 0.1% to 0.5% by weight/vol., led to a significant increase in the concentration of ethanol in young beer by 17–46% compared to the control sample [26].

While baking may not be directly related to the fermentation industry, one of the crucial stages in bread production involves fermenting dough using *Saccharomyces cerevisiae* yeast. Therefore, the findings of studies on the activation of baking yeast with fruit and berry extracts are of interest.

Biologically active components and acids found in fruit and berry extracts can either inhibit or activate *Saccharomyces cerevisiae* yeast. The impact of various berry extracts on the activity of S. cerevisiae baking yeast has been investigated. These experiments were conducted using "Extra" baking yeast and dry extracts of raspberry, chokeberry, sea buckthorn, and rosehip. Raspberry extract (3–4%) suppressed the growth and propagation of yeast. After 1 hour of exposure, the number of yeast cells decreased by 1.5–2 times, when compared to the control sample. Sea buckthorn extract had a stimulating effect: the growth rate of yeast cells increased up to 40% when compared to the control sample. Chokeberry and rosehip extracts had minimal effect on yeast cell growth rate. However, 2–3% chokeberry extract enhanced dough fermentation. The extracts of fruit and berries naturally increased the dough's acidity, which influenced the growth rate of yeast cells. Sea buckthorn extracts significantly elevated the acidity (up to 4.24 pH units), resembling acid stress, leading to increased

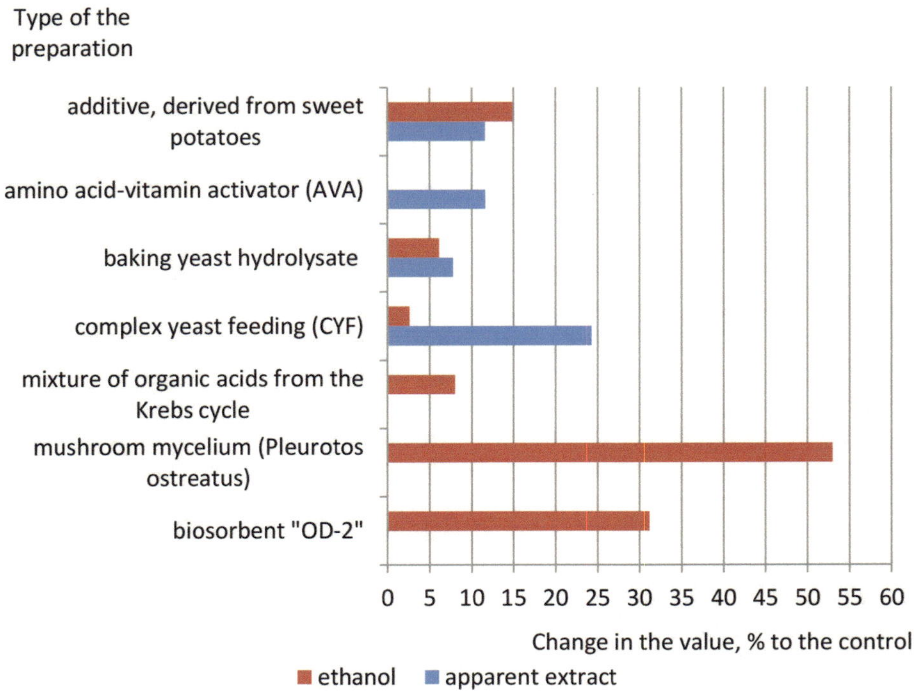

Figure 1.
Increase in ethyl alcohol content and decrease in apparent extract (%) in test samples of finished beer compared to similar values in control samples.

yeast cell growth rate of (1.53×10^6–1.55×10^6 compared to 1.10×10^6 in 1 cm^3 of the control sample). On the other hand, samples with raspberry extract exhibited the slowest growth rate, as raspberry extract is known for its pronounced fungistatic effect, resulting in a 1.5–2 times decrease in the number of yeast cells after an hour of fermentation [27].

A method [28] has been proposed for the preliminary activation of baking yeast to enhance their resistance to acid stress. It was observed that yeast treated with solutions of hydrogen peroxide in concentrations ranging from 0.5 to 3.0 mM retained higher fermentation activity when subjected to stressful conditions (such as a 2% lactic acid solution) compared to the control sample.

The effectiveness of several chemical activation methods was compared by increasing in the concentration of ethyl alcohol and decreasing in the content of the apparent extract (**Figure 1**).

3. Physical activation methods

Activation methods in this group are primarily based on direct or indirect wave/field effects, with the exception of the use of mechanical vibrations (rotary pulsation action). This will be discussed below when examining the combined yeast processing method [23].

Based on the analysis of the literature, it can be concluded that the greatest interest is in the study of the effectiveness of ultrasonic treatment. Examples of such an approach are provided at the beginning of this section.

3.1 Ultrasound and audible range sound processing

Ultrasonic treatment is widely employed for the destruction or inactivation of treated objects, including yeast cells [29]. However, altering the parameters of such treatment can yield the opposite effect and activate yeast cell population, as confirmed by the data presented below.

The same physical and mechanical effects utilized in sonochemistry, such as strong shear forces, particle fragmentation, and increased mass and heat transfer, are also applied in the food industry. Powerful ultrasound is employed to influence the development of living cells, aiming to boost sterilization efficiency, impact enzyme activity [30], and enhance the overall quality of food products [31].

Fermentation processes involve enzymes produced by microorganisms' cells to carry out chemical transformations. Ultrasound can be employed in such processes for monitoring the course of fermentation or influencing its. High-frequency ultrasound (>2 mHz) is well-known as a tool for measuring changes in chemical composition during the fermentation process, offering real-time insights into the reaction's progress. Low-frequency ultrasound (20–50 kHz) can affect the fermentation process by improving mass transfer and cell permeability. This leads to increased process efficiency and productivity. Additionally, this type of ultrasound can be used to eliminate microorganisms that may disrupt the fermentation process [32].

The activation method of Safale T-58 brewer's top fermentation yeast *Saccharomyces cerevisiae* was examined, employing ultrasound at an oscillation frequency of 44 kHz. Ultrasonic treatment significantly intensified the fermentation process and improved the quality of final products. The yeast suspension was placed in an ultrasound machine and processed at the oscillation frequency of 44 kHz. During the exposure, the temperature was measured every minute from the 1st to 20th, as well as at the 25th, 30th, 35th minutes. The survival of yeast cells was recorded by defining the percentage of dead cells. It was found that under the influence of ultrasound, the medium was heated by 1° C for 1 min, and by the 40th minute of treatment, the temperature of the medium reached 57° C. The yeast treated with ultrasound served as an inoculum at the stage of the fermentation of beer wort. The fermentation was carried out at a temperature of 22° C. The fermentation activity increased by 36% in the sample after treating seed yeast with ultrasound for 2 min. The remaining samples differed slightly from the control sample during the first 72 hours of fermentation and lost their fermentation activity after 80 hours of fermentation [33].

In their later work, the same group of authors [34] researched the activating and disintegrating effect of ultrasound at a frequency of 44 kHz and intensity of 1.0 W/cm^2 on brewer's yeast. It was found that a 10-minute ultrasonic treatment of yeast is sufficient to achieve a stimulating effect. Further ultrasound treatment is impractical, since the percentage of dead cells in the yeast suspension exceeds the permissible level (more than 10%). The experiment showed that two-minute ultrasound treatment improved the physiological activity of the seed yeast. The beer obtained using this technique had a higher quality.

The effect of ultrasound on the results of fermentation of beer wort from Korean six-row barley was also studied. Beer samples were processed in an ultrasonic bath for 4 days during the primary fermentation. The ultrasound frequency was 40 kHz, and the input power was 120, 160 and 200 watts. The ultrasound treatment was carried out for 2, 6, and 12 hours for each input power. The physicochemical and organoleptic properties were measured, and the quality of the beer was evaluated. Ultrasound with a power of 160 W increased the yield of ethanol by 13.18% [35].

The study examined the impact of ultrasound at a frequency of 22 kHz and an intensity of 1.0 W/cm^2 on the outcomes of water–heat treatment of wort derived from winter triticale grain [36]. It was established that ultrasonic exposure results in increased activity of grain α-amylases enzymes. The wort obtained was used for cultivating seed yeast with simultaneous ultrasound treatment. It was also observed that such treatment leads to an augmentation of biomass and intensification of the seed yeast growth process. The processed wort was then fermented with activated yeast to produce alcohol with a yield of 67.3 dal/t of conditional starch and a reduced content of toxic impurities.

The potential for increasing ethanol yield in alcohol production through the treatment of *Saccharomyces cerevisiae* yeast using low-intensity ultrasonic irradiation at fixed frequencies of 20, 23, 25, 25, 28, 33, and 40 kHz was considered [37]. Under optimal ultrasonic treatment conditions (ultrasound frequency - 28 kHz, power density - 180 W/l, processing time - 24 h), the maximum ethanol yield increased by 34.87% compared to the control sample. Transcriptome sequencing revealed that ultrasound treatment regulated the expression of genes involved in pyruvate metabolism, glycolysis, the pentose phosphate pathway, and glucose transport. Quantitative polymerase chain reaction in real time also confirmed changes in gene expression. Metabolomics showed that ultrasound treatment increased the intracellular glucose and nicotinamide adenine dinucleotide (NADH) content, crucial metabolites for ethanol synthesis. Additionally, ultrasound treatment reduced the levels of acetate and its derivatives, leading to a decrease in the reverse consumption of pyruvate, thereby facilitating ethanol synthesis. These alterations in gene expression and metabolite content are likely the primary reasons for the increase in ethanol yield in *Saccharomyces cerevisiae* after ultrasound irradiation [37].

The study investigated the impact of ultrasonic treatment with different operating modes and different frequencies on the accumulation of cells and metabolites of *Saccharomyces cerevisiae* yeast. The results demonstrated that *in situ* ultrasound treatment can enhance the controlled parameters, and the effect of ultrasound with a fixed frequency was greater than that of wide-frequency ultrasound. When exposed to ultrasound with a fixed frequency, the concentration of metabolites in the culture medium increased, whereas it decreased after treatment with wide-frequency ultrasound. Conversely, when the ultrasound frequency exceeded 33 kHz, the accumulation of *S. cerevisiae* biomass weakened, accompanied by an increase in the proportion of nonviable cells in the culture medium. At a frequency of 23 kHz and a fermentation duration of 48 hours, the ethanol content increased by 19.33%, as well the content of β-phenylethanol and other volatile metabolites, such as esters [38].

Another study examined the effect of ultrasound treatment on the growth of *S. cerevisiae* yeast, utilizing ultrasonic frequencies of 20, 25, 45, and 130 kHz, with processing duration ranging from of 2 to 30 min. The kinetics of yeast biomass accumulation following ultrasonic treatment was assessed using modeling techniques and scanning electron microscopy (SEM). It was revealed that ultrasound at frequencies of 45 and 130 kHz had no adverse impact on the lag phase and growth rate of yeast populations, unlike the 20 kHz frequency. Ultrasound treatment at 20 kHz resulted in a significant reduction in yeast concentration (by 1.3 log CFU/cm^3) and caused significant damage to the external structures of *S. cerevisiae* cells [39].

The effect of low-intensity ultrasound, varying in frequencies, processing time, and ultrasound power, on *Saccharomyces cerevisiae* in different growth phases was assessed by measuring biomass growth. Additionally, the permeability of cell membranes and the ethanol tolerance of ultrasound-treated *Saccharomyces cerevisiae* cells were studied. It was discovered that under optimal processing conditions during cultivation (ultrasound frequency: 28 kHz, power: 140 W/dm^3, treatment duration: 1 hour), yeast biomass increased by 127.03%. Ultrasound exposure improved the permeability of cell membranes, resulting in increased extracellular protein, nucleic acid, and fructose-1.6-diphosphate content. Furthermore, ultrasound treatment could increase damage to yeast cells when exposed to high concentration of alcohol. However, this effect did not significantly impact yeast tolerance to ethanol [40].

It was also observed that the fermentation process could be activated due to the indirect impact of ultrasound: it does not process the yeast biomass at one stage or another, but the fermentation medium before pitching with yeast.

Consequently, the influence of ultrasound and thermal pretreatment on ethanol yield from cassava chips was investigated. Cassava suspensions were treated with ultrasound for 10 and 30 seconds at amplitudes of 80, 160, and 320 microns/s, corresponding to low, medium, and high power levels, respectively. Processed and untreated (control) ultrasound samples were then subjected to simultaneous liquefaction-saccharification and fermentation. Based on the efficiency of converting cassava starch into ethanol, it was concluded that higher ethanol yields are directly related to the duration of ultrasound treatment, rather than to its power level. The ethanol yield from the ultrasound-treated sample was 2.7 times higher than that from the control sample. The fermentation rate was also significantly higher, with the fermentation duration being shortened by nearly 24 hours for samples treated with ultrasound to achieve the same ethanol yield as in control samples. Thus, ultrasound pretreatment increased both the total ethanol yield and the fermentation rate. Compared to the heat-treated samples, the ethanol yield in ultrasound-treated samples was almost 29% higher. The combined heat and ultrasound treatment did not significantly affect the overall ethanol yield from cassava chips. Ultrasound was also preferred over preheat treatment due to its lower energy requirements [41].

Similar findings were reported in a study [42] where pretreatment with ultrasound during the liquefaction stage was tested in the subsequent combined saccharification/fermentation of corn flour using *Saccharomyces cerevisiae* var. *ellipsoideus* for bioethanol production through a periodic method. Preliminary ultrasonic treatment (at a frequency of 40 kHz) was conducted for various durations at different temperatures before adding a liquefying enzyme. Based on glucose concentration, the optimal duration of ultrasound treatment was determined to be 5 minutes at a treatment temperature of 60° C. This resulted in an increase in glucose concentration by 6.82% compared to the untreated control sample. Additionally, the kinetics of combined saccharification/fermentation was evaluated, and the impact of ultrasound pretreatment on increasing ethanol yield was studied. Ultrasound pretreatment led to an 11.15% increase in ethanol concentration (compared to the control sample) after 32 hours of saccharification/fermentation. The study found that ultrasound's effect, driven by cavitation and acoustic flow, stimulated the breakdown of starch granules and the release of glucose, thereby accelerating starch hydrolysis.

Furthermore, research conducted at our university has focused on the acoustic effects on raw materials and intermediates in brewing production over several years. *Saccharomyces cerevisiae* yeast from various strains was subjected to audible sound exposure at fixed frequencies. It has been established that sound processing could either enhance or impair the technological characteristics of brewer's yeast [43].

This depends primarily on the processing parameters and the frequency of the sound. The following parameters were applied: sound frequency of 2765 Hz, duration of exposure - 30 min, oscillation amplitude - 100%, source power - 2 W, and distance from the source to the treated object - 5 cm. Such preliminary acoustic treatment of dry seeded yeast that were applied to the fermentation of the model nutrient medium (5% sucrose solution) led to a utilization rate increase in nutrient compared to the control sample by 66%, along with a 5% reduction in the proportion of nonviable cells [44].

3.2 Exposure to pulsed electromagnetic field

There is an opinion [45] that a pulsed electromagnetic field (PEMF) has a positive impact on yeast metabolism, cell biomass growth, and ethanol production, especially with low-frequency electromagnetic waves, which are known to have biological effects. An electromagnetic pulsed generator was used to investigate the influence of electromagnetic pulsed fields on yeast cell reproduction and ethanol accumulation in a fermenting medium, with a focus on intensifying the technological process. The fermentation of wort was carried out in a two-stage mode with treatment using a pulsed electromagnetic field (frequency - 4 Hz, power - 1 μT). The study employed beer wort with a density of 15–35% (15% unmalted raw material) and *Saccharomyces cerevisiae* top fermenting yeast. A positive effect of the pulsed electromagnetic field was shown on yeast cell proliferation and the fermentation process. This approach also revealed the potential for accelerating alcoholic fermentation and increasing ethanol yield by 15%. The physicochemical evaluation of fermented materials showed a significant increase in foaming and foam resistance. The fermented wort underwent distillation; gas chromatography analysis of the distillate indicated variations in the content of trace impurities, such as higher alcohols (n-butyl, isobutyl, and isoamyl). The proposed technology has the potential to optimize the fermentation process, adjust the quality of volatile components, and improve the sensory characteristics of fermented materials and distillates for alcoholic beverage production (beer, whiskey wort, brandy, alcoholic beverages, etc.).

The impact of extremely low-frequency magnetic fields on ethanol accumulation by *Saccharomyces cerevisiae* yeast during the fermentation of sugarcane molasses in a periodic fermentation process was explored. The cell suspension from the fermenter was directed through two magnetic field generators in recirculation mode. The recirculation rate and magnetic field strength varied within the range of 0.6–1.4 $m{\bullet}s^{-1}$ and 5–20 μT, respectively. The best results were achieved when treated with a magnetic field of 20 μT at a recirculation rate of 0.9–1.2 $m{\bullet}\ s^{-1}$, resulting in a yield of ethanol approximately 17% higher than in the control sample. Moreover, the fermentation duration was reduced by 2 hours [46].

3.3 Electron-ion processing (EIP)

The potential for activating brewer's yeast through EIP to enhance membrane permeability and increase nutrient and oxygen availability for cells has been confirmed. These activated cells were found to maintain their viability for 3–5 cycles after EIP, attributed to the activation of the permease system. A correlation was established between the EIP modes and glycogen content of yeast cells. When using yeast that has undergone EIP, the main fermentation cycle of beer can be shortened by 15–40%. These activated cells facilitate vigorous wort fermentation without the need for additional fermentation activators. Implementing the proposed EIP method for brewer's yeast activation prior to its introduction into the fermentation apparatus results in beer with a fermentation degree exceeding 80%, enhancing the stability of the final product. EIP-treated yeast can be utilized for 10–11 generations, extending their operational lifespan by 1.5 times compared to untreated yeast. This also reduces the cost associated with maintaining a pure yeast culture. Production tests confirmed the effectiveness of EIP, showing that processing low-quality yeast led to a 10–60% increase in viability and a 3–8% boost in the final fermentation degree while maintaining high beer quality standards. The beer produced using EIP-treated yeast met all standards and exhibited superior physicochemical and organoleptic properties compared to beer samples produced using traditional methods [47, 48].

Glushhenko [49] provides a description of the equipment used for electron-ion processing of yeast. The impact of electron-ion treatment on yeast cells viability was investigated, revealing a reduction in nonviable yeast cells numbers The most pronounced effect was observed when yeast was processed in a "yeast + wort" medium. The efficiency of electron-ion processing is contingent on the intensity of the non-uniform electric field and the exposure time. Implementing electron-ion processing results in a significant enhancement of the yeast's physiological state.

3.4 Processing with visible light and laser radiation in the infrared range

Another approach for yeast-saccharomycete activation is based on exposure to visible spectrum light wavelengths.

The earliest publication we found on this subject dates back to 1984 [50]. This doctoral dissertation research demonstrates that exposure to shortwave optical radiation (410–520 nm) with an intensity of 0.12–0.20 W/m^2 leads to a substantial activation of metabolite transformations within the glycolytic system. Visible light exposure triggers the decarboxylation of pyruvate, facilitating the generation of a substrate for the operation of the di- and tricarboxylic acid cycle, thus alleviating substrate deficiencies in the Krebs cycle. In addition, it increases the activity of alcohol dehydrogenase, the final enzyme of alcoholic fermentation, and some respiratory enzymes. It also enhances the activity of isocitrate dehydrogenase, succinate dehydrogenase, aldehyde dehydrogenase, and the pyruvate dehydrogenase complex. Research also revealed that exposure to 410 nm light wavelength increases the rate of protein biosynthesis by 1.6 times compared to the control sample. Furthermore, an optimal parameter for applying optical radiation to influence the technological characteristics of beer seeding yeast was established. Exposure to 410 nm light at an intensity of 0.12–0.20 W/m^2 for 6 hours prior to inoculation into the wort was found to boost yeast fermentation activity by 20–25%. This lead to a reduction of 1 day in the main fermentation duration without compromising the quality of the final product. Additionally, beer produced with photoactivated yeast displayed enhanced resistance to colloidal turbidity, attributed to a significant reduction in the concentration of high-molecular-weight protein substances compared to the control sample. Regrettably, in our view, this research, although promising, did not receive further attention at that time and was only resumed three decades later.

Thus Kobelev and Bagaeva [51], *Saccharomyces cerevisiae* baker's yeast was used. The initial yeast was grown on a solid nutrient Reader medium until the beginning of the stationary phase of population growth (72 hours). Then, they were transferred to a liquid medium, where the cells were exposed to light of various spectra. The experiments were carried out in three light-isolated blocks: 1st - white light (the source of illumination is fluorescent lamps LDS–40); 2nd – blue light (the source is fluorescent lamps LG-40, the transmission area is 420–460 nm); and 3rd – red light (the source is fluorescent lamps LC–40, the transmission area is 620–640 nm). The exposure time was 6 hours. The control was a yeast culture grown in the dark. The research results indicated a notable enhancement in yeast biomass when exposed to both blue and red light. Specifically, the biomass increase under blue light exposure was found to be 10% higher compared to the control sample, while under red light, this increase amounted to 5%. The rise in yeast biomass under white light conditions was consistent with the growth observed in the dark. This suggests that the effect of blue light is associated with the direct absorption of blue light quanta by components of the mitochondrial energy system, leading to an increased synthesis of high-energy compounds. Red light also affects yeast cells, but unlike blue light, its impact appears to influence a broader spectrum of cellular systems associated with overall metabolism.

For several years, our university has also been engaged in researching the effects of monochromatic light exposure within the visible spectrum, as well as near-ultraviolet and infrared spectra, on brewer's yeast. Previously, a positive impact of such treatment on top-fermentation yeast [52] was demonstrated. This research is currently ongoing, utilizing bottom fermentation *Saccharomyces cerevisiae* Saflager S-189 yeast. Light with a certain wavelength is applied to dry yeast for 60 minutes, after which seeding is performed. Cultivation is carried out for 3 days at 28° C in a model nutrient medium, which is a sterilized 5% sucrose solution. The controlled parameters and their values (expressed as a percentage relative to the control samples) after 3 days of cultivation are presented in **Figure 2**.

Based on this data, it can be concluded that light with such wavelengths has a multidirectional effect on the state of the yeast population. The processing effect is less pronounced than most of the activation methods described earlier. Nevertheless, light treatment with a wavelength of 650 nm provided a higher degree of utilization of nutrients, a smaller proportion of nonviable cells, and a significantly lower increase in biomass than in the control sample. This can be an advantage in the brewing industry, as it helps in reducing the quantity surplus yeast, which, in compliance with environmental regulations, must be disposed of in an environmentally responsible manner. We have decided to continue the research, in order to consider the possibility of optimizing the processing parameters.

The effectiveness of low-intensity laser radiation (LILR) in the infrared range on brewer's yeast of top fermentation was evaluated as a means of intensification of fermentation processes [53]. A treatment duration equal to 2 minutes was chosen as rational. The impact of radiation with pulse frequencies was tested at 300, 600, 760, 1200, 1500, and 3000 Hz. The object of treatment was *Saccharomyces cerevisiae* top fermentation dry yeast. It was found that when processing experimental samples with LILR at a frequency of 600 Hz, the number of nonviable yeast decreases by 75% and when processing at a frequency of 300 and 3000 Hz, by 48% compared to the control sample. In all samples, with the exception of treatment at a frequency of 3000 Hz, an increase in the number of cells with glycogen was observed. The greatest extent was observed in the case of treatment at frequencies of 300 and 600 Hz—by 42%—compared to the control sample. The maximum accumulation of biomass was revealed in the experimental sample after treatment at 600 Hz. In the same sample, on the 7th day of fermentation of 12% beer wort, the visible extract was 4.6%, while in the control sample only on the 9th day, it was 5.8%. This proves the possibility of reducing the duration of the main fermentation stage by 2 days. In the given sample, the amount of diacetyl and 2-butanol (0.6 and 0.2 mg/dm^3) was

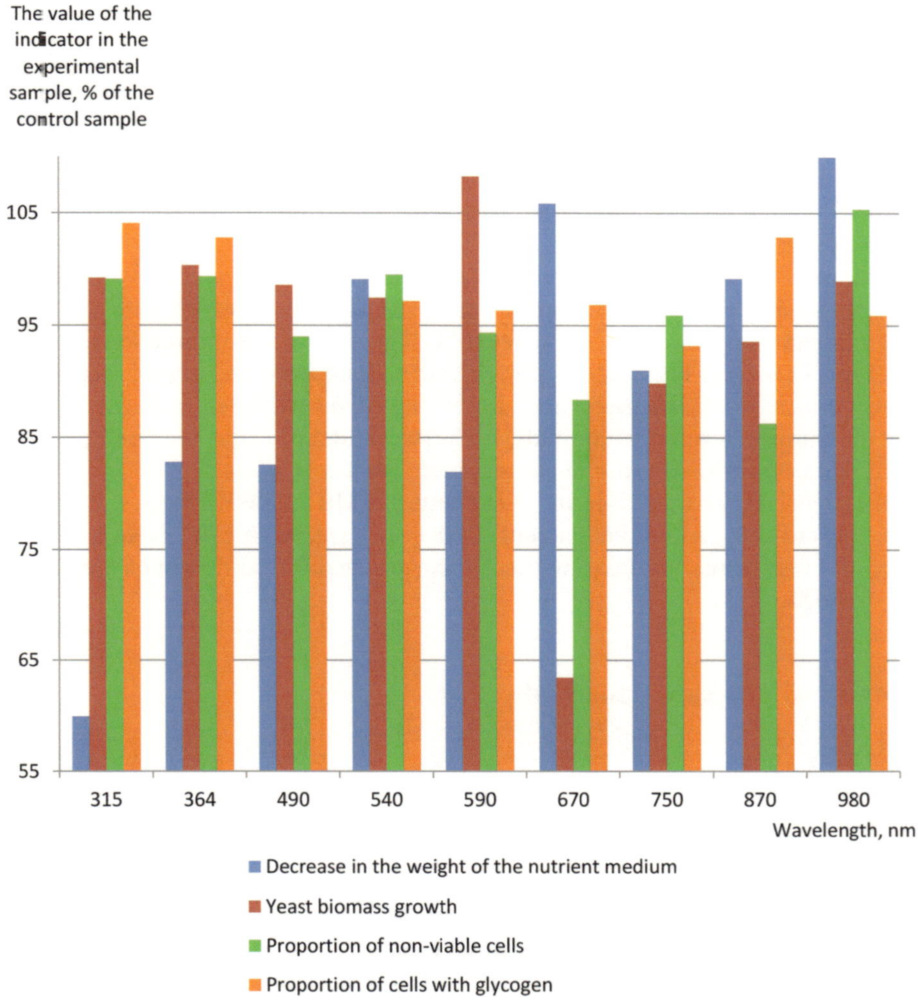

Figure 2.
The effect of fixed wavelength light treatment on the indicators of bottom fermenting brewer's yeast population.

lower than in the control sample (0.9 and 0.4 mg/dm^3). The beer obtained with the use of top fermentation yeast treated with low frequencies (300, 600, 760 Hz) has a mild full taste and floral aroma. At the same time, it was found that processing by LILR at frequencies of 1500 and 3000 Hz led to a deterioration in most of the indicators monitored.

3.5 Other physical methods of yeast activation

The effect of a magnetic field on alcoholic fermentation using *Saccharomyces cerevisiae* yeast strain DAUFPE-1012 was studied. The yeast culture was exposed to a constant magnetic field of 220 microTesla generated by NdFeB (Neodymium-Iron-Boron) rod magnets for 24 hours at a temperature of - 23 +/−1°C. The magnets were placed diametrically opposite (N to S) on a cylindrical tubular reactor. Biomass growth in the culture medium of the reactor (yeast extract + glucose of 2%) was monitored over 24 hours by measuring changes in optical density, which correlated with the dry mass of cells. Ethanol and glucose concentrations were measured every 2 hours. As a result of this treatment, the biomass (g/dm^3) increased by 2.5 times, and the ethanol concentration increased by 3.4 times compared to the control sample. Glucose consumption was higher in the magnetized cultures, which was in line with increased ethanol production [54].

However, an opposite viewpoint was also published, suggesting that constant or variable magnetic fields have no influence on cellular processes in *Saccharomyces cerevisiae* yeast [55].

The impact of electric current on yeast characteristics has also been studied. When a direct current of 10 mA or alternating current of 100 mA was applied to the culture medium, a significant increase in the rate of cell growth and accumulation of ethyl alcohol was observed. The content of higher alcohols, esters, and organic acids in culture media treated with direct and alternating current differed from that in the untreated sample. Several compounds, such as acetaldehyde and acetic acid, were formed from ethanol as a result of the electrode reaction [56].

Pulsating electromagnetic-induced currents (PEMIC, PEMF, PMF), simple alternating currents, and direct currents have been explored for their effects on cells, tissues, and organisms, stimulating membrane permeability and various metabolic processes [57].

4. Combined processing methods

A method for activating brewer's yeast was investigated [5] and patented [58]. This method provides for the treatment of a suspension of yeast cells with an activation medium (whey, beer wort, or a mixture of whey and beer wort) in a ratio of 1:0.5 through acoustic exposure to low-frequency vibrations (20–2·10^4 Hz) in a rotary pulsating machine for 2 minutes. The rotor speed and the inter-cylinder clearance were 2000 min^{-1} (209.33 c^{-1}) and 0.2–0.3 mm, respectively. The processing parameters were selected to avoid an excessive number of dead cells, which should be kept under 10%. This processing method significantly improved the quality of brewer's yeast. Compared to the control sample, the total concentration of cells increased by 3.9–8.1%, the percentage of dead cells decreased by 1.7–2.3 times, and the number of budding cells increased almost twofold. After 7 days of fermentation of 11% wort in the control sample, the actual extract was 4.9%, while in the experimental samples, similar concentrations of dry substances (4.6–4.5) were achieved after 5–6 days of the main fermentation. The volume fraction of ethyl alcohol in the finished beer of the control sample was 4.58%, whereas in the experimental variants, it ranged from 5.27 to 5.37% of the volume, with the actual extract at 3.5% and 2.3–2.4%, respectively.

A patent [59] was obtained for a yeast activation method designed for the alcohol industry. In this method, a suspension of alcoholic yeast in the production grain wort is subjected to an acoustic field with an oscillation frequency of 22 kHz and an oscillation intensity of 1.0 W/cm^2 for 3.5–4.5 minutes at a temperature of 28–34°C. A fraction of 5–10% of the total volume of the processed yeast suspension is then selected and undergoes a secondary treatment with an acoustic field at an oscillation frequency of 22 kHz and an oscillation intensity of 1.0 W/cm^2 for 35–45 minutes at a temperature of 48–52°C. This secondary treatment is aimed at destroying yeast shells and obtaining yeast extract. The resulting extract is subsequently mixed with the suspension of processed yeast. This process leads to a significant increase in yeast

DOI: http://dx.doi.org/10.5772/intechopen.1003283

Activation method	The field of yeast use	Increasing the degree of fermentation, %	Increase in ethanol concentration, %	Increase in cell titer/yeast biomass growth, %	Shortening the duration of the stage	Increase in the proportion of budding cells, %	Increase in the proportion of cells with glycogen, %	Reduction of the proportion of non-viable cells	A source
Chemical									
Sweet potato supplement	Brewing	5.8	3.6 in green beer	44	Lagering - for 2–3 days				7
Succinic acid	Alcohol production			14*		71*	11.5*		10
Amino acid-vitamin activator 0.2%	Brewing	6.6		87	The main fermentation - for 1 day	42	33.9		11
Complex yeast feeding	Brewing					50–100	60–70	8–30	18
Milk whey	Brewing				The main fermentation - for 1 day	100	30		23
Physical									
Ultrasound 23 kHz	Alcohol production		19.3						38
Sound at frequency of 2765 Hz	Brewing	66						5	44
Light with a wavelength of 410 nm	Brewing	20–25			The main fermentation - for 1 day				50

Activation method	The field of yeast use	Increasing the degree of fermentation, %	Increase in ethanol concentration, %	Increase in cell titer/yeast biomass growth, %	Shortening the duration of the stage	Increase in the proportion of budding cells, %	Increase in the proportion of cells with glycogen, %	Reduction of the proportion of non-viable cells	A source
Low-intensity laser radiation 600 Hz	Brewing				The main fermentation - for 2 days		42	75	53
Magnetic field			240	150					54
Combined									
Rotary pulsation treatment + whey	Brewing		15–17	3.9–8.1	The main fermentation - for 1-2 days	100		70–130	58
Ultrasound + yeast extract	Alcohol production			100–200	Fermentation of alcoholic wort - for 10–12 hours				59

**After 24 hours of cultivation.*

Table 1.
The effectiveness of various methods of activation of Saccharomyces cerevisiae yeast (change in the value of the indicator compared to the control sample).

biomass by 2–3 times, faster start of wort fermentation at the first day of fermentation, and a reduced fermentation duration by 10–12 hours.

5. Comparing the effectiveness of various methods of activation of *Saccharomyces cerevisiae* yeast

From the information provided, it is evident that various methods are employed for different types of yeast, and their effectiveness is evaluated using diverse criteria. Nevertheless, we have chosen to conduct a comparative analysis of the effectiveness of selected methods by consolidating the data from in the literature into a single table (**Table 1**).

6. Conclusion

It can be seen that the proposed methods for improving the technological characteristics of *Saccharomyces cerevisiae* yeast are very diverse and aim at improving various properties of the cells of the population. At the same time, different authors recommend various parameters of the same type of exposure, in order to ensure an activating effect. Separate groups of researchers express sometimes opposing opinions about the effect of one and the same type of treatment. In our perspective, this suggests that the efficacy of treatment depends not only on the nature and methods of exposure but also on a large number of other variables, including the initial characteristics of yeast cells, yeast strain, the environmental conditions in which yeast develops within the culture medium, and various other factors. Therefore, it is advisable to continue the research and obtain more detailed information about the effectiveness of yeast activation methods. A number of the approaches herein discussed seem promising when applied on an industrial scale in the aims of intensifying the various technological stages in fermentation industries, ranging from yeast generation to wort fermentation.

Conflict of interest

The authors declare no conflict of interest.

Author details

Dmitry Karpenko* and Artem Grishin
Federal State Budgetary Educational Institution of Higher Education "Russian Biotechnological University (ROSBIOTECH)", Moscow, The Russian Federation

*Address all correspondence to: doka.65@mail.ru

References

[1] Galbraith S, Bhatia H, Liu H, Yoon S. Media formulation optimization: Current and future opportunities. Current Opinion in Chemical Engineering. 2018;**22**:42-47. DOI: 10.1016/j.coche.2018.08.004

[2] D'amore T. Cambridge prize lecture improving yeast fermentation performance. Journal of The Institute of Brewing. 1992;**98**:375-382

[3] Dequin S. The potential of genetic engineering for improving brewing, wine-making and baking yeasts. Applied Microbiology and Biotechnology. 2001;**56**:577-588. DOI: 10.1007/s002530100700

[4] Permjakova LV. Классификация стимуляторов жизненной активности дрожжей (classification of preparations to promote yeast vital activity). Техника и технология пищевых производств. 2016;**42**(3):46-55. (in Russ.)

[5] Pomozova VA, Permjakova LV, Safonova EA, Artemasov VV. Активация пивных дрожжей (Brewer's yeast activation). Пиво и напитки. 2002;**2**:26-27. (In Russ.)

[6] Meledina TV. Сырье и вспомогательные материалы в пивоварении (Raw materials and auxiliary materials in brewing). Санкт-Петербург: Издательство «Профессия»; 2003. 304 с. (In Russ.)

[7] Gagieva LC, Cugkiev BG, Dzantieva LB, Makiev ON. Технологические аспекты использования растительного сырья в качестве активаторов бродильных процессов (Aspects of technology use vegetable raw materials as activators of fermentation processes). Пиво и напитки. 2011;**2**:28-29. (in Russ.)

[8] Palagina MV, Drekko KA, Plehova NG. Влияние биостимуляторов дальневосточных растений на физиологическую активность пивоваренных дрожжей (Influence of biostimulators Far East plants on the physiological activity of brewing yeast). Пиво и напитки. 2011;**2**:33-35. (in Russ.)

[9] Grebenchikov VA, Gernet MV. Использование активаторов дрожжей при производстве кваса (the use of activators of yeast in the production of kvass). Пиво и напитки. 2003;**3**:34-37. (in Russ.)

[10] Krikunova LN, Rjabova SM, Peschanskaja VA, Urusova LM. Влияние янтарной кислоты на метаболизм дрожжей *Saccharomyces cerevisiae*. (effect of succinic acid on the metabolism of the yeast *Saccharomyces cerevisiae*). Пиво и напитки. 2015;**1**:36-38. (in Russ.)

[11] Hnykin AM, Sadova AI, Timaev AM. Разработка метода активации сухих пивоваренных дрожжей для заводов малой мощности (development of a method of activation of dry brewing yeast for small power plants). Пиво и напитки. 2012;**2**:12-16. (in Russ.)

[12] Gernet MV, Gribkova IN, Borisenko OA. Исследование возможности активации дрожжей при приготовлении ферментированных напитков (the yeast activation possibility study in the fermented beverages preparation). Пищевая промышленность. 2020;**8**:56-59. (In Russ.). DOI: 10.24411/0235-2486-2020-10087

[13] Bidihova MJ, Lavrova VL, Gernet AM, Gruzdeva AE. Повышение

жизнеспособности пивоваренных дрожжей с использованием Спирулины платенсис (Increasing the viability of brewing yeast using Spirulina platensis). Пиво и напитки. 2002;**6**:10-12. (In Russ.)

[14] Kiseleva IV, Puchkova EA, Kobelev KV, Gernet MV, Lavrova VL. Способ интенсификации процесса сбраживания сусла (Method for intensifying the wort fermentation process). Пиво и напитки. 2004;**2**:38-39. (In Russ.)

[15] Gorelov SS. Влияние экстракта дрожжей на процесс сбраживания плотного пивного сусла (the influence of yeast extract on the fermentation process of high gravity beer wort). Пиво и напитки. 2008;**5**:38-40. (In Russ.)

[16] Krechetnikova AN, Il'jashenko NG, Shaburova LN, Bodrova OJ. Активирующий эффект воздействия дрожжевого экстракта на клетки *Saccharomyces cerevisiae* (activating effect of yeast extract on *Saccharomyces cerevisia*e cells). Производство спирта и ликероводочных изделий. 2006;**3**:29-30. (In Russ.)

[17] Gorelov SS, Il'jashenko NG, Krechetnikova AN, Beteva EA, Gernet MV. Влияние дрожжевых экстрактов на процесс сбраживания сусла из крахмалсодержащего сырья (the influence of yeast extracts on the process of fermentation of wort from starch-containing raw materials). Производство спирта и ликероводочных изделий. 2005;**1**:22-23. (In Russ.)

[18] Permjakova LV, Pomozova VA, Apenova DS, Russkih RV. Активация сухих пивных дрожжей с помощью комплексной дрожжевой подкормки (activation of dry brewer's yeast using complex yeast feeding). Пиво и напитки. 2012;**1**:18-21. (in Russ.)

[19] Permjakova LV, Pomozova VA, Pavlov AA, Horunzhina SI. Применение новых видов пищевых подкормок для дрожжей в производстве пива (application of new types of nutritional supplements for yeast in beer production). Техника и технология пищевых производств. 2013;**2**(29):46-52. (In Russ.)

[20] JeA H, Abramov SA, Kotenko SC, JeA I. Влияние стимулятора биосинтеза этанола – геотермальной воды на морфологические особенности дрожжей *Saccharomyces cerevisiae* в различных условиях культивирования (the influence of the stimulator of ethanol biosynthesis - geothermal water on the morphological characteristics of the yeast *Saccharomyces cerevisiae* under various cultivation conditions). Хранение и переработка сельхозсырья. 2010;**8**:44-46. (In Russ.)

[21] Abramov SA, Vlasova OK, Kotenko SC. Морфофизиологические свойства дрожжей рода *saccharomyces* на кремнийсодержащих средах (morphophysiological properties of yeasts of the genus *saccharomyces* on silicon-containing media). Виноделие и виноградарство. 2008;**4**:14-15. (In Russ.)

[22] Pavlov AA, Pomozova VA, PermjakovaLV, VereshhaginAL. Активация пивных дрожжей смесью органических кислот (activation of brewer's yeast with a mixture of organic acids). Современные проблемы науки и образования. 2013;**5**:127-134. (In Russ.)

[23] Kozlov SG. Исследование и разработка способов активации дрожжей с использованием молочной сыворотки (Research and development of methods for activating yeast using whey) [thesis]. Кемерово: Кемеровский технологический институт пищевой промышленности; 2002. (In Russ.)

[24] Permjakova LV. Регулирование биотехнологических свойств пивных

дрожжей путем корректировки состава питательной среды : монография (Regulation of biotechnological properties of brewer's yeast by adjusting the composition of the nutrient medium: monograph). Кемерово: КемГУ; 2017 248 с. (In Russ.)

[25] Druzhinina ES, Gernet MV, Kolesnikova VF. Интенсификация брожения с использованием биомассы гриба *Pleurotos ostreatus* (вешенка) (Intensification of fermentation using biomass of the fungus *Pleurotos ostreatus* (oyster mushroom)). Пиво и напитки. 2003;**1**:26-28 (in Russ.)

[26] Karpenko DV. Разработка технологии получения биосрбентов на основе осадочных пивных дрожжей и их применение для производства пива, этилового спирта и других пищевых продуктов (Development of Technology for Producing Biosrbents Based on Sedimentary brewer's Yeast and their Use for the Production of Beer, Ethyl Alcohol and Other Food Products) [Thesis]. Москва: Московский государственный университет пищевых производств; 2005. (In Russ.)

[27] Kuz'mina SS, Kozubaeva LA, Egorova EJ, Kulushtaeva BM, Smol'nikova FH. Активность дрожжей *Saccharomyces cerevisiae* в условиях стресс-провокации плодово-ягодными экстрактами (Effect of berry extracts on *Saccharomyces cerevisiae* yeast). Техника и технология пищевых производств. 2021;**51**(4):819-831. DOI: 10.21603/2074-9414-2021-4-819-831. (In Russ.)

[28] Jamashev TA, Reshetnik OA. Влияние предварительной активации дрожжей пероксидом водорода на их адаптацию к осмотическому стрессу (Effect of preliminary activation of yeast with hydrogen peroxide on their adaptation to osmotic stress). Вестник Казанского технологического университета. 2010;**11**:312-316. (In Russ.)

[29] Tsukamoto I, Constantinoiu E, Furuta M, Nishimura R, Maeda Y. Inactivation effect of sonication and chlorination on *Saccharomyces cerevisiae*. Calorimetric analysis. Ultrasonic Sonochemistry. 2004;**11**(3-4):167-172. DOI: 10.1016/j.ultsonch.2004.01.014

[30] Mason TJ, Paniwnyk L, Lorimer JP. The uses of ultrasound in food technology. Ultrasonics Sonochemistry. 1996;**3**(3):S253-S260. DOI: 10.1016/S1350-4177(96)00034-X

[31] Chandrapala J, Oliver C, Kentish S, Ashok-kumar M. Ultrasonics in food processing – Food quality assurance and food safety. Food Science and Technology. 2012;**26**(2):88-98. DOI: 10.1016/j.tifs.2012.01.010

[32] Shikha Ojha K, Mason TJ, O'Donnell CP, Kerry JP, Tiwari BK. Ultrasound technology for food fermentation applications. Ultrasonics Sonochemistry. 2017;**34**:410-417. DOI: 10.1016/j.ultsonch.2016.06.001

[33] Kaluzhina OJ, Jakovleva KS, Kashapova RA, Chernenkov EN, Chernenkova AA, Bodrov AJ. Влияние ультразвука на пивоваренные дрожжи (the effect of ultrasound on brewing yeast). Вестник ВГУИТ. 2020;**82**:103-109. (In Russ.). DOI: 10.20914/2310-1202-2020-1-103-109

[34] Kalugina O, Nafikova A, Chernenkov E, Leonova S, Chernenkova A, Badamshina E, et al. Application of ultrasound for enhancing fermentation rates in brewing technology. Acta Scientiarum Polonorum, Technologia Alimentaria. 2021;**20**(3):301-312. DOI: 10.17306/J.AFS.2021.0950

[35] Choi EJ, Ahn H, Kim M, Han H, Kim WJ. Effect of ultrasonication on fermentation kinetics of beer using six-row barley cultivated in Korea. Journal of The Institute of Brewing. 2015;**121**(4):510-517. DOI: 10.1002/jib.262

[36] Kaluzhina OY, Krechetnikova AN, Smirnova IV, Gusev AN, Nafikova AR. Alcohol technology intensification with the application of ultrasound. Bulgarian Journal of Agricultural Science. 2019;**19**(Suppl.2):98-104

[37] Yang Y, Ren W, Xu H, Cheng L, Dapaah M, He R, et al. Incorporating transcriptomic-metabolomic analysis reveal the effect of ultrasound on ethanol production in *Saccharomyces cerevisiae*. Ultrasonics Sonochemistry. 2021;**79**:105791. DOI: 10.1016/j.ultsonch.2021.105791

[38] Zhang Z, Xiong F, Wang Y, Dai C, Xing Z, Dabbour M, et al. Fermentation of *Saccharomyces cerevisiae* in a one liter flask coupled with an external circulation ultrasonic irradiation slot: Influence of ultrasonic mode and frequency on the bacterial growth and metabolism yield. Ultrasonics Sonochemistry. 2019;**54**:39-47. DOI: 10.1016/j.ultsonch.2019.02.017

[39] Soro AB, Oliveira M, O'Donnell CP, Tiwari BK. Ultrasound assisted modulation of yeast growth and inactivation kinetics. Ultrasonics Sonochemistry. 2021;**80**:105819. DOI: 10.1016/j.ultsonch.2021.105819

[40] Dai C, Xiong F, He R, Zhang W, Ma H. Effects of low-intensity ultrasound on the growth, cell membrane permeability and ethanol tolerance of *Saccharomyces cerevisiae*. Ultrasonics Sonochemistry. 2017;**36**:191-197. DOI: 10.1016/j.ultsonch.2016.11.035

[41] Nitayavardhana S, Shrestha P, Rasmussen ML, Lamsal BP, van Leeuwen J, Khanal SK. Ultrasound improved ethanol fermentation from cassava chips in cassava-based ethanol plants. Bioresource Technology. 2010;**101**(8):2741-2747. DOI: 10.1016/j.biortech.2009.10.075

[42] Nikolić S, Mojović L, Rakin M, Pejin D, Pejin J. Ultrasound-assisted production of bioethanol by simultaneous saccharification and fermentation of corn meal. Food Chemistry. 2010;**122**(1):216-222. DOI: 10.1016/j.foodchem.2010.02.063

[43] Karpenko DV, Gernet MV, Krjukova EV, Gribkova IN, Nurmukhanbetova DE, Assembayeva EK. Acoustic vibration effect on genus *saccharomyces* yeast population development. News of the Academy of Sciences of the Republic of Kazakhstan. Series of Geology and Technical Sciences. 2019;**4**(436):103-112. DOI: 10.32014/2019.2518-170X.103

[44] Karpenko DV. Определение рациональных параметров акустической обработки с целью активации пивных дрожжей. Здоровье, питание и биотехнологии (Determination of rational parameters of acoustic processing in order to activate brewer's yeast). 2020;**2**(1):140-152. DOI: 10.36107/hfb.2020.i1.s290. (In Russ.)

[45] Mamarasulov BD, Nasirova OA, Mirzarahmetova DT. Интенсификация процесса сбраживания пивного сусла (intensification of the fermentation process of beer wort). Пиво и напитки. 2017;**5**:24-27. (In Russ.)

[46] Perez VH, Reyes AF, Justo OR, Alvarez DC, Alegre RM. Bioreactor coupled with electromagnetic field generator: Effects of extremely low frequency electromagnetic fields on ethanol production by *Saccharomyces cerevisiae*. Biotechnology Progress.

2007;**23**(5):1091-1094. DOI: 10.1021/bp070078k

[47] Osipova MV. Интенсификация процесса брожения методом электронно-ионной обработки (ЭИО) пивных дрожжей (Intensification of the Fermentation Process by Electron-Ion Processing (EIP) of brewer's Yeast) [Thesis]. Москва: Новгородский Государственный Университет имени Ярослава Мудрого; 2007. (In Russ.)

[48] Osipova MV, Glushhenko LF. Интенсификация брожения пива посредством электронно-ионной обработки (ЭИО) пивных дрожжей (intensification of beer fermentation through electron-ion processing (EIP) of brewer's yeast). Пиво и напитки. 2006;**5**:22-24. (In Russ.)

[49] Glushhenko NA. О некоторых эффектах влияния электронно-ионной обработки на дрожжевые микроорганизмы (About some effects of electron-ion treatment on yeast microorganisms). Вестник Новгородского государственного университета им. Ярослава Мудрого. 2013;**2**(71):36-40

[50] Shaburova GV. Интенсификация производства пива путем фотостимуляции метаболизма дрожжей (Intensification of beer production by photostimulation of yeast metabolism) [thesis]. Кемерово: Кемеровский технологический институт пищевой промышленности; 1984. (In Russ.)

[51] Kobelev AV, Bagaeva TV. Влияние разного светового спектра на рост дрожжей *Saccharomyces cerevisiae* (effect of different light spectrum on the growth of yeast *Saccharomyces cerevisiae*). Ученые записки Казанского университета: Естественные науки. 2012;**154**:98-102. (In Russ.)

[52] Suprunjuk AJ, Karpenko DV. Влияние обработки монохроматическим светом на характеристики пивных дрожжей (Effect of monochromatic light treatment on the characteristics of brewer's yeast). In: В сборнике: Общеуниверситетская научная конференция молодых учёных и специалистов «День Науки». Часть II. 16 апреля 2016. Москва: МГУПП; 2016. pp. 134-138. (In Russ.)

[53] Shaburova LN, Danilova AN, Ponomareva MS, Gernet MV. Действие импульсной частоты лазерного излучения на дрожжи верхового брожения (Effect of pulsed frequency laser radiation on top-fermenting yeast). Пиво и напитки. 2019;**2**:16-19. (In Russ.)

[54] da Motta MA, Muniz JB, Schuler A, Da Motta M. Static magnetic fields enhancement of *saccharomyces cerevisae* ethanolic fermentation. Biotechnology Progress. 2004;**20**(1):393-396. DOI: 10.1021/bp034263j

[55] Anton-Leberre V, Haanappel E, Marsaud N, Trouilh L, Benbadis L, Boucherie H, et al. Exposure to high static or pulsed magnetic fields does not affect cellular processes in the yeast *Saccharomyces cerevisiae*. Bioelectromagnetics. 2010;**31**(1):28-38. DOI: 10.1002/bem.20523

[56] Nakanishi K, Tokuda H, Soga T, Yoshinaga T, Takeda M. Effect of electric current on growth and alcohol production by yeast cells. Journal of Fermentation and Bioengineering. 1998;**85**(2):250-253

[57] Grosse H-H, Bauer E, Berg H. Electrostimulation during fermentation. Bioelectrochemistry and Bioenergetics. 1988;**20**(1-3):279-285. DOI: 10.1016/S0302-4598(98)80024-X

[58] Pomozova VA, Permjakova LV, Plotnikov VA, Safonova EA, Kozlov SG, Artemasov VV, et al. Способ активации пивных дрожжей (Brewer's yeast activation method). Патент РФ (Patent) RU 2 234 529 C2. 2004. 8 p (In Russ.)

[59] Bodrova OJ, Krechetnikova AN, Il'jashenko NG, Shaburova LN, Gernet MV. Способ активации спиртовых дрожжей (Method for activating alcoholic yeast). Патент на изобретение (Patent) RU 2288262 C1. 2006. 5 p (In Russ.)

Section 2

Wine Biotechnology

Chapter 4

Distinctive Role of Yeast Strain on Aromatic Profile of Wines Made from Minority Grape Cultivars: Chemical and Sensory Characterization of Aroma Components

José Pérez-Navarro, Adela Mena-Morales, Sergio Gómez-Alonso, Esteban García-Romero and Pedro Miguel Izquierdo-Cañas

Abstract

This chapter synthetizes the main results that our research group has obtained about the specific influence of a commercial *Saccharomyces cerevisiae* strain on the aromatic profile of fermented musts from four minority grape varieties (*Vitis vinifera* L.) cultivated in Castilla-La Mancha (Spain), that is, Moribel, Tinto Fragoso, Albillo Dorado and Montonera del Casar. In addition, wines made from the grape cultivars Tempranillo and Airén were evaluated. To determine the main yeast-derived odor relevant in these grape varieties, the aromatic profiles of grape cultivars and the resulting wines were studied by gas chromatography coupled to mass spectrometry and wines were subjected to Napping, a rapid sensory evaluation method. The results revealed wine sensory differences which are consequence of different aromatic profiles of wines produced with these grape cultivars. The combination of quantitative chemical analysis of volatile compounds together with sensory analysis of wines point out different patterns of aroma compound formation and release. Thus, the yeast strain used in the fermentation step is one of the main factors that affect the sensory properties of wines.

Keywords: *Saccharomyces cerevisiae*, yeast, grape, wine, volatile compounds, aroma, gas chromatography, sensory analysis

1. Introduction

Wine is a beverage obtained usually from grapes via the alcoholic fermentation process carried out by yeasts. The use of commercial *Saccharomyces cerevisiae* strains has been widespread in wine industry due to its good fermentative properties and the ability to produce quality wines, particularly through modifications of volatile compound profile [1]. The distinctive aroma of wine is determined by several factors such as grape variety and maturity, viticultural and winemaking practices, and storage conditions [2]. Several volatile compounds are produced in grapes, increasing their concentration during the vegetative growing period although this tendency changes during grape ripening [3]. Among aroma compounds present in grapes, a significant part comes from specific glycosidically linked forms (odorless precursors) which are transformed into free forms (odor compounds) during the winemaking process [4]. In wine, a large number of volatile compounds can be a part of its aromatic profile. However, around 70 odor compounds play major roles in the aromatic characteristics of wines [5]. Most of these volatile compounds belonging to different chemical classes are formed from precursors during fermentation by *Saccharomyces cerevisiae* or grape-derived compounds released and/or modified by the yeast action [6]. Fermentation-derived volatiles including higher alcohols, acids, ethyl esters and acetates are the most abundant in total aroma composition of wine. The modulation of the levels of these compounds by the yeast strain in charge of fermentation leads to sensory differences [7].

Spain is the country with the biggest vineyard surface in the world and has a great varietal biodiversity [8], despite the disappearance of many grape varieties (*Vitis vinifera* L.) in Europe was caused by the attack of the phylloxera in the late nineteenth century [9]. In the last decades, prospecting and recovery works started in the wine-growing areas of the world due to the great interest of wine sector in recovering ancient varieties [10], some of which are minoritarian and in danger of disappearing. In this way, grapevine prospecting work allowed to preserve the old vine heritage of the Castilla-La Mancha Spanish region, which resulted in the identification of more than 40 new grape genotypes not previously registered in any database [11]. Notable among these minority grape varieties are Moribel, Tinto Fragoso, Albillo Dorado and Montonera del Casar (**Figure 1**). These grape genotypes are grown in the Collection

Figure 1.
Leaf and bunch morphology in the Vitis vinifera L. grape cultivars: Moribel (a), Tinto Fragoso (b), Albillo Dorado (c) and Montonera del Casar (d).

of Grapevine Varieties from Castilla-La Mancha (CGVCLM) created to preserve the local grape diversity of this winemaking region.

The aim of the chapter is to synthesize the main results that our research group has obtained on the impact of a commercial *Saccharomyces cerevisiae* strain on the volatile composition and aroma profile of wines produced by these minority grape cultivars to elucidate the chemical changes in the components potentially responsible for aroma sensory properties. This work was carried out by studying the aromatic profile of these grape varieties and resulting wines by gas chromatography coupled to mass spectrometry (GC-MS) and the sensory properties of wines were evaluated by Napping. These were compared to wines made from Airén and Tempranillo, the most cultivated grape varieties in this winemaking region. To our knowledge, this is the first comprehensive study about the aroma components of these minority grape cultivars and the resulting wines.

2. Materials and methods

2.1 Chemicals

All solvents and reagents were of analytical grade (>99%) and a Milli-Q purification system (Merck-Millipore, Darmstadt, Germany) was used to obtain the pure water. Several commercial standards were employed for volatile compound identification and quantification purposes, which were purchased from Extrasynthese (Genay, France), Fluka (Buchs, Germany), Merck (Darmstard, Germany), and Sigma-Aldrich Chemie (Steinheim, Germany). 4-nonanol and 4-methyl-2-pentanol were used as internal standards and were supplied by Sigma-Aldrich Chemie (Steinheim, Germany).

2.2 Grape samples

Grape samples of four minority varieties (*Vitis vinifera* L.), two red (Moribel and Tinto Fragoso), and two white (Albillo Dorado and Montonera del Casar), were evaluated in this chapter. Grapes were harvested at the optimal ripening stage and in good sanitary conditions from an experimental vineyard of the Instituto Regional de Investigación y Desarrollo Agroalimentario y Forestal from Castilla-La Mancha (IRIAF) located in Tomelloso (Spain). The minority varieties were compared with two of the most traditionally cultivated grapes in this winemaking region, that is, Tempranillo and Airén. For each variety, a total of three batches of grape clusters were sampled from grapevines in several zones of the vineyard. Sampling was made by selecting randomly berries from the top, central, and bottom of grape bunches in the lab, choosing 500 g of grapes from each batch to evaluate the aromatic potential.

2.3 Winemaking process

Winemaking process was carried out in the winery of IRIAF and all produced wines were dry, i.e. containing less than 5 g/L of residual sugar. White wines were elaborated from 75 kg of grapes. The grapes were destemmed and crushed, with the addition of 80 mg/L of SO_2. Cold pre-fermentation maceration was developed at 5°C for 24 h. Subsequently, the resultant grape must was separated from the solid phase by pressing, and total acidity was adjusted to 4.5 g/L in the case of grape must with lower values. In addition, ascorbic acid (100 mg/L) and lysozyme (100 mg/L) were added to avoid oxidation and malolactic fermentation respectively. Fermentation was carried out at 17°C in 100 L fermentation vessels using the commercial yeast Uvaferm VN® (Lallemand Inc., Zug, Switzerland) at 20 g/hL. Alcoholic fermentation was controlled by measures of the density. When a relative density of 1.010 was reached, the alcoholic fermentation continued at 20°C. When the glucose + fructose concentration was below 5 g/L, the alcoholic fermentation was terminated. Then, the wines were racked and sulphited (25 mg/L of free SO_2). For 1 month, the fermentation lees were stirred regularly (*Bâtonnage* technique). Wines were subjected to cold treatment at −5°C for 15 days, and the free SO_2 concentration was corrected up to 25 mg/L. Wines were then filtered through 0.2 μm membranes, bottled, and stored at 16–18°C.

Red wines were elaborated from 75 kg of grapes in vats of 100 L, with skin maceration until the alcoholic fermentation ended. After stemming and crushing, a concentration of 50 mg/L of SO_2 was added before the inoculation with *S. cerevisiae* selected yeast Uvaferm VN® (Lallemand Inc., Zug, Switzerland) at 20 g/hL. The alcoholic fermentation temperature was maintained at 22°C and its development was monitored daily by measuring the density until 0.995. After alcoholic fermentation, the wines were racked and malolactic fermentation was carried out at a temperature of 22°C using *Oenococcus oeni* lactic acid bacteria VP41® (Lallemand Inc., Zug, Switzerland) at 1 g/hL. Malolactic fermentation was monitored by the quantification of L-malic acid in wines. When the malic acid content reached values below 0.2 g/L, the wines were sulphited to 25 mg/L of free SO_2. Subsequently, wines were left for 30 days in vats at 5°C and then underwent cold treatment at −5°C for 15 days. After, the wines were racked and sulphited to obtain 25 mg/L of free SO_2. Finally, wines were filtered through 0.2 μm membranes, bottled, and stored in controlled conditions at 16–18°C. All wines were carried out in duplicate for each grape cultivar.

2.4 Volatile compound extraction

For each variety, an amount of 500 g of grapes was selected and crushed for 3 min with an ULTRA-TURRAX digital T50 crusher and subsequently centrifuged at 3500 rpm for 15 min. The obtained supernatant was then filtered, disregarding the first wort fraction. A volume of 25 mL was collected and 500 μL of 4-nonanol 0.1 g/L (internal standard) was added. The extraction of the free fraction of volatile compounds was performed following the procedure described in the literature [12], using SPE cartridges (LiChrolut EN, Merck, 0.2 g phase for wine, 0.3 g phase for grape must). A volume of 25 mL of grape must or wine was selected and passed through the SPE resin with the internal standard. Subsequently, 25 mL of Milli-Q water was added to remove polar compounds and sugars. The free fraction of volatile compounds was recovered by passing 15 mL of a pentane-dichloromethane solution (2:1, v/v). The eluate obtained was concentrated by distillation on a Vigreux column and then under a nitrogen stream to 150 μL, keeping at −20°C until analysis. The glycosidically bound volatile fraction of grapes was obtained by passing 25 mL of ethyl acetate-methanol (9:1, v/v) after recovering the aroma-free portion from the SPE resin. Subsequently, the obtained fraction was evaporated to dryness in a rotary evaporator and reconstituted with 5 mL of 0.2 M citrate buffer (pH = 5.00). Glycosidic aromatic precursors were released by enzymatic hydrolysis. The reconstituted sample was thawed and 250 mg of enzyme Lallzyme® BETA (Lallemand Inc., Zug, Switzerland) was added, remaining tightly covered in an oven at 40°C for 18 hours. Then it was left to cool and 500 μL of 4-nonanol 0.1 g/L was added. The process of volatile compound extraction described above was repeated once again, using SPE cartridges of 0.2 g of phase and 5 mL capacity.

2.5 GC-MS analysis of volatile compounds

Volatile compound analysis was performed with a Focus GC gas chromatograph system coupled to a mass spectrometer (ISQ, ThermoQuest, Waltham, MA, USA). A BP21, polyethylene glycol treated with nitroterephthalic acid, capillary column (50 m × 0.32 mm i.d.; 0.25 µm thick of Free Fatty Acid Phase (FFAP)) was used. For major volatile compounds, 0.8 µL of a mix solution (100 µL of wine sample, 100 µL of 4-methyl-2-pentanol (50 mg/L) as internal standard, and 1 mL of Milli-Q water) were injected in split mode. The gas chromatographic conditions were as follows: helium was used as carrier gas with a constant flow of 1.2 mL/min, the injector temperature was 195°C and oven temperature program was: 32°C (2 min), 5°C/min to 120°C, 75°C/min to 190°C, and 18 min at 190°C. Minor volatile compounds were analyzed by injecting 1 µL of sample in splitless mode after solid phase extraction (SPE). Operating conditions were as follows: carrier gas was helium (1 mL/min); injector temperature, 220°C; and oven temperature program was: 40°C for 15 min, 2°C/min to 100°C, 1°C/min to 150°C, 4°C/min to 210°C and 55 min at 210°C. The mass spectrometer operated in the electron impact mode with an electron energy of 70 eV, ion source temperature 250°C, and the global run time was recoded in full scan mode (mass scanning range, 40–250 amu). The identification of volatile compounds was performed by chromatographic retention times and mass spectra using commercial standards. These compounds were quantified by analyzing the characteristic m/z fragment for each compound following the internal standard method. When the commercial standards were not available, the volatile compound concentration was expressed as internal standard equivalents obtained by normalizing the compound peak area to that of the internal standard, multiplying by the internal standard concentration.

2.6 Odor activity values

The Odor Activity Values (OAVs) were determined to evaluate the contribution of a chemical compound to the wine aroma, providing the importance of a specific component to the sample odor. OAVs were calculated as the ratio between the individual compound concentration and the perception threshold found in literature [13, 14]. A possible contribution to the aroma of wine was considered when OAV was higher than 0.1.

2.7 Wine sensory analysis

Wines were sensorial evaluated by a trained panel of 11 experienced tasters from the IRIAF. The assessment was carried out under ISO standards related to taster selection and training [15], methodology and vocabulary [16], and tasting room [17], following the Napping technique [18]. Wine-taster members were asked to smell and taste the samples and to place them on white sheet of 40 × 60 cm, according to their similarities and dissimilarities. Samples close together on the sheet had similar sensory properties but if tasters perceived samples very differently, they had to place them far from each other. A second session was carried out providing a list of attributes imposed by the judge-in-chief and previously chosen by an expert panel. Napping sensory maps were obtained by Multiple Factorial Analysis (MFA) which provides useful information on the general perception of evaluated wines.

2.8 Statistical analysis

The statistical treatments were performed using the SPSS software version 23.0 (SPSS Inc., Chicago, Illinois, USA) and XLSTAT 2017 statistical software (Addinsoft, Paris, France). To determine statistically significant differences in the concentration of volatile compounds of wine aroma, a one-way analysis of variance (ANOVA, Student–Newman–Keuls/Tukey test, $p < 0.05$) was made. Sensory data were subjected to MFA to obtain the Napping sensory maps of wines.

3. Results and discussion

3.1 Aroma composition of grapes

The study of the aroma composition of these grapes allowed us to identify and quantify the free and bound volatile fractions usually located in berry skins by GC-MS (**Tables 1** and **2**).

Different families of volatiles described the varietal composition of these minority cultivars, that is, C6 compounds, terpenes, norisoprenoids, benzenic compounds, and alcohols. C6 compounds are derived from linoleic and lignoceric acids via enzymatic reactions and are responsible for herbaceous aroma in grapes and wines [20]. Among the free C6 compounds, 1-hexanol, *cis*-3-hexenol, *trans*-2-hexenol, hexanal, and *trans*-2-hexenal were present at relatively high concentrations in the six grape varieties, which is in agreement with previous results from *V. vinifera* grapes [21].

The bound fraction of C6 compounds was dominated by 1-hexanol, hexanal, and *trans*-2-hexenal. In all cases, total concentrations of bound C6 compounds were lower than those found in the free forms. Red grapes were characterized by a higher total concentration of free C6 compounds than other grape varieties grown in Spain [4]. Montonera del Casar was the grape genotype with a higher content of the free C6 compounds, reaching 2.72 g/L. However, the most abundant bound fraction of C6 compounds was determined in Tempranillo grapes (0.54 g/L). Terpenes have characteristic fruity and floral notes and are considered a positive quality factor because of their contribution to the varietal aroma although they are present at low levels [22]. Linalool, nerol, and geraniol were the terpenoids identified in these *V. vinifera* grapes. Among the terpenoids, geraniol was predominant in the terpene composition, mainly in Tinto Fragoso genotype. The concentration of these compounds observed in the glycosidically-bound forms was remarkable. Total concentrations of bound terpenes were 7–17 times higher than those observed in the free forms of the six grape cultivars. Similar results were found in a study on Godello and Agudelo grapes [23]. Norisopenoids are formed by carotenoid degradation and have been identified as potential impact odorants in wine [24]. These compounds were only identified in the bound form and significant differences were found between grape cultivars, showing the highest values in Tempranillo grapes. Total concentrations of norisoprenoids ranged from 0.05 to 0.17 mg/L.

Benzaldehyde, benzyl alcohol, 2-phenylethanol, phenylacetaldehyde, vanillin, and acetovanillone were the benzenic compounds found in the six grape varieties. Eugenol was only present in bound form. The concentration of each bound benzenic compound was, in general, higher than those observed in the free one, mainly benzyl alcohol and 2-phenylethanol which have been described as a floral aroma component [25]. Eugenol contributes to the spicy and smoky notes and has a low perception

DOI: http://dx.doi.org/10.5772/intechopen.1003666

Compounds	Tempranillo		Moribel		Tinto Fragoso	
	Free	Bound	Free	Bound	Free	Bound
1-Hexanol[1]	0.31 ± 0.02 b	0.15 ± 0.03 a	0.34 ± 0.03 b	0.08 ± 0.01 a	0.61 ± 0.09 c	0.14 ± 0.02 a
trans-3-Hexenol[2]	5.72 ± 0.27 a	ND	5.95 ± 0.45 a	ND	11.51 ± 2.70 b	ND
cis-3-Hexenol[1]	0.21 ± 0.02 d	0.08 ± 0.00 b	0.36 ± 0.03 e	0.05 ± 0.01 a	0.14 ± 0.02 c	0.03 ± 0.00 a
trans-2-Hexenol[1]	0.29 ± 0.03 d	0.03 ± 0.00 a	0.24 ± 0.02 c	0.01 ± 0.00 a	0.52 ± 0.03 e	0.07 ± 0.01 b
cis-2-Hexenol[2]	7.45 ± 0.50 d	2.55 ± 0.19 b	4.67 ± 0.38 c	1.45 ± 0.45 a	8.48 ± 0.84 e	2.26 ± 0.37 ab
Hexanal[1]	1.50 ± 0.07 d	0.12 ± 0.01 a	1.34 ± 0.11 c	0.08 ± 0.00 a	0.96 ± 0.11 b	0.06 ± 0.01 a
trans-3-Hexenal[2]	9.96 ± 1.00 a	ND	8.88 ± 0.96 a	ND	10.28 ± 1.26 a	ND
trans-2-Hexenal[1]	0.12 ± 0.01 ab	0.16 ± 0.02 c	0.17 ± 0.01 c	0.09 ± 0.01 a	0.13 ± 0.01 b	0.11 ± 0.02 ab
cis-2-Hexenal[2]	82.61 ± 4.95 d	0.95 ± 0.21 a	66.07 ± 12.07 c	0.68 ± 0.07 a	34.77 ± 7.63 b	0.68 ± 0.13 a
C6 compounds[1]	2.54 ± 0.15 b	0.54 ± 0.07 a	2.52 ± 0.13 b	0.30 ± 0.01 a	2.42 ± 0.27 b	0.41 ± 0.06 a
Linalool[2]	0.60 ± 0.15 ab	0.81 ± 0.24 b	0.25 ± 0.05 a	0.67 ± 0.35 ab	0.57 ± 0.01 ab	0.49 ± 0.14 ab
Nerol[2]	0.43 ± 0.07 a	6.88 ± 0.56 c	0.33 ± 0.05 a	3.02 ± 0.36 b	0.81 ± 0.10 a	7.40 ± 0.96 c
Geraniol[2]	1.35 ± 0.28 a	28.23 ± 2.48 b	1.57 ± 0.31 a	12.66 ± 1.89 a	6.85 ± 0.61 a	51.20 ± 11.15 c
Terpenes[2]	2.37 ± 0.34 a	35.91 ± 3.26 c	2.15 ± 0.37 a	16.35 ± 2.20 b	8.22 ± 0.71 ab	59.09 ± 11.87 d
3-Oxo-α-ionol[1]	ND	0.15 ± 0.02 b	ND	0.08 ± 0.01 a	ND	0.08 ± 0.01 a
β-Ionol[2]	ND	16.51 ± 1.29 c	ND	2.96 ± 0.20 a	ND	6.80 ± 0.83 b
Norisoprenoids[1]	ND	0.17 ± 0.02 b	ND	0.09 ± 0.01 a	ND	0.08 ± 0.02 a
Benzaldehyde[2]	5.41 ± 0.77 a	31.71 ± 6.41 bc	6.77 ± 0.39 a	36.18 ± 3.46 c	5.51 ± 1.06 a	27.87 ± 2.21 b
Benzyl alcohol[1]	0.01 ± 0.00 a	1.12 ± 0.22 c	0.08 ± 0.00 a	1.50 ± 0.20 d	0.02 ± 0.00 a	0.72 ± 0.04 b
Phenylacetaldehyde[2]	3.63 ± 1.08 ab	4.32 ± 0.51 ab	6.99 ± 0.53 c	3.17 ± 0.54 a	5.72 ± 1.96 bc	2.74 ± 0.31 a
2-Phenylethanol[1]	0.02 ± 0.00 a	0.57 ± 0.07 c	0.19 ± 0.03 b	0.58 ± 0.03 c	0.01 ± 0.00 a	0.19 ± 0.01 b

Compounds	Tempranillo		Moribel		Tinto Fragoso	
	Free	Bound	Free	Bound	Free	Bound
Eugenol[2]	ND	14.80 ± 1.88 b	ND	2.84 ± 0.62 a	ND	26.30 ± 1.83 c
Vanillin[1]	0.01 ± 0.00 a	0.05 ± 0.00 b	0.11 ± 0.04 c	0.04 ± 0.00 ab	0.01 ± 0.00 a	0.02 ± 0.00 ab
Acetovanillone[1]	0.01 ± 0.00 a	0.15 ± 0.03 c	0.01 ± 0.00 a	0.08 ± 0.00 b	0.00 ± 0.00 a	0.15 ± 0.01 c
Benzenic compounds[1]	0.04 ± 0.00 a	1.94 ± 0.31 d	0.40 ± 0.06 b	2.24 ± 0.24 e	0.05 ± 0.01 a	1.13 ± 0.04 c
1-Butanol[2]	12.23 ± 1.25 a	98.88 ± 4.09 e	20.17 ± 1.73 b	32.81 ± 1.74 c	30.76 ± 2.62 c	38.28 ± 4.27 d
3-Methyl-2-butanol[1]	ND	0.08 ± 0.01 b	ND	0.03 ± 0.00 a	ND	0.03 ± 0.00 a
3-Methyl-3-butanol[1]	1.79 ± 0.12 b	0.25 ± 0.02 a	4.53 ± 0.39 c	0.11 ± 0.01 a	2.48 ± 0.73 b	0.05 ± 0.01 a
1-Pentanol[2]	6.63 ± 0.86 a	40.70 ± 1.82 c	9.78 ± 0.93 a	25.48 ± 0.84 b	7.98 ± 0.54 a	26.45 ± 3.73 b
2-Ethyl-1-hexanol[1]	0.01 ± 0.00 bc	0.04 ± 0.00 e	0.02 ± 0.00 cd	0.02 ± 0.00 d	0.01 ± 0.00 e	0.01 ± 0.00 a
1-Octanol[2]	1.07 ± 0.11 a	11.97 ± 1.31 d	1.40 ± 0.26 a	5.68 ± 0.42 c	1.32 ± 0.19 a	3.37 ± 0.44 b
3-Octanol[2]	2.58 ± 0.09 b	2.51 ± 0.11 b	2.66 ± 0.05 b	2.49 ± 0.16 b	2.59 ± 0.02 b	2.12 ± 0.17 a
Alcohols[1]	0.04 ± 0.00 a	0.52 ± 0.03 d	0.05 ± 0.00 a	0.23 ± 0.01 c	0.06 ± 0.01 a	0.15 ± 0.02 b
Total[1]	2.62 ± 0.16 bc	3.20 ± 0.43 c	2.98 ± 0.08 bc	2.88 ± 0.26 bc	2.53 ± 0.27 b	1.83 ± 0.11 a

[1]*mg/L.*
[2]*µg/L.*
Abbreviation: ND = not detected.
Different letters in the same row indicate significant differences (ANOVA, Student–Newman–Keuls test, $p < 0.05$).
Mean value ± standard deviation, $n = 3$.
Adapted from Pérez-Navarro et al. [19].

Table 1.
Volatile composition of Tempranillo, Moribel, and Tinto Fragoso grape musts.

Compounds	Airén		Albillo Dorado		Montonera del Casar	
	Free	Bound	Free	Bound	Free	Bound
1-Hexanol[1]	0.17 ± 0.01 c	0.03 ± 0.00 a	0.27 ± 0.06 d	0.10 ± 0.02 b	0.46 ± 0.04 e	0.04 ± 0.00 a
trans-3-Hexenol[2]	5.28 ± 0.26 a	ND	4.49 ± 0.84 a	ND	31.17 ± 2.79 b	ND
cis-3-Hexenol[1]	0.45 ± 0.02 e	0.03 ± 0.00 a	0.37 ± 0.03 d	0.08 ± 0.01 b	0.16 ± 0.02 c	0.01 ± 0.00 a
trans-2-Hexenol[1]	0.25 ± 0.03 b	0.01 ± 0.00 a	0.27 ± 0.01 b	0.02 ± 0.00 a	0.60 ± 0.04 c	0.02 ± 0.00 a
cis-2-Hexenol[2]	3.80 ± 0.22 b	0.93 ± 0.35 a	6.17 ± 1.56 c	1.86 ± 0.93 a	12.08 ± 1.15 d	0.94 ± 0.32 a
Hexanal[1]	0.84 ± 0.04 b	0.06 ± 0.00 a	1.10 ± 0.14 c	0.06 ± 0.00 a	1.24 ± 0.12 d	0.06 ± 0.00 a
trans-3-Hexenal[2]	7.29 ± 0.15 a	ND	10.89 ± 0.27 c	ND	10.26 ± 0.18 b	ND
trans-2-Hexenal[1]	0.19 ± 0.01 b	0.08 ± 0.01 a	0.15 ± 0.00 c	0.19 ± 0.02 d	0.11 ± 0.02 b	0.17 ± 0.02 c
cis-2-Hexenal[2]	56.35 ± 1.83 b	0.51 ± 0.11 a	79.79 ± 8.79 c	0.44 ± 0.12 a	84.79 ± 8.34 c	0.46 ± 0.06 a
C6 compounds[1]	1.90 ± 0.11 c	0.21 ± 0.02 a	2.27 ± 0.13 d	0.45 ± 0.03 b	2.72 ± 0.18 e	0.30 ± 0.03 ab
Linalool[2]	0.26 ± 0.02 a	0.62 ± 0.11 a	0.24 ± 0.03 a	0.55 ± 0.35 a	0.25 ± 0.02 a	0.62 ± 0.18 a
Nerol[2]	0.39 ± 0.04 a	6.27 ± 0.55 d	0.36 ± 0.06 a	2.81 ± 0.23 b	0.65 ± 0.08 a	4.51 ± 0.44 c
Geraniol[2]	1.39 ± 0.26 a	19.14 ± 2.16 b	3.16 ± 0.38 a	21.27 ± 7.77 b	2.09 ± 0.63 a	38.65 ± 4.35 c
Terpenes[2]	2.04 ± 0.26 a	26.02 ± 1.91 b	3.76 ± 0.41 a	24.63 ± 7.93 b	3.83 ± 0.98 a	43.79 ± 4.83 c
3-Oxo-α-ionol[1]	ND	0.04 ± 0.01 a	ND	0.13 ± 0.01 c	ND	0.07 ± 0.01 b
β-Ionol[2]	ND	5.20 ± 0.78 b	ND	14.91 ± 0.87 c	ND	1.71 ± 0.20 a
Norisoprenoids[1]	ND	0.05 ± 0.02 a	ND	0.14 ± 0.01 c	ND	0.07 ± 0.01 b
Benzaldehyde[2]	4.40 ± 0.50 a	16.39 ± 1.78 a	7.76 ± 0.69 a	25.49 ± 1.32 a	17.76 ± 0.87 a	498.96 ± 32.79 b
Benzyl alcohol[1]	0.01 ± 0.00 a	0.49 ± 0.03 b	0.01 ± 0.00 a	0.61 ± 0.03 c	0.02 ± 0.00 a	0.84 ± 0.14 d
Phenylacetaldehyde[2]	4.76 ± 0.34 b	2.65 ± 0.40 a	5.11 ± 0.08 b	2.81 ± 0.26 a	12.27 ± 1.99 c	3.79 ± 0.27 ab

Compounds	Airén		Albillo Dorado		Montonera del Casar	
	Free	Bound	Free	Bound	Free	Bound
2-Phenylethanol[1]	0.01 ± 0.00 a	0.25 ± 0.03 b	0.02 ± 0.00 a	0.46 ± 0.04 d	0.03 ± 0.00 a	0.34 ± 0.03 c
Eugenol[2]	ND	5.50 ± 1.56 b	ND	2.09 ± 0.36 a	ND	1.00 ± 0.22 a
Vanillin[2]	6.75 ± 2.39 a	24.71 ± 6.55 b	27.68 ± 8.64 b	9.93 ± 1.05 a	66.42 ± 11.29 c	9.37 ± 0.65 a
Acetovanillone[2]	0.89 ± 0.58 a	79.25 ± 10.71 c	4.16 ± 0.69 a	62.71 ± 3.57 b	5.27 ± 1.45 a	55.52 ± 5.86 b
Benzenic compounds[1]	0.03 ± 0.00 a	0.87 ± 0.06 b	0.08 ± 0.01 a	1.17 ± 0.07 c	0.15 ± 0.01 a	1.74 ± 0.17 d
1-Butanol[2]	2.72 ± 0.27 a	6.92 ± 1.08 b	23.10 ± 2.22 d	45.67 ± 1.36 e	6.35 ± 0.29 b	10.65 ± 1.23 c
3-Methyl-2-butanol[2]	ND	48.19 ± 3.81 a	ND	44.88 ± 4.71 a	ND	47.28 ± 1.87 a
3-Methyl-3-butanol[2]	1.74 ± 0.24 a	94.26 ± 9.15 b	2.05 ± 0.12 a	104.86 ± 14.51 b	2.41 ± 0.17 a	92.79 ± 10.81 b
1-Pentanol[2]	5.96 ± 0.88 a	15.88 ± 2.65 c	9.80 ± 1.12 b	28.06 ± 1.94 e	9.12 ± 0.25 b	19.28 ± 2.00d
2-Ethyl-1-hexanol[2]	9.94 ± 0.39 a	22.60 ± 1.69 c	11.58 ± 0.73 a	27.08 ± 3.29 d	14.82 ± 0.63 b	22.60 ± 0.59 c
1-Octanol[2]	1.15 ± 0.24 a	3.42 ± 0.77 b	1.37 ± 0.20 a	3.43 ± 1.07 b	1.15 ± 0.14 a	2.52 ± 0.37 ab
3-Octanol[2]	2.14 ± 0.05 a	2.38 ± 0.10 b	2.61 ± 0.12 b	2.40 ± 0.16 b	2.61 ± 0.07 b	2.02 ± 0.13 a
Alcohols[1]	0.02 ± 0.00 a	0.19 ± 0.01 c	0.05 ± 0.00 b	0.25 ± 0.02 d	0.04 ± 0.00 ab	0.20 ± 0.02 c
Total[1]	1.96 ± 0.11 b	1.34 ± 0.07 a	2.40 ± 0.91 c	2.03 ± 0.09 b	2.91 ± 0.17 d	2.35 ± 0.21 c

[1]*mg/L.*
[2]*µg/L.*
Abbreviation: ND = not detected.
Different letters in the same row indicate significant differences (ANOVA, Student–Newman–Keuls test, $p < 0.05$).
Mean value ± standard deviation, $n = 3$.
Adapted from Pérez-Navarro et al. [19].

Table 2.
Volatile composition of Airén, Albillo Dorando, and Montonera del Casar grape musts.

threshold (6 μg/L) [26], thus may have a significant impact on wine aroma. In all grapes evaluated, the concentrations of this compound exceeded its odor threshold with the highest concentration in Tinto Fragoso grapes (26.30 μg/L). In this study, red grape cultivars were characterized by a higher eugenol concentration than white ones. Several alcohols in both free and bound forms were identified in the studied grapes. However, 3-methyl-2-butanol was only detected in the free form. The six grape cultivars had greater levels of bound alcohols than the free fraction. The higher amounts of alcohol in the glycosidic form were 5 to 13 times higher than those in the free forms. These compounds derived from grapes do have not a significant impact on wine aroma due to the number of alcohols formed during alcoholic fermentation as secondary yeast metabolites [6]. The total concentration of free volatiles varied according to grape cultivar, with values between 1.96 and 2.98 mg/L. In general, total concentration of the bound fraction of volatile compounds was lower than those observed in the free forms. However, the proportion of glycosylated fraction present in all grapes may have a relevant role in the aromatic potential of these grapes, with a concentration ranging from 1.34 to 3.20 mg/L. The bound volatile compounds can be converted into free form by hydrolysis, modifying the aromatic profile of wines and enhancing the varietal character [27].

3.2 Aroma composition of wines

Wines fermented by the commercial *Saccharomyces cerevisiae* strain used were analyzed by gas chromatography and mass spectrometry to evaluate the influence of the yeast strain on aromatic profile of wines made from Airén, Tempranillo, and the minority grape varieties. The results of the GC-MS analysis of wines are given in **Table 3**. Differences in the aroma components of wines were noticeable even the typical varietal aroma was preserved.

Volatile compounds affecting the secondary aroma of wine are produced *via* yeast metabolism during alcoholic fermentation and have a significant impact on sensory properties of wine [30]. Some of these compounds are alcohols and esters which are present at high concentrations in the evaluated wines. Alcohols are secondary metabolites of yeast and optimal levels of them provide fruity odors. However, an excessive amount of these compounds can have negative impact on aroma and flavor of wine, resulting in pungent smell and taste [31].

In wine, variations in the alcohol concentration can occur using different yeast strains during alcoholic fermentation [32]. Total alcohol concentrations ranged from 284 to 519 mg/L in the evaluated samples, showing greater values in red wines. Isoamyl alcohol is synthesized in the yeast cell through the Ehrlich pathway [6]. In all wines, this compound was the most abundant alcohol produced (173–305 mg/L). Most of ethyl esters of carboxylic acids are biosynthesized by the yeast metabolism during fermentation. All samples contained ethyl lactate which was produced at high levels in red wines due to the malolactic fermentation. Ethyl acetate is a compound related to fruity aromas and was the most abundant ester determined in wines, with concentration values between 50 and 85 mg/L. Other significant esters are isoamyl acetate (banana odor) and 2-phenyltehyl acetate (honey, fruity and flowery aromas).

Acetaldehyde is also produced by the metabolism of yeasts and its production is affected by SO_2 content of the media [33]. Significant differences in acetaldehyde content were observed among wines, showing lower levels in red wines due to the least amount of SO_2 used in red winemaking. An important group of volatile compounds in wines are carboxylic acids and their lactones which are products of the lipid

Compounds	Tempranillo	Moribel	Tinto Fragoso	Airén	Albillo Dorado	Montonera del Casar
1-Hexanol[1]	2.50 ± 0.02 d	1.82 ± 0.04 c	3.16 ± 0.09 e	0.80 ± 0.18 b	0.66 ± 0.03 b	0.39 ± 0.01 a
trans-3-Hexenol[1]	0.07 ± 0.01 a	0.15 ± 0.01 b	0.09 ± 0.01 a	0.17 ± 0.04 b	0.16 ± 0.00 b	0.08 ± 0.00 a
cis-3-Hexenol[1]	0.99 ± 0.07 b	1.25 ± 0.08 b	1.11 ± 0.01 b	1.86 ± 0.37 c	1.01 ± 0.01 b	0.05 ± 0.00 a
C6 compounds[1]	3.56 ± 0.05 c	3.21 ± 0.13 c	4.36 ± 0.11 d	2.83 ± 0.59 c	1.82 ± 0.09 b	0.52 ± 0.01 a
Linalool[2]	2.16 ± 0.04 ab	1.40 ± 0.17 a	4.76 ± 0.54 c	2.98 ± 0.58 b	2.78 ± 0.20 b	3.17 ± 0.19 b
Citronellol[2]	2.88 ± 0.38 b	1.60 ± 0.09 a	4.25 ± 0.08 c	2.11 ± 0.29 ab	2.21 ± 0.35 ab	2.49 ± 0.04 b
Geraniol[2]	12.18 ± 0.84 c	14.64 ± 0.77 d	18.12 ± 0.43 e	4.94 ± 0.91 a	4.12 ± 0.60 a	6.75 ± 0.43 b
Terpenes[2]	17.22 ± 0.42 c	17.64 ± 0.69 c	27.13 ± 0.89 d	10.02 ± 1.78 ab	9.11 ± 1.14 a	12.41 ± 0.66 b
β-Damascenone[2]	4.73 ± 0.90 ab	2.30 ± 0.65 a	6.41 ± 0.22 ab	3.50 ± 0.52 a	10.57 ± 2.98 b	7.76 ± 2.67 ab
β-Ionone[2]	0.52 ± 0.01 b	0.47 ± 0.05 b	0.58 ± 0.06 b	0.10 ± 0.02 a	0.08 ± 0.01 a	0.02 ± 0.00 a
Norisoprenoids[2]	5.24 ± 0.91 ab	2.78 ± 0.70 a	6.99 ± 0.16 ab	3.60 ± 0.54 ab	10.66 ± 3.00 b	7.78 ± 2.67 ab
Benzaldehyde[2]	ND	ND	ND	2.32 ± 0.79 b	0.98 ± 0.15 a	44.13 ± 0.31 c
Eugenol[2]	4.17 ± 0.45 a	13.39 ± 1.41 b	59.24 ± 5.82 c	2.86 ± 0.15 a	0.93 ± 0.17 a	0.82 ± 0.18 a
Guaiacol[2]	10.89 ± 0.28 b	7.38 ± 0.13 ab	9.93 ± 4.07 b	2.14 ± 0.47 a	2.30 ± 0.66 a	11.82 ± 1.43 b
2-Phenyletanol[1]	28.62 ± 3.09 b	20.93 ± 1.97 a	39.69 ± 3.15 c	17.70 ± 5.85 a	17.86 ± 1.19 a	11.74 ± 0.29 a
Vanillin[2]	22.82 ± 1.51 b	28.61 ± 5.97 b	31.97 ± 3.82 b	7.33 ± 1.26 a	9.44 ± 2.51 a	5.75 ± 0.40 a
Acetovainillone[1]	0.09 ± 0.00 a	0.09 ± 0.00 a	0.44 ± 0.03 b	ND	ND	ND
Benzenic compounds[1]	28.74 ± 3.09 b	21.08 ± 1.97 ab	40.24 ± 3.19 c	17.71 ± 5.85 a	17.87 ± 1.19 a	11.80 ± 0.29 a
Methanol[1]	80.97 ± 29.53 a	84.51 ± 5.42 a	118.16 ± 40.50 a	48.95 ± 1.90 a	40.53 ± 1.24 a	120.37 ± 3.28 a
1-Propanol[1]	32.35 ± 5.66 a	38.36 ± 1.99 a	46.68 ± 11.50 a	ND	ND	ND
Isoamyl alcohol[1]	262.11 ± 24.31 c	242.84 ± 8.16 bc	305.28 ± 7.17 d	200.17 ± 0.75 ab	231.18 ± 21.49 bc	173.87 ± 0.71 a
3-Ethoxy-propanol[2]	27.36 ± 10.92 a	35.23 ± 4.76 a	115.55 ± 70.69 a	3.55 ± 1.63 a	2.22 ± 1.59 a	14.58 ± 1.90 a
3-Methylthio-propanol[1]	0.26 ± 0.06 a	0.23 ± 0.05 a	0.85 ± 0.06 a	0.25 ± 0.08 a	0.22 ± 0.10 a	0.25 ± 0.04 a

Compounds	Tempranillo	Moribel	Tinto Fragoso	Airén	Albillo Dorado	Montonera del Casar
Isobutanol[1]	41.13 ± 1.25 a	36.92 ± 3.12 a	48.34 ± 4.24 a	34.88 ± 3.54 a	52.19 ± 11.76 a	50.95 ± 1.34 a
Alcohols[1]	416.84 ± 49.48 a	402.90 ± 2.43 a	519.41 ± 55.06 b	284.26 ± 2.73 a	324.12 ± 34.58 a	345.45 ± 1.27 a
Isobutyric acid[1]	0.23 ± 0.01 a	0.23 ± 0.05 a	0.29 ± 0.16 b	ND	ND	ND
Butyric acid[1]	0.48 ± 0.07 a	0.50 ± 0.07 a	0.28 ± 0.17 a	0.49 ± 0.07 a	0.32 ± 0.09 a	0.58 ± 0.04 a
Isovaleric acid[1]	1.06 ± 0.06 a	1.20 ± 0.23 a	1.27 ± 0.83 a	1.08 ± 0.30 a	0.81 ± 0.22 a	0.91 ± 0.11 a
Hexanoic acid[1]	12.15 ± 1.37 b	14.01 ± 0.47 b	5.75 ± 3.42 a	12.29 ± 1.83 b	10.87 ± 0.52 ab	10.16 ± 0.39 ab
Octanoic acid[1]	19.57 ± 2.67 ab	26.39 ± 2.06 b	10.42 ± 6.06 a	48.48 ± 7.08 c	38.65 ± 1.28 c	43.93 ± 4.87 c
Decanoic acid[1]	4.18 ± 0.98 a	6.46 ± 1.15 a	2.90 ± 1.68 a	12.26 ± 0.35 b	12.91 ± 0.38 b	19.99 ± 3.71 c
Acids[1]	37.68 ± 5.16 ab	48.79 ± 4.03 bc	20.91 ± 12.32 a	74.61 ± 8.93 c	63.56 ± 1.47 c	75.58 ± 9.13 c
Acetaldehyde[1]	10.69 ± 1.30 a	8.72 ± 0.43 a	11.16 ± 1.36 a	76.06 ± 6.43 d	41.43 ± 3.46 c	32.17 ± 3.58 b
Aldehydes[1]	10.69 ± 1.30 a	8.72 ± 0.43 a	11.16 ± 1.36 a	76.06 ± 6.43 d	41.43 ± 3.46 c	32.17 ± 3.58 b
Ethyl acetate[1]	63.42 ± 12.04 a	85.49 ± 1.91 b	52.09 ± 3.42 a	43.35 ± 5.13 a	50.21 ± 5.49 a	60.28 ± 2.18 a
Isoamyl acetate[1]	7.17 ± 2.38 b	10.34 ± 1.34 b	2.13 ± 0.27 a	7.08 ± 1.13 b	6.97 ± 0.60 b	6.32 ± 0.41 b
2-Phenylethyl acetate[1]	0.28 ± 0.04 b	0.22 ± 0.06 b	0.13 ± 0.01 a	0.26 ± 0.03 b	0.26 ± 0.02 b	0.26 ± 0.02 b
Ethyl lactate[1]	36.35 ± 0.43 c	24.34 ± 1.08 b	38.87 ± 4.22 c	5.06 ± 0.52 a	7.41 ± 0.72 a	6.47 ± 0.44 a
Ethyl butyrate[1]	0.33 ± 0.03 a	0.47 ± 0.03 b	0.22 ± 0.01 a	0.25 ± 0.06 a	0.30 ± 0.04 a	0.26 ± 0.01 a
Ethyl hexanoate[1]	0.57 ± 0.15 a	0.61 ± 0.01 a	0.59 ± 0.04 a	1.03 ± 0.12 b	0.89 ± 0.06 ab	0.85 ± 0.00 ab
Ethyl octanoate[1]	0.68 ± 0.10 a	0.66 ± 0.01 a	0.55 ± 0.06 a	1.47 ± 0.19 b	1.34 ± 0.17 b	1.39 ± 0.04 b
Ethyl decanoate[1]	0.12 ± 0.02 a	0.12 ± 0.00 a	0.11 ± 0.02 a	0.27 ± 0.01 b	0.36 ± 0.07 c	0.41 ± 0.01 c
Esters[1]	108.94 ± 15.17 bc	122.24 ± 1.53 c	94.68 ± 5.04 b	58.75 ± 5.45 a	67.73 ± 5.70 a	76.24 ± 3.01 a
Furaneol[2]	12.30 ± 0.78 a	23.54 ± 0.05 a	14.18 ± 4.07 a	27.34 ± 4.71 a	15.24 ± 8.36 a	17.40 ± 3.93 a
Furanic compounds[2]	12.30 ± 0.78 a	23.54 ± 0.05 a	14.18 ± 4.07 a	27.34 ± 4.71 a	15.24 ± 8.36 a	17.40 ± 3.93 a

Compounds	Tempranillo	Moribel	Tinto Fragoso	Airén	Albillo Dorado	Montonera del Casar
γ-Butyrolactone[1]	0.07 ± 0.01 a	0.05 ± 0.00 a	0.16 ± 0.04 b	0.04 ± 0.01 a	0.07 ± 0.02 a	0.11 ± 0.00 a
γ-Nonalactone[2]	10.13 ± 0.02 c	12.85 ± 1.16 d	23.38 ± 2.09 e	3.29 ± 0.68 a	7.19 ± 0.50 b	6.29 ± 0.13 b
δ-Dodecalactone[2]	46.11 ± 5.28 a	45.73 ± 3.79 a	51.46 ± 0.70 a	108.12 ± 28.01 b	70.54 ± 17.90 a	63.00 ± 0.07 a
Lactones[1]	0.13 ± 0.00 a	0.12 ± 0.01 a	0.24 ± 0.04 b	149.15 ± 37.65 ab	144.67 ± 39.34 ab	175.37 ± 3.37 ab

[1]*mg/L.*
[2]*μg/L.*
Abbreviation: ND = not detected.
Different letters in the same row indicate significant differences (ANOVA, Student–Newman–Keuls test, $p < 0.05$).
Mean value ± standard deviation, $n = 2$.
Adapted from Pérez-Navarro et al. [19, 28, 29].

Table 3.
Volatile composition of red and white wines.

metabolism of yeast and provide fatty and wax-like smell [34]. The concentrations of these compounds were under the odor threshold and did not influence final aroma of wines.

Green and grassy odors are important for the final aroma of wines and are represented by C6 compounds. In all wine samples, 1-hexanol and *trans*-3-hexenol were quantified in a concentration below the odor threshold value [35], except for *cis*-3-hexenol which is the predominant isomer of 3-hexenol in wines [36].

The production of monoterpenes from different precursors depends on *S. cerevisiae* strain, although its hydrolysis action on bonded terpene fraction in the first fermentation stages is one of the most relevant contributions [37]. Linalool, citronellol, and geraniol were the terpenes identified and quantified in the evaluated wines, which are associated with floral odors [38]. In general, geraniol was the most abundant terpene determined in all samples (4.12–18.12 μg/L). These values were close to those reported previously in literature [39]. The concentrations of these compounds were under the odor thresholds so the contribution of terpenes to resulting wine aroma was minimal.

The concentrations of β-damascenone were higher than those of β-ionone in all wines. The odor threshold of β-damascenone was exceeded, contributing to wine aroma profile [40]. Among benzenic compounds, benzaldehyde was only determined in white wines with a concentration that did not exceed the odor threshold (2000 μg/L) [38] although it could have a synergic effect on wine aroma, providing fruity notes. Tinto Fragoso wine was characterized by a volatile composition rich in eugenol. This was also observed in grapes so this compound may be considered as a varietal market for this grape cultivar. The 2-phenylethanol was determined at significantly higher concentration than the odor threshold in red wines, supporting rose odors of terpenes like geraniol. Red wines showed 2-phenylethanol concentrations similar to those reported for wines made from Bobal grapes [41].

A large number of volatile compounds were determined in the evaluated wines. However, the impact of these compounds on the overall aroma character is different. To evaluate the influence of each volatile compound on wine aroma, Odor Activity Values (OAVs) were calculated as the ratio between the concentration of the aroma compound and its odor threshold concentration (**Table 4**).

The OAVs higher than 0.1 were grouped into several aromatic series to estimate the overall wine aroma and each compound was assigned to one or several aromatic series (fruity, floral, green/fresh, sweet, spicy, fatty, and other odors) based on similar odor descriptors.

A plot of the values obtained for the aromatic series in the wines made from the six grape cultivars is shown in **Figure 2**, indicating the mean values of each series. The main aroma categories were series I (fruity), III (green, fresh), and VII (other odors) which contributed most to the aroma profile of wines. A great impact of odor compounds on the aromatic profile was observed in white wines, mainly in series I and VII. Tinto Fragoso was characterized by higher OAVs in the spicy aromatic series, which may be explained by its eugenol concentration.

3.3 Sensory profiles

The Napping technique was employed to evaluate the sensory attributes identified in the produced wines from the grape cultivars studied. **Figure 3A** shows Napping data obtained from the sensory analysis of red wines.

A clear separation between red wines was observed in the plot, which indicates that wines made from Tempranillo, Moribel, and Tinto Fragoso showed a distinctive sensory

Compounds	Odor descriptors	Odor threshold (μg/L)	Aromatic series
1-Hexanol	Flower, green, cut grass	8000 [35]	II, III
trans-3-Hexenol	Green	400 [38]	III
cis-3-Hexenol	Green, cut grass	400 [35]	III
Linalool	Floral	15 [35]	II
Citronellol	Floral	100 [35]	II
Geraniol	Roses, geranium	30 [35]	II
β-Damascenone	Sweet, fruit	0.05 [35]	I, IV
β-Ionone	Floral, violet	0.09 [26]	II
Benzaldehyde	Sweet, cherry, almond	2000 [38]	I, IV
Eugenol	Spices, clove, honey	6 [26]	IV, V, VII
Guaiacol	Medicine, caramel, smoke	10 [26]	IV, VI
2-Phenyletanol	Floral, roses	10,000 [35]	II
Vanillin	Vanilla	60 [42]	V, VII
Acetovainillone	Vanilla, clove	1000 [42]	VI
Methanol	Chemical, medicinal	668,000 [38]	VI
1-Propanol	Ripe fruit, alcohol	830,000 [38]	I, VI
Isoamyl alcohol	Solvent, fusel	30,000 [35]	VII
3-Ethoxy-propanol	Fruity	100 [43]	I
3-Methylthio-propanol	Cooked, vegetable	500 [35]	VI
Isobutanol	Bitter, green	40,000 [35]	III, VI
Isobutyric acid	Rancid, butter, cheese	2300 [26]	VI
Butyric acid	Rancid, cheese, sweat	173 [26]	VI
Isovaleric acid	Acid, rancid	33 [26]	IV, VI
Hexanoic acid	Sweat	420 [26]	VI
Octanoic acid	Sweat, cheese	500 [26]	VI
Decanoic acid	Rancid fat	1000 [26]	VI
Acetaldehyde	Pungent, ripe apple	500 [35]	I, VI
Ethyl acetate	Fruity, solvent	7500 [35]	I, VI
Isoamyl acetate	Banana	30 [26]	I
2-Phenylethyl acetate	Floral, roses	250 [35]	II
Ethyl lactate	Acid, medicine	154,636 [42]	VI
Ethyl butyrate	Fruity	20 [35]	I
Ethyl hexanoate	Green apple	14 [26]	I
Ethyl octanoate	Sweet, fruity	5 [26]	I, II, IV
Ethyl decanoate	Sweet, fruity	200 [26]	I, IV
Furaneol	Burnt sugar, caramel, maple	5 [44]	IV
γ-Butyrolactone	Sweet, toast, caramel	35,000 [40]	IV

Compounds	Odor descriptors	Odor threshold (µg/L)	Aromatic series
γ-Nonalactone	Coconut	30 [26]	IV
δ-Dodecalactone	Coconut	88 [38]	IV

Aromatic series: I (fruity), II (floral), III (green, fresh), IV (sweet), V (spicy), VI (fatty), and VII (other odors).

Table 4.
Aromatic series, odor descriptors, and thresholds of some volatile compounds in wines.

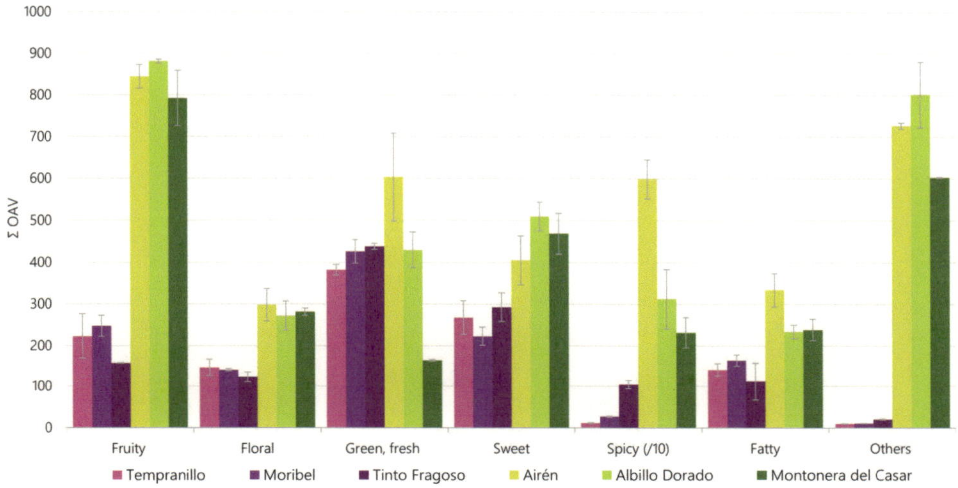

Figure 2.
Aromatic series in red and white wines (mean value and standard deviation, n = 2). Adapted from Pérez-Navarro et al. [19, 28, 29].

profile. Tinto Fragoso wine was characterized by its color with more intensity and purplish nuances. This wine had the most aromatic profile of evaluated red wines, providing spicy notes and persistence in mouth. Wines made from Tempranillo and Moribel grapes showed different odor descriptors. Tempranillo was characterized by red fruity odors, while forest berry and ripe fruit notes defined the aromatic profile of Moribel wine.

The sensory evaluation of white wines by a panel of experienced wine tasters is shown in **Figure 3B**. Airén wine was characterized by green apple odors that are well-known for this grape cultivar [39], and it was located close to Albillo Dorado wine in the plot, which had floral and fruity aromas. More herbaceous odors were determined in Montonera del Casar wine.

4. Conclusions

This chapter gathers the first comprehensive characterization of the chemical aroma composition and sensory properties of wines made from several minority grape cultivars from Castilla-La Mancha (Spain), showing the importance of a proper yeast stain for the development of the desired aroma components in wine. The strong impact exerted by the strain of *Saccharomyces cerevisiae* on wine aroma composition becomes in many cases only from the presence of aromatic fractions from grapes,

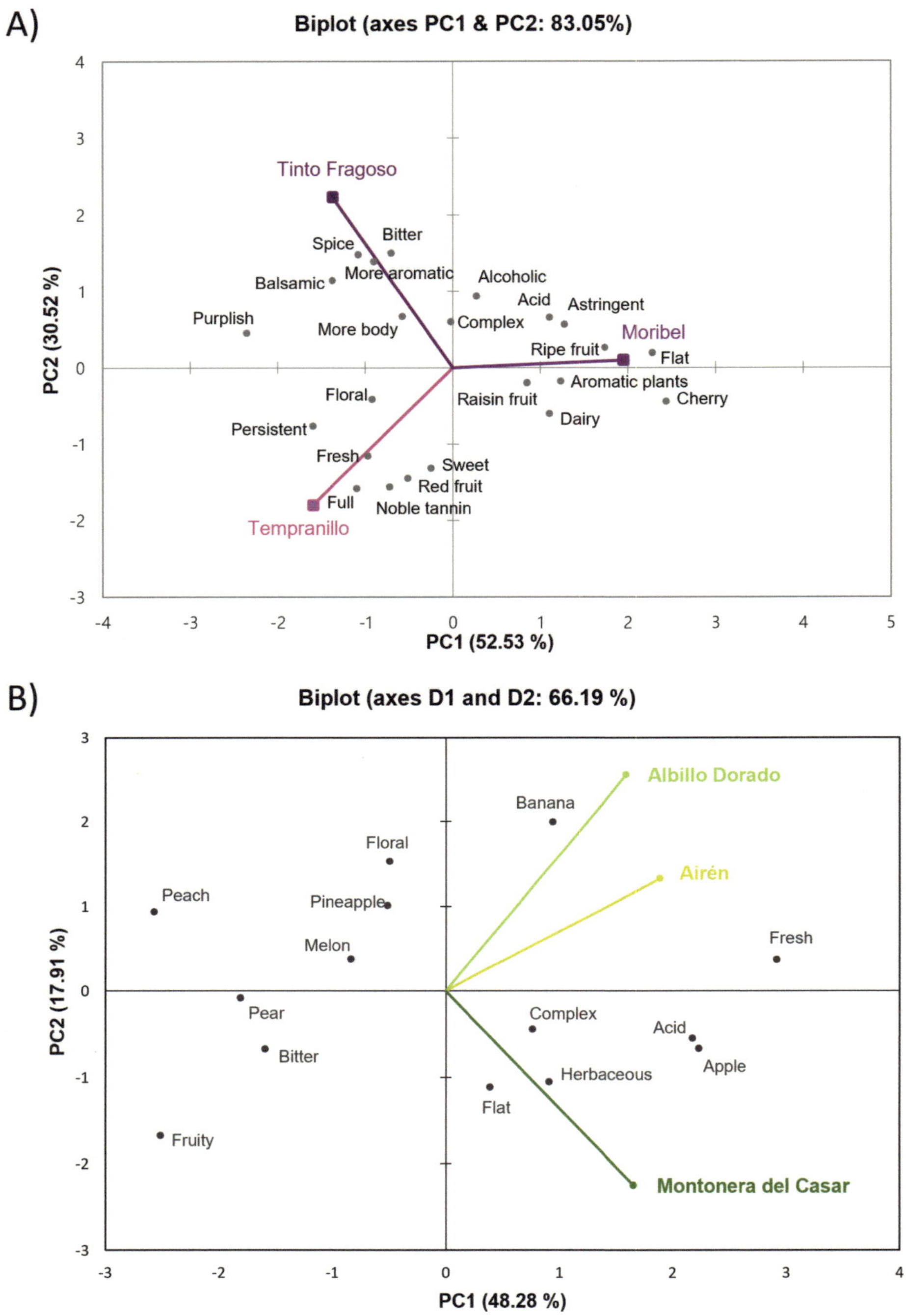

Figure 3.
Sensory characterization of red (A) and white (B) wines. Adapted from Pérez-Navarro et al. [19, 28, 29].

having a notable influence on yeast metabolism. Based on the results obtained, two patterns of aroma formation and release from the yeast employed in the vinification of the six grape cultivars are the production of ethyl esters and acetates that provide fruit and fresh attributes and the release of aroma components from grape precursors which enhance the aromatic complexity of wines.

Therefore, the yeast strain selection is pivotal in grape must fermentation due to its significant impact on wine sensory properties, affecting consumer preferences. In addition, this work displayed that these minority grape varieties present an aroma profile characteristic and may be considered a viable alternative to well-known grape varieties used in wine production in this region.

Acknowledgements

This research was funded by the Castilla-La Mancha Regional Government (project POII-2014-008-P). J. P-N also thanks the European Union for financial support through the European Social Fund Plus (ESF+).

Conflict of interest

The authors declare no conflict of interest.

Author details

José Pérez-Navarro[1,2]*, Adela Mena-Morales[3], Sergio Gómez-Alonso[1,4], Esteban García-Romero[3] and Pedro Miguel Izquierdo-Cañas[3]

1 Regional Institute for Applied Scientific Research (IRICA), University of Castilla-La Mancha, Ciudad Real, Spain

2 Higher Technical School of Agronomic Engineering, University of Castilla-La Mancha, Ciudad Real, Spain

3 Instituto Regional de Investigación y Desarrollo Agroalimentario y Forestal de Castilla-La Mancha (IRIAF), Centro de Investigación de la Vid y el Vino de Castilla-La Mancha (IVICAM), Tomelloso, Spain

4 Faculty of Chemical Sciences and Technologies, University of Castilla-La Mancha, Ciudad Real, Spain

*Address all correspondence to: jose.pnavarro@uclm.es

References

[1] Karabegović I, Malićanin M, Danilović B, Stanojević J, Stojanović SS, Nikolić N, et al. Potential of non-Saccharomyces yeast for improving the aroma and sensory profile of Prokupac red wine. Oeno One. 2021;**55**(2):181-195

[2] Gambetta JM, Bastian SE, Cozzolino D, Jeffery DW. Factors influencing the aroma composition of chardonnay wines. Journal of Agricultural and Food Chemistry. 2014;**62**(28):6512-6534

[3] Hashizume K, Samuta T. Grape maturity and light exposure affect berry methoxypyrazine concentration. American Journal of Enology and Viticulture. 1999;**50**(2):194-198

[4] Canosa P, Oliveira JM, Masa A, Vilanova M. Study of the volatile and glycosidically bound compounds of minority Vitis vinifera red cultivars from NW Spain. Journal of the Institute of Brewing. 2011;**117**(3):462-471

[5] Ferreira V, De-la-Fuente-Blanco A, Sáenz-Navajas M. Wine aroma vectors and sensory attributed. In: Reynolds AG, editor. Managing Wine Quality. 2nd ed. Cambridge, UK: Woodhead Publishing Limited; 2022. pp. 3-39

[6] Swiegers J, Bartowsky E, Henschke P, Pretorius I. Yeast and bacterial modulation of wine aroma and flavour. Australian Journal of Grape and Wine Research. 2005;**11**(2):139-173

[7] Denat M, Ferreira V, Pérez D, Sáenz-Navajas MP, Heras J. Impact of two yeast strains on Tempranillo red wine aroma profiles throughout accelerated ageing. OENO One. 2021;**55**(4):181-195

[8] Roca P. State of the World Vine and Wine Sector. OIV Press Conference; 2022. DOI: 10.1787/9789264244047-44-en

[9] Wilmink J, Breuer M, Forneck A. Effects of grape phylloxera leaf infestation on grapevine growth and yield parameters in commercial vineyards: A pilot study. OENO One. 2022;**56**(1):197-208

[10] Robinson J, Harding J, Vouillamoz J. Wine Grapes: A Complete Guide to 1,368 Vine Varieties, Including their Origins and Flavours. Harmondsworth, UK: Allen Lane/Penguin; 2012

[11] Mena A, Martínez J, Fernández-González M. Recovery, identification and relationships by microsatellite analysis of ancient grapevine cultivars from Castilla-La Mancha: The largest wine growing region in the world. Genetic Resources and Crop Evolution. 2014;**61**:625-637

[12] Ibarz MJ, Ferreira V, Hernández-Orte P, Loscos N, Cacho J. Optimization and evaluation of a procedure for the gas chromatographic–mass spectrometric analysis of the aromas generated by fast acid hydrolysis of flavor precursors extracted from grapes. Journal of Chromatography A. 2006;**1116**(1-2):217-229

[13] Francis I, Newton J. Determining wine aroma from compositional data. Australian Journal of Grape and Wine Research. 2005;**11**(2):114-126

[14] Vilanova M, Masa A, Tardaguila J. Evaluation of the aroma descriptors variability in Spanish grape cultivars by a quantitative descriptive analysis. Euphytica. 2009;**165**:383-389

[15] International Organization for Standardization. ISO 8586-1:1993 - Sensory Analysis. General Guidance for the Selection, Training and Monitoring of Assessors. Part 1: Selected Assessors; 1993

[16] International Organization for Standardization. ISO 11035:1994 - Sensory Analysis. Identification and Selection of Descriptors for Establishing a Sensory Profile by a Multidimensional Approach; 1994

[17] International Organization for Standardization. ISO 8589:2007 - Sensory Analysis. General Guidance for the Design of Test Rooms; 2007

[18] Pagès J. Recueil direct de distances sensorielles: Application à l'évaluation de dix vins blancs du Val-de-Loire. Sciences Des Aliments. 2003;**23**(5-6):679-688

[19] Pérez-Navarro J. Calidad enológica de nuevos genotipos de uva identificados en el Banco de Germoplasma de Vid de Castilla-La Mancha [thesis]. Ciudad Real: University of Castilla-La Mancha; 2019

[20] Kalua CM, Boss PK. Evolution of volatile compounds during the development of Cabernet Sauvignon grapes (*Vitis vinifera* L.). Journal of Agricultural and Food Chemistry. 2009;**57**(9):3818-3830

[21] Qian X, Sun L, Xu X-Q, Zhu B-Q, Xu H-Y. Differential expression of VvLOXA diversifies C6 volatile profiles in some Vitis vinifera table grape cultivars. International Journal of Molecular Sciences. 2017;**18**(12):2705

[22] Mateo J, Jiménez M. Monoterpenes in grape juice and wines. Journal of Chromatography A. 2000;**881**(1-2):557-567

[23] Vilanova M, Genisheva Z, Bescansa L, Masa A, Oliveira JM. Changes in free and bound fractions of aroma compounds of four *Vitis vinifera* cultivars at the last ripening stages. Phytochemistry. 2012;**74**:196-205

[24] Vilanova M, Martínez C. First study of determination of aromatic compounds of red wine from Vitis vinifera cv. Castañal grown in Galicia (NW Spain). European Food Research and Technology. 2007;**224**:431-436

[25] Lamikanra O, Grimm CC, Inyang ID. Formation and occurrence of flavor components in Noble muscadine wine. Food Chemistry. 1996;**56**(4):373-376

[26] Ferreira V, López R, Cacho JF. Quantitative determination of the odorants of young red wines from different grape varieties. Journal of the Science of Food and Agriculture. 2000;**80**(11):1659-1667

[27] Mendes Ferreira A, Clímaco M, Mendes FA. The role of non-Saccharomyces species in releasing glycosidic bound fraction of grape aroma components—A preliminary study. Journal of Applied Microbiology. 2001;**91**(1):67-71

[28] Pérez-Navarro J, Izquierdo-Canas PM, Mena-Morales A, Martinez-Gascuena J, Chacon-Vozmediano JL, Garcia-Romero E, et al. First chemical and sensory characterization of Moribel and Tinto Fragoso wines using HPLC-DAD-ESI-MS/MS, GC-MS, and napping techniques: Comparison with Tempranillo. Journal of the Science of Food and Agriculture. 2019;**99**(5):2108-2123

[29] Pérez-Navarro J, Izquierdo-Cañas PM, Mena-Morales A, Chacón-Vozmediano JL, Martínez-Gascueña J, García-Romero E, et al. Comprehensive chemical and sensory assessment of wines made from white grapes of *Vitis vinifera* cultivars Albillo Dorado and Montonera del Casar: A comparative study with Airén. Food. 2020;**9**(9):1282

[30] Ferreira V, Fernández P, Peña C, Escudero A, Cacho JF. Investigation on the role played by fermentation esters in the aroma of young Spanish wines

by multivariate analysis. Journal of the Science of Food and Agriculture. 1995;**67**(3):381-392

[31] Swiegers JH, Pretorius IS. Yeast modulation of wine flavor. Advances in Applied Microbiology. 2005;**57**:131-175

[32] Giudici P, Romano P, Zambonelli C. A biometric study of higher alcohol production in Saccharomyces cerevisiae. Canadian Journal of Microbiology. 1990;**36**(1):61-64

[33] Herraiz T, Martin-Alvarez PJ, Reglero G, Herraiz M, Cabezudo MD. Differences between wines fermented with and without Sulphur dioxide using various selected yeasts. Journal of the Science of Food and Agriculture. 1989;**49**(2):249-258

[34] Vilanova M, Oliveira JM. Application of gas chromatography on the evaluation of grape and wine aroma in Atlantic viticulture (NW Iberian Peninsula). In: Salih B, Celikbicak O, editors. Gas Chromatography in Plant Science, Wine Technology, Toxicology and Some Specific Applications. London, UK: IntechOpen; 2012. pp. 109-146

[35] Guth H. Identification of character impact odorants of different white wine varieties. Journal of Agricultural and Food Chemistry. 1997;**45**(8):3022-3026

[36] Boido E. Modificaciones producidas por la fermentación maloláctica en la fracción aromática de los vinos Tannat. Sección Enología. 2002

[37] Ugliano M, Bartowsky EJ, McCarthy J, Moio L, Henschke PA. Hydrolysis and transformation of grape glycosidically bound volatile compounds during fermentation with three Saccharomyces yeast strains. Journal of Agricultural and Food Chemistry. 2006;**54**(17):6322-6331

[38] Etievant P. Volatile Compounds in Food and Beverages: Wine. New York: Marcel Dekker; 1991

[39] Palomo ES, Hidalgo MD, Gonzalez-Vinas M, Perez-Coello M. Aroma enhancement in wines from different grape varieties using exogenous glycosidases. Food Chemistry. 2005;**92**(4):627-635

[40] Escudero A, Campo E, Fariña L, Cacho J, Ferreira V. Analytical characterization of the aroma of five premium red wines. Insights into the role of odor families and the concept of fruitiness of wines. Journal of Agricultural and Food Chemistry. 2007;**55**(11):4501-4510

[41] García-Carpintero EG, Sánchez-Palomo E, González-Viñas M. Aroma characterization of red wines from cv. Bobal grape variety grown in La Mancha region. Food Research International. 2011;**44**(1):61-70

[42] Gómez-Míguez MJ, Cacho JF, Ferreira V, Vicario IM, Heredia FJ. Volatile components of Zalema white wines. Food Chemistry. 2007;**100**(4):1464-1473

[43] Peinado R, Moreno J, Medina M, Mauricio J. Changes in volatile compounds and aromatic series in sherry wine with high gluconic acid levels subjected to aging by submerged flor yeast cultures. Biotechnology Letters. 2004;**26**:757-762

[44] Escudero A, Gogorza B, Melus M, Ortin N, Cacho J, Ferreira V. Characterization of the aroma of a wine from Maccabeo. Key role played by compounds with low odor activity values. Journal of Agricultural and Food Chemistry. 2004;**52**(11):3516-3524

Chapter 5

Use of UHPH to Sterilize Grape Juices and to Facilitate the Implantation of *Saccharomyces* and Other Emerging Fermentation Biotechnologies in Wines

Antonio Morata, Carlos Escott, Juan Manuel del Fresno, Buenaventura Guamis, Iris Loira, María Antonia Bañuelos, Carmen López, Felipe Palomero and Carmen González

Abstract

Ultra-High Pressure Homogenization (UHPH) is an emerging non-thermal technology that uses continuous pumping at 200–600 MPa (generally 300), and then the fluid is depressurized against a special valve made with highly resistant alleys and coated with diamond powder. The UHPH process is able to sterilize grape juice or whatever food fluid and at the same time is extremely efficient in inactivating oxidative enzymes. The total elimination of wild microorganisms in grape juice allows the perfect implantation of the selected *Saccharomyces cerevisiae* starter facilitating the expression of particular metabolic profiles or the expression of specific extracellular activities. Furthermore, the use of complex emerging biotechnologies such as the use of co-inoculations with *S. cerevisiae* and lactic acid bacteria to perform simultaneously alcoholic and malolactic fermentations works much better if the initial juice has a very low load of wild microorganisms. Additionally, the development of spoilage processes by wild microorganisms that can remain in the wine after fermentation is practically impossible when the initial juice is treated by UHPH. Thus, UHPH helps to obtain wines more stable from a microbiological perspective and therefore safer and more hygienic.

Keywords: UHPH, wine, *Saccharomyces cerevisiae*, non-*Saccharomyces*, *Lachancea thermotolerans*, *Metschnikowia pulcherrima*, *Hanseniaspora* spp., SO_2, clean labels, emerging biotechnologies

1. Introduction

Ultra-High Pressure Homogenization (UHPH) is an emerging non-thermal technology with many interesting advantages in food processing, including: powerful antimicrobial effect, high inactivation of oxidative enzymes, nanofragmentation, formation of stable colloidal dispersions, nanocoating, and nanoencapsulation [1–4]. The liquid must be pumped at ultra-high pressure (200–600 MPa, normally 300 MPa), reaching a speed of Mach 3 in the capillary pipeline upstream of the valve and a very high intense kinetic energy that is released against the valve with intense impact forces and shear stresses. The temperature at the valve can reach 70–80°C (**Figure 1**) when the fluid is at room temperature or refrigerated (4–20°C), or even much higher if the fluid is preheated. Downstream of the valve, subsequent depressurization leads to cooling of the liquid that can be also enhanced by the use of heat exchangers (**Figure 1**).

After passing through the valve, the colloidal particles in the fluid and the microorganisms are nanofragmented, resulting in inactivation of the microorganisms. It can be observed how the initial juice contains a lot of colloidal particles from the cell walls of the pulp but also tartrate crystals and wild microbial cells (**Figure 2a**). From optical microscopic observations, it can be seen that after UHPH treatment, the colloidal structure is thinner in the juice, which makes it highly stable and difficult to settle, giving the appearance of a haze in the beaker and of a small granulation under optical microscopy (**Figure 2b**). The colloidal particle size of a pressed juice is usually in the range of a few micrometers and after UHPH processing, is reduced to nanometric scale. We have measured by AFM and laser light scattering a particle size range of 0.8–3.2 μm which is decreased by nanofragmentation after UHPH to 235–744 nm [5, 6].

Yeast counts of 6-log cfu/mL can be easily reached within a few hours in fresh grape must after grape pressing due to the growth and development of wild microorganisms and that population can be completely eliminated with a single pass through UHPH at 300 MPa [7]. Under the same conditions, 4-log bacteria and even spores can also be inactivated depending on the in-valve temperature [5, 7].

At the same time, the UHPH process is very gentle and protective of sensitive molecules with high impact on sensory quality, such as anthocyanins [6], terpenes [5],

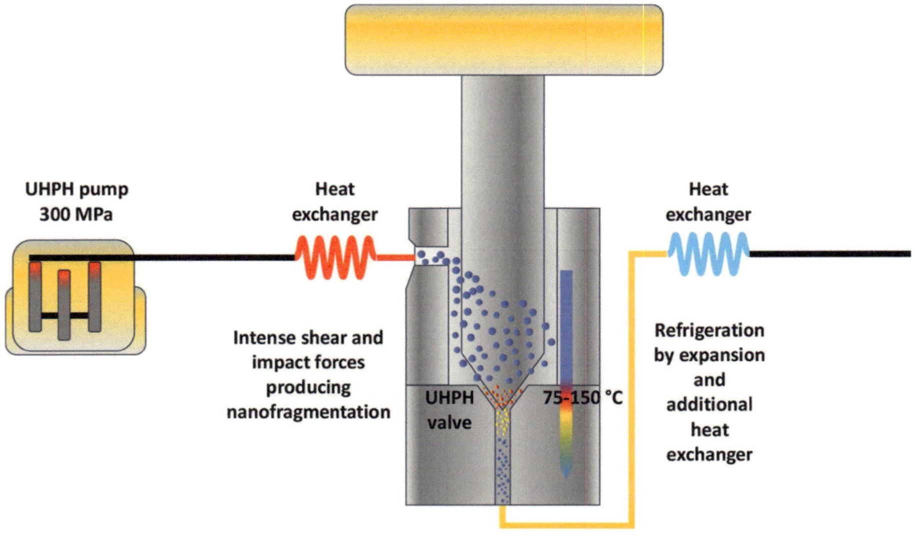

Figure 1.
Main components of the UHPH system (not at scale), including the design of the UHPH valve, the nanofragmentation process, and the temperature change of the colloidal particles.

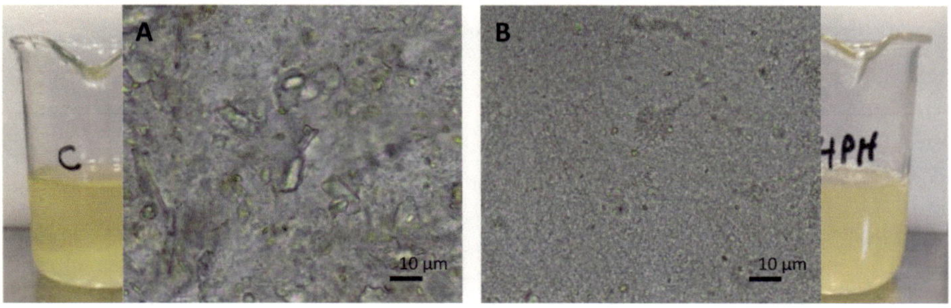

Figure 2.
Macroscopical appearance and optical microscopy of the colloids in a white grape juice Vitis vinifera L Verdejo variety (A) after pressing and settling, (B) processed by UHPH at 300 MPa.

or thiols [8]. Even when high temperatures (75–150°C) can be reached in the valve, the amount of thermal markers as HMF [5] or furosine [9] is very low or not detectable. Furthermore, the formation of the carcinogenic molecule ethyl carbamate in sake after pasteurization and its absence after UHPH processing has recently been observed [10].

2. Antimicrobial effect and implantation of *Saccharomyces*

UHPH is a highly effective technology to eliminate vegetative cells of yeasts and bacteria working at 300 MPa. Yeasts are particularly sensitive to the effect of UHPH and can be easily controlled (**Table 1**), even considering very high populations that are abnormal in healthy grapes. The same degree of inactivation can be reached with vegetative cells of aerobic bacteria.

The degree of inactivation can be comparable to HTST (High Temperature Short Time) pasteurization when processed without thermal assistance at the valve, and similar to UHT (Ultra-High Temperature) sterilization, but with less thermal damage, when the temperature at the valve reaches 140–160°C to inactivate sporulated bacteria [1, 3]. A sporulated bacterium that causes significant problems in juices and especially in concentrated juices is *Alicyclobacillus acidoterrestris*, but it can be easily controlled by UHPH at 300 MPa with mild inlet temperatures of 80–90°C. This bacterium is aerobic, acidophilic, thermoresistant, capable of sporulating, and responsible for the production of off-flavors in juices, such as 2,6-dibromophenol and guaiacol.

The initial yeast population in wine grapes normally ranges between 3 and 4-log CFU/mL depending on maturity and environmental conditions. It consists mainly of low fermentative apiculate yeasts, such as *Hanseniaspora* spp., *Metschnikowia* spp., *Candida* spp., *Pichia* spp., *Lachancea* spp., *Kluyveromyces*, and *Saccharomyces* [11–14], and other yeast-like saprophytic fungi, such as *Aureobasidium pullulans*. *Saccharomyces cerevisiae* is found in very low counts and is, therefore, hardly representative of the initial grape population, although, due to its peculiar physiology, it will predominate during alcoholic fermentation.

No clear resistance to UHPH processing has been observed among the different non-*Saccharomyces* species, all of them can be easily eliminated at populations ranging from 3 to 5-log CFU/mL (**Table 1**) [5–7]. Furthermore, we have observed an even higher inactivation in UHPH processing of yeast biomasses. Therefore, the use of UHPH is a powerful tool to eliminate initial wild yeast populations from grape juices

Grape variety	UHPH conditions	Microorganisms	Initial count (CFU/mL)	Count after UHPH (CFU/mL)	Reference
Vitis vinifera L. variety Hondarribi zuri	300 ± 3 MPa, inlet temperature 20°C, in-valve 98°C (0.02s), outlet 25°C	Total yeasts	>5-log	nd	[7]
		Non-*Saccharomyces* yeasts	>5-log	nd	
		Total aerobic bacteria	≈4-log	nd	
		Lactic acid bacteria	≈4-log	nd	
Vitis vinifera L. variety Muscat of Alexandria	307 ± 3 MPa, inlet temperature 23–25°C, in-valve 78–65°C (0.02 s), outlet 13–15°C	Total yeasts	3.62-log	nd	[5]
		Non-*Saccharomyces* yeasts	3.52-log	nd	
		Total aerobic bacteria	3.88-log	nd	
		Lactic acid bacteria	3.26-log	nd	
Vitis vinifera L. variety Cabernet sauvignon	300 ± 3 MPa, inlet temperature 4°C, in-valve 78 ± 2 °C (<0.2 s), outlet 15°C	Total yeasts	≈4-log	nd	[6]
		Non-*Saccharomyces* yeasts	4.08-log	nd	
		Total aerobic bacteria	3.23-log	nd	

nd: not detected.

Table 1.
Effect of UHPH on the control of wild yeast and bacteria.

and to facilitate the implantation of *Saccharomyces* starters inoculated at even lower doses that those recommended by yeast producers (often 20–40 g/Hl of active dry yeast). For those *Saccharomyces* strains with low fermentative power or slower fermentation kinetics, or non-*Saccharomyces* yeasts that are weak in terms of competition for nutrients and colonization of the medium, pre-treatment of the juice with UHPH greatly increases their chances of implantation, growth, and development of their metabolomic profile [15].

The selection of *Saccharomyces cerevisiae* strains looks for specific properties that can help to improve fermentation, to produce low volatile acidity and off-flavors (some sulfur compounds, volatile acidity, ethyl acetate, and others), to decrease alcoholic strength, or to develop suitable fermentations under stress conditions such as high sugar concentrations, low temperature, or limited yeast assimilable nitrogen [16–18]. Special features with impact on aroma, such as the production of acetate esters of higher alcohols or ethanol esters of medium chain fatty acids, the expression of beta-glucosidase enzymes to release volatile terpenes, or carbon lyase activities to release thiols, are also of interest [16–17]. Similarly, specific properties with effect on color stability, such as the production of pyruvate and acetaldehyde to form stable vitisin pigments or the expression of hydroxy cinnamate decarboxylase activity to form vinylphenolic pyranoanthocyanins during fermentation or the formation of polymeric pigments, are also studied for the selection of *Saccharomyces cerevisiae* strains [17]. The influence of the yeast on the structure, body, and softening of tannins during fermentation is also considered by selecting appropriate strains that can produce more glycerol or release cell wall polysaccharides [16–18].

3. Emerging fermentation biotechnologies and UHPH

Currently, several non-*Saccharomyces* yeasts are widely used in enology to control pH by acidification with *Lachancea thermotolerans* [19] or deacidification with *Schizosaccharomyces pombe* [20] or *japonicus* [21], and to improve aroma, flavor, and structure with *Torulaspora delbrueckii* [22], *Hanseniaspora* spp. [23] or *Metschnikowia pulcherrima* [24]. All these non-*Saccharomyces* yeast species are usually less competitive than wild or inoculated *Saccharomyces* and their implantation and development are much better in juices processed by UHPH [15, 25]. *Lachancea thermotolerans* produces effective acidification during alcoholic fermentation by metabolizing sugars to lactic acid which can decrease the pH under enological conditions by up to 0.5 units and, at the same time, slightly reduce the alcohol content. Furthermore, it produces low volatile acidity and has positive effects on the formation of esters and the release of volatile thiols [19].

Schizosaccharomyces pombe and *S. japonicus* have been used for their ability to degrade malic acid to ethanol by the maloalcoholic fermentation and also have shown a good impact on wine palatability by releasing high amounts of cell wall polysaccharides [20, 21]. Some strains can reach 13–15% v/v ethanol and can be used for complete fermentation in the absence of *S. cerevisiae* [20]. Moreover, *S. pombe* has shown a high release of pyruvate, which influences the formation of vitisin A-type pigments [20]. Also, some strains have a positive effect on aroma, but the main problem can be the excessive production of volatile acidity [20]. *Torulaspora delbrueckii* was the first non-*Saccharomyces* produced at industrial level as dry yeast due to its positive impact on structure and aroma; juices fermented with this yeast species increase fruitiness and complexity [22]. Some specific markers of *Torulaspora delbrueckii* metabolism, such as 3-ethoxy propanol, have been reported. *Hanseniaspora* spp. are normally described as high producers of acetate esters and therefore with a significant impact on fruity and floral aroma perception. Some of them are especially potent in the formation of bencenoids, such as *H. vineae* which also has a high fermentative power and low volatile acidity formation [23]. This species also has a positive role in the release of free terpenes from non-volatile glycosides. *Metschnikowia pulcherrima* is another key non-*Saccharomyces* in wine biotechnology for its ability to improve the aromatic complexity, express enzymatic activities and its interesting effects on bioprotection [24]. Recently, its high compatibility with *Lachancea thermotolerans* and a synergistic effect on acidification have also been reported [26].

Most of these non-*Saccharomyces* species have low fermentative power ranging from 1 to 10% v/v ethanol [19, 22–24, 27], and are unable to finish alcoholic fermentation in still wines, with a probable alcohol ranging from 12 to 16% v/v ethanol. Therefore, they must be used together with *S. cerevisiae* to completely dry the musts and obtain wines without residual sugars. At industrial level, two typical strategies are used: sequential fermentations [28] or mixed fermentations (co-inoculations) [29, 30]. The first option facilitates the control of fermentation and metabolomic influence of the non-*Saccharomyces* species by just deciding the inoculation time of the *Saccharomyces* yeast strain. However, it requires two inoculations. The second one is easier to carry out at industrial level, because a single inoculation is enough, but the ratio of non-*Saccharomyces*/*Saccharomyces* populations in the inoculum must be carefully evaluated to obtain good implantation and suitable metabolomic expression of each yeast species. In all situations, the use of juice previously processed by UHPH increases the possibilities of achieving good implantation without the competitiveness of wild yeasts [5–7, 15].

Additionally, most of these non-*Saccharomyces* yeasts are sensitive or highly sensitive to SO_2 [19, 22–24, 27]. Actually, in the past, and also now, sulfites are used to control the development of wild non-*Saccharomyces* yeasts, to facilitate fermentation by *S. cerevisiae,* and to control some parameters, such as volatile acidity or ethyl acetate that can decrease wine quality and can be promoted by some non-*Saccharomyces* yeasts. The use of UHPH avoids the deleterious effects of sulfites on non-*Saccharomyces* yeasts while improving their implantation during fermentation.

Currently, it is also frequent to develop alcoholic (AF) and malolactic (MLF) fermentations, at the same time to reduce the total fermentation time with safer fermentations [31–35] and often with higher contents of ethyl and acetate esters [31]. For this particular biotechnology, it is important to avoid unsuitable wild yeast and bacteria species and to achieve a good implantation of the inoculated bacteria-yeast starter. It is usually recommended to make yeast-bacteria co-inoculations on healthy grapes with pH equal or lower than 3.5 and with a low wild population, to avoid undesired developments of wild bacteria and the increase of volatile acidity and other metabolites that can behave as off-flavors [32].

Therefore, UHPH is a powerful tool to facilitate the use of emerging biotechnologies in wine fermentation. Nowadays, the use of more complex mixtures of microorganisms with special growth conditions and delicate fermentative metabolisms makes it more essential to eliminate competitive wild species to better implant these inoculants and favor the expression of their metabolisms to improve the sensory quality of the wine [26, 36–42]. In this framework, UHPH produces sterilized or pasteurized juices in which all the competitiveness can be avoided while preserving the sensory quality of the juice [5–8].

4. Control of polyphenol oxidase enzymes, antioxidant capacity, and SO_2 reduction

Oxidative enzymes are very detrimental to wine quality and rapidly affect color, reducing paleness in white wines and increasing browning in whites (**Figure 3**), rosés, and reds. The main oxidative enzymes in white wines are polyphenol oxidase enzymes (PPOs), among which the most important are tyrosinase in healthy grapes, and laccase in grapes affected by *Botrytis cinerea*. PPOs catalyze the oxidation of *ortho*-diphenols to quinones that polymerize to relatively insoluble brown melanoidins [43]. This browning decreases the color quality of most wines, usually related to evolution, oxidation, but also to the degradation of the fruitiness and varietal aroma.

The protection of juice and wine from oxidation by PPOs is done by adding antioxidants such as SO_2, or ascorbic acid. SO_2 is highly effective, but currently under discussion due to its toxicity for the general public, but especially for sensitive people. SO_2 has a double role: anti-oxidase by inhibiting the effect of PPOs and antioxidant by protecting *ortho*-diphenols from oxidation by PPOs [44]. Considering the potential

Figure 3.
Effect on color of PPOs after a few hours of air exposition in the absence of antioxidants in a control (left) and in a juice processed by UHPH (right) at 300 MPa.

toxicity of SO_2, new molecules, such as glutathione (GSH), are being studied and used to control oxidation and browning in juices and wines [45, 46].

UHPH causes the fragmentation or unfold and denaturation of several enzymes by impact and shear efforts, including oxidative PPOs. In white varieties fermented in the absence of SO_2, it can be observed how musts processed by UHPH remain pale and without browning even under high aeration conditions and with sensitive varieties, such as Muscat (**Figure 3**). We have also observed that oxidation continues after fermentation and if we clarify the wines with fining agents, the wines obtained from the non-processed juices continue to form brown compounds, but this does not occur with wines obtained from musts treated by UHPH. The effects on PPO inactivation, antioxidant activity, and anthocyanin contents are summarized (**Table 2**) for 3 grape varieties (2 whites and 1 red). UHPH control of PPO is very useful when it is desired to apply emerging biotechnologies, such as the use of non-*Saccharomyces* yeasts that are very sensitive to SO_2. By using UHPH, it is possible to preserve the aromatic quality and prevent oxidation in musts of delicate aromatic white varieties, and at the same time, maintain suitable conditions for the implantation of non-*Saccharomyces* yeasts by the total elimination of wild yeasts and the absence of toxic sulfites.

In addition to the improved sensory quality and stability of the juices processed by UHPH, higher antioxidant activity was observed, probably due to the inactivation of oxidative enzymes (**Table 2**). These effects made the UHPH juices healthier and with

Grape variety	UHPH conditions	Inactivation of PPOs and browning	Antioxidant activity	Anthocyanins	Reference
Vitis vinifera L. variety Hondarribi zuri	300 ± 3 MPa, inlet temperature 20°C, in-valve 98°C (0.02 s), outlet 25°C	Strong inactivation (>90%) Paler juices (10%)	+45%	White variety	[7]
Vitis vinifera L. variety Muscat of Alexandria	307 ± 3 MPa, inlet temperature 23–25°C, in-valve 78–65°C (0.02 s), outlet 13–15°C	Strong inactivation (>90%) Paler juices that remain stable more than 3 days in the absence of SO2	+56%	White variety	[5]
Vitis vinifera L. variety Cabernet sauvignon	300 ± 3 MPa, inlet temperature 4°C, in-valve 78 ± 2°C (<0.2 s), outlet 15°C	Strong inactivation Lower browning of anthocyanins	+6%	Protection of anthocyanins especially acylated forms (+9%)	[6]

Table 2.
Effect of UHPH on the control of PPOs and oxidation.

better nutraceutical properties. These characteristics are maintained in the subsequent wines after fermentation. The inactivation of enzymes by UHPH probably by unfold and denaturation could also have a positive effect on protein haze in white wines [5]. Therefore, UHPH musts can have fewer proteins able to produce turbidity or even the unfolding produced by the UHPH can facilitate the removal of unstable proteins by the use of proteases without the need for thermal pre-treatments.

5. Conclusion

UHPH is a powerful emerging non-thermal technology able to eliminate wild microorganisms facilitating the use of new biotechnologies with non-competitive non-*Saccharomyces* yeasts and non-*Oenococcus* lactic acid bacteria. At the same time, UHPH is very effective in controlling oxidative enzymes and can therefore be used to produce fermentations in the absence or with very low levels of SO_2, what favors the development of sensitive yeasts such as many non-*Saccharomyces* species. Moreover, UHPH is a gentle technique with a protective effect on sensitive molecules, such as anthocyanins, terpenes, thiols, and others that are strongly affected by other conventional treatments. Additionally, UHPH can be considered an environmentally sustainable technique with low water and energy consumption.

Acknowledgements

This research was funded by MINISTERIO DE CIENCIA E INNOVACIÓN, project ENOINNOVAPRESS – PID2021-124250OB-I00 Ministerio de Ciencia e Innovación. This work was partially supported by the State Program to Promote

Scientific-Technical Research and its Transfer of the Spanish Ministry of Science and Innovation through the MALTA CONSOLIDER TEAM Research Network (RED2022-134388-T).

Conflict of interest

The authors declare no conflict of interest.

Appendices and nomenclature

AFM	Atomic Force Microscopy
CFU	Colony Forming Units
GSH	glutathione
HMF	hydroxy methyl furfural
MLF	malolactic fermentation
MPa	mega pascal
PPOs	poly phenol oxidases
UHPH	Ultra-High Pressure Homogenization

Author details

Antonio Morata[1*], Carlos Escott[1], Juan Manuel del Fresno[1], Buenaventura Guamis[2], Iris Loira[1], María Antonia Bañuelos[1], Carmen López[1], Felipe Palomero[1] and Carmen González[1]

1 MALTA CONSOLIDER TEAM Research Network (RED2022-134388-T), enotecUPM, ETSIAAB, Universidad Politécnica de Madrid, Madrid, Spain

2 Departament de Ciència Animal I Dels Aliments, Facultat de Veterinària, Centre d'Innovació, Recerca I Transferència en Tecnologia Dels Aliments (CIRTTA), TECNIO, XaRTA, Universitat Autònoma de Barcelona, Bellaterra, Spain

*Address all correspondence to: antonio.morata@upm.es

References

[1] Zamora A, Guamis B. Opportunities for ultra-high-pressure homogenization (UHPH) for the food industry. Food Engineering Reviews. 2015;**7**:130-142. DOI: 10.1007/s12393-014-9097-4

[2] Patrignani F, Lanciotti R. Applications of high and ultra high pressure homogenization for food safety. Frontiers in Microbiology. 2016;**7**:1132. DOI: 10.3389/fmicb.2016.01132

[3] Morata A, Guamis B. Use of UHPH to obtain juices with better nutritional quality and healthier wines with low levels of SO_2. Frontiers in Nutrition. 2020;**7**:598286. DOI: 10.3389/fnut.2020.598286

[4] Morata A, Loira I, Guamis B, Raso J, et al. Emerging technologies to increase extraction, control microorganisms, and reduce SO_2. In: Cosme F, Nunes FM, Filipe-Ribeiro L, editors. Chemistry and Biochemistry of Winemaking, Wine Stabilization and Aging. London, UK: IntechOpen; 2021. DOI: 10.5772/intechopen.92035

[5] Bañuelos MA, Loira I, Guamis B, Escott C, Del Fresno JM, Codina-Torrella I, et al. White wine processing by UHPH without SO_2. Elimination of microbial populations and effect in oxidative enzymes, colloidal stability and sensory quality. Food Chemistry. 2020;**332**:127417. DOI: 10.1016/j.foodchem.2020.127417

[6] Vaquero C, Escott C, Loira I, Guamis B del Fresno JM, et al. Cabernet sauvignon red must processing by UHPH to produce wine without SO_2: The colloidal structure, microbial and oxidation control, colour protection and sensory quality of the wine. Food and Bioprocess Technology. 2022;**15**:620-634. DOI: 10.1007/s11947-022-02766-8

[7] Loira I, Morata A, Bañuelos MA, Puig-Pujol A, Guamis B, González C, et al. Use of ultra-high pressure homogenization processing in winemaking: Control of microbial populations in grape musts and effects in sensory quality. Innovative Food Science and Emerging Technologies. 2018;**50**:50-56. DOI: 10.1016/j.ifset.2018.10.005

[8] Morata A, Escott C, Vaquero C, del Fresno JM, Guamis B, Loira I, et al. Improving the implantation of non-*Saccharomyces* yeasts in winemaking by UHPH processing. Oral communication. In: Proceedings of the 44th World Congress of Vine and Wine; 5-9 June 2023; Cadiz-Jerez, Spain. BIO Web of Conferences. Vol. 68. Dijon, France: OIV; 2023. p. 02001

[9] Ferragut V, Hernández-Herrero M, Veciana-Nogués MT, Borras-Suarez M, González-Linares J, Vidal-Carou MC, et al. Ultra-high-pressure homogenization (UHPH) system for producing high-quality vegetable-based beverages: Physicochemical, microbiological, nutritional and toxicological characteristics. Journal of the Science of Food and Agriculture. 2015;**95**:953-961. DOI: 10.1002/jsfa.6769

[10] Arakawa G, Yokoi K. Application of multiple ultra-high-pressure homogenization to the pasteurization process of Japanese rice wine, sake. Journal of Bioscience and Bioengineering. 2023;**136**:117-122. DOI: 10.1016/j.jbiosc.2023.05.003

[11] Cocolin L, Bisson LF, Mills DA. Direct profiling of the yeast dynamics in wine fermentations. FEMS Microbiology

Letters. 2000;**189**(1):81-87. DOI: 10.1111/j.1574-6968.2000.tb09210.x

[12] Mills DA, Johannsen EA, Cocolin L. Yeast diversity and persistence in Botrytis-affected wine fermentations. Applied and Environmental Microbiology. 2002;**68**(10):4884-4893. DOI: 10.1128/AEM.68.10.4884-4893.2002

[13] Fleet GH. Wine yeasts for the future. FEMS Yeast Research. Nov 2008;**8**(7):979-995. DOI: 10.1111/j.1567-1364.2008.00427.x

[14] Anagnostopoulos D, Kamilari E, Tsaltas D. Contribution of the microbiome as a tool for estimating Wine's fermentation output and authentication. In: Morata A, Loira I, editors. Advances in Grape and Wine Biotechnology. London, UK: IntechOpen; 2019. DOI: 10.5772/intechopen.85692

[15] Loira I, Escott C, del Fresno JM, Bañuelos MA, González C, Guamis B, et al. Chapter 5 – White must preservation by ultra-high pressure homogenization without SO_2. In: Morata A, editor. White Wine Technology. London, UK: Academic Press; 2022. pp. 49-59. DOI: 10.1016/B978-0-12-823497-6.00014-4

[16] Pretorius IS. Tailoring wine yeast for the new millennium: Novel approaches to the ancient art of winemaking. Yeast. 2000;**16**:675-729. DOI: 10.1002/1097-0061(20000615)16:8<675::AID-YEA585>3.0.CO;2-B

[17] Suárez-Lepe JA, Morata A. New trends in yeast selection for winemaking. Trends in Food Science & Technology. 2012;**23**:39-50. DOI: 10.1016/j.tifs.2011.08.005

[18] Loira I, Morata A, Bañuelos MA, Suárez-Lepe JA. Chapter 15 – Isolation, selection, and identification techniques for non-*Saccharomyces* yeasts of oenological interest. In: Grumezescu AM, Holban AM, editors. Biotechnological Progress and Beverage Consumption. London, UK: Academic Press; 2020. pp. 467-508. DOI: 10.1016/B978-0-12-816678-9.00015-1

[19] Morata A, Loira I, Tesfaye W, Bañuelos MA, González C, Suárez Lepe JA. *Lachancea thermotolerans* applications in wine technology. Fermentation. 2018;**4**:53. DOI: 10.3390/fermentation4030053

[20] Loira I, Morata A, Palomero F, González C, Suárez-Lepe JA. *Schizosaccharomyces pombe*: A promising biotechnology for modulating wine composition. Fermentation. 2018;**4**:70. DOI: 10.3390/fermentation4030070

[21] Portaro L, Maioli F, Canuti V, Picchi M, Lencioni L, Mannazzu I, et al. *Schizosaccharomyces japonicus/Saccharomyces cerevisiae* mixed starter cultures: New perspectives for the improvement of Sangiovese aroma, taste, and color stability. LWT. 2022;**156**:113009. DOI: 10.1016/j.lwt.2021.113009

[22] Ramírez M, Velázquez R. The yeast *Torulaspora delbrueckii*: An interesting but difficult-to-use tool for winemaking. Fermentation. 2018;**4**:94. DOI: 10.3390/fermentation4040094

[23] Martin V, Valera MJ, Medina K, Boido E, Carrau F. Oenological impact of the *Hanseniaspora/Kloeckera* yeast genus on wines—A review. Fermentation. 2018;**4**:76. DOI: 10.3390/fermentation4030076

[24] Morata A, Loira I, Escott C, del Fresno JM, Bañuelos MA, Suárez-Lepe JA. Applications of *Metschnikowia pulcherrima* in wine biotechnology. Fermentation. 2019;**5**:63. DOI: 10.3390/fermentation5030063

[25] Morata A, Escott C, Bañuelos MA, Loira I, del Fresno JM, González C, et al. Contribution of non-*Saccharomyces* yeasts to wine freshness: A review. Biomolecules. 2020;**10**:34. DOI: 10.3390/biom10010034

[26] Escott C, Vaquero C, Loira I, López C, González C, Morata A. Synergetic effect of *Metschnikowia pulcherrima* and *Lachancea thermotolerans* in acidification and aroma compounds in airén wines. Food. 2022;**11**:3734. DOI: 10.3390/foods11223734

[27] Jolly NP, Varela C, Pretorius IS. Not your ordinary yeast: Non-*Saccharomyces* yeasts in wine production uncovered. FEMS Yeast Research. 2014;**14**:215-237. DOI: 10.1111/1567-1364.12111

[28] Gobbi M, Comitini F, Domizio P, Romani C, Lencioni L, Mannazzu I, et al. *Lachancea thermotolerans* and *Saccharomyces cerevisiae* in simultaneous and sequential co-fermentation: A strategy to enhance acidity and improve the overall quality of wine. Food Microbiology. 2013;**33**:271-281. DOI: 10.1016/j.fm.2012.10.004

[29] Ciani M, Comitini F, Mannazzu I, Domizio P. Controlled mixed culture fermentation: A new perspective on the use of non-*Saccharomyces* yeasts in winemaking. FEMS Yeast Research. 2010;**10**:123-133. DOI: 10.1111/j.1567-1364.2009.00579.x

[30] Comitini F, Gobbi M, Domizio P, Romani C, Lencioni L, Mannazzu I, et al. Selected non-*Saccharomyces* wine yeasts in controlled multistarter fermentations with *Saccharomyces cerevisiae*. Food Microbiology. 2011;**28**:873-882. DOI: 10.1016/j.fm.2010.12.001

[31] Knoll C, Fritsch S, Schnell S, Grossmann M, Krieger-Weber S, du Toit M, et al. Impact of different malolactic fermentation inoculation scenarios on Riesling wine aroma. World Journal of Microbiology and Biotechnology. 2012;**28**:1143-1153. DOI: 10.1007/s11274-011-0917-x

[32] Pardo I, Ferrer S. Chapter 7 - Yeast-bacteria coinoculation. In: Morata A, editor. Red Wine Technology. London, UK: Academic Press; 2019. pp. 99-114. DOI: 10.1016/B978-0-12-814399-5.00007-4

[33] Krieger-Weber S. Application of yeast and bacteria as starter cultures. In: König H, Unden G, Fröhlich J, editors. Biology of Microorganisms on Grapes, in Must and in Wine. 2nd ed. Cham: Springer International Publishing; 2017. pp. 605-634. DOI: 10.1007/978-3-319-60021-5_25

[34] Bartowsky EJ, Costello PJ, Chambers PJ. Emerging trends in the application of malolactic fermentation. Australian Journal of Grape and Wine Research. 2015;**21**:663-669. DOI: 10.1111/ajgw.12185

[35] Krieger-Weber S, Heras JM, Suarez C. *Lactobacillus plantarum*, a new biological tool to control malolactic fermentation: A review and an outlook. Beverages. 2020;**6**:23. DOI: 10.3390/beverages6020023

[36] Vaquero C, Loira I, Heras JM, Carrau F, González C, Morata A. Biocompatibility in ternary fermentations with *Lachancea thermotolerans*, other non-*Saccharomyces* and *Saccharomyces cerevisiae* to control pH and improve the sensory profile of wines from warm areas. Frontiers in Microbiology. 2021;**12**:656262. DOI: 10.3389/fmicb.2021.656262

[37] Morata A, Arroyo T, Bañuelos MA, Blanco P, Briones A, Cantoral JM, et al. Wine yeast selection in the Iberian

Peninsula: *Saccharomyces* and non-*Saccharomyces* as drivers of innovation in Spanish and Portuguese wine industries. Critical Reviews in Food Science and Nutrition. 2022;**10**:1-29. DOI: 10.1080/10408398.2022.2083574

[38] Vaquero C, Escott C, Heras JM, Carrau F, Morata A. Co-inoculations of *Lachancea thermotolerans* with different *Hanseniaspora* spp.: Acidification, aroma, biocompatibility, and effects of nutrients in wine. Food Research International. 2022;**161**:111891. DOI: 10.1016/j.foodres.2022.111891

[39] Del Fresno JM, Escott C, Loira I, Herbert-Pucheta JE, Schneider R, Carrau F, et al. Impact of *Hanseniaspora vineae* in alcoholic fermentation and ageing on lees of high-quality white wine. Fermentation. 2020;**6**:66. DOI: 10.3390/fermentation6030066

[40] Vaquero C, Izquierdo-Cañas PM, Mena-Morales A, Marchante-Cuevas L, Heras JM, Morata A. Use of *Lachancea thermotolerans* for biological vs. chemical acidification at pilot-scale in white wines from warm areas. Fermentation. 2021;**7**:193. DOI: 10.3390/fermentation7030193

[41] Del Fresno JM, Escott C, Loira I, Carrau F, Cuerda R, Schneider R, et al. The impact of *Hanseniaspora vineae* fermentation and ageing on lees on the terpenic aromatic profile of white wines of the Albillo variety. International Journal of Molecular Sciences. 2021;**22**:2195. DOI: 10.3390/ijms22042195

[42] Vaquero C, Loira I, Bañuelos MA, Heras JM, Cuerda R, Morata A. Industrial performance of several *Lachancea thermotolerans* strains for pH control in white wines from warm areas. Microorganisms. 2020;**8**:830. DOI: 10.3390/microorganisms8060830

[43] Martinez MV, Whitaker JR. The biochemistry and control of enzymatic browning. Trends in Food Science & Technology. 1995;**6**:195-200. DOI: 10.1016/S0924-2244(00)89054-8

[44] Giacosa S, Río Segade S, Cagnasso E, Caudana A, Rolle L, Gerbi V. Chapter 21 – SO_2 in wines: Rational use and possible alternatives. In: Morata A, editor. Red Wine Technology. London, UK: Academic Press; 2019. pp. 309-321. DOI: 10.1016/B978-0-12-814399-5.00021-9

[45] Martínez J, González-Arenzana L. Chapter 3 – Use of glutathione in the winemaking of white grape varieties. In: Morata A, editor. White Wine Technology. London, UK: Academic Press; 2022. pp. 29-38. DOI: 10.1016/B978-0-12-823497-6.00021-1

[46] Kilmartin P. Chapter 23 - Impacts of phenolics and prefermentation antioxidant additions on wine aroma. In: Morata A, editor. White Wine Technology. London, UK: Academic Press; 2022. pp. 301-312. DOI: 10.1016/B978-0-12-823497-6.00029-6

Chapter 6

Effect of Aging on Lees on the Volatile Profile of Malvasia *aromatica* Wines Fermented with *Saccharomyces* Native Yeasts in PDO "Vinos de Madrid"

Julia Crespo, Margarita García, Teresa Arroyo, Valeria Romero and Juan M. Cabellos

Abstract

The loss of aromatic and sensory quality in wines because of climate change in traditional winemaking areas is a challenge for winemakers. Aging on lees of the wine fermented with Saccharomyces native yeasts has been tested as a technique to try to improve the sensory characteristics of *Malvasia aromatica* white wines in PDO "Vinos de Madrid." The grapes were pre-cold macerated and fermented with *S. cerevisiae* CLI 271 and CLI 889 (native yeast strains). Then, commercial lees were used for aging of wines for a five-month at low temperature in order to compare with the effect of *S. cerevisiae* CLI 271 and CLI 889 without lees treatment. Aromatic and organoleptic properties of wines aged on lees were studied using GC-FID and HS-SPME/GC–MS to quantify volatile compounds and a taster panel to sensorial analysis. There was a significant decrease in the ester family in wines aged on lees being more pronounced in CLI 889 wines. The treatment contributed to enhance the fruity and floral aromatic properties in CLI 271 and CLI 889 wines, respectively according to tasting panel, which showed a hedonic preference for CLI 271 wines without lees treatment and CLI 889 wines aged on lees.

Keywords: lees, native yeast, aroma, climate change, Malvasia *aromatica*

1. Introduction

Aging on lees is a technique used after fermentation associated with the improvement of the sensory properties of wine through yeast autolysis via enzymatic self-degradation of cellular constituents that begins after the death of the yeast. It is a technique used for decades mainly in the production of white wines. Its main objective is to reduce astringency and bitterness, increase body, structure and roundness in the mouth and improve aromatic persistence and complexity of wines [1–3] in addition to contributing to the reduction of undesirable flavors [4]. The main conditioning and essential factor to allow cell autolysis is the time. It is estimated that the process begins as soon as the cell dies and increases significantly after the second month. The autolysis process is slow and involves risks of microbiological spoilage, reduction and organoleptic off-flavors [1, 5]. Therefore, aging on lees is combined with battonnages to improve the contact of lees with the wine and help the faster release of polysaccharides from the yeasts [6] and compounds such as amino acids and lipids that can be aroma precursors.

Aroma is one of the main quality attributes of white wines and a very important aspect for consumers. The volatile compounds provided by the grapes, more recognized as varietal aromas, are responsible for their aromatic typicity and their presence is influenced by factors such as variety, cultural practices, terroir, geographical location and climate [7]. The compounds released in the course of alcoholic fermentation also have a decisive influence on the volatile composition of wines and their synthesis depends mainly on the microorganisms involved.

PDO "Vinos de Madrid" is located in the center of the Iberian Peninsula with a prevalence of hot summers, cold winters and low levels of precipitation. Climate predictions indicate a gradual increase in temperatures, a decrease in precipitation and a greater frequency of severe phenomena such as frosts, storms and heat waves with greater incidence in the center of Spain [8]. These events could compromise the correct development of the technological and aromatic ripening of the grapes, preventing the production of intense aromatic white wines. On the other hand, the loss of aromatic and sensory typicity attached to a particular region is also a challenge for winemakers. The yeast activity during a spontaneous fermentation could contribute to fewer desirable attributes to the wine and the quality can be variable between seasons. Thus, the use of commercial yeasts could be employed to obtain a product of uniform quality [9, 10]. However, the typical character of the harvest as well as its aromatic properties could be lost.

In order to adapt to the climate change-related effects, as well as the preservation of the typical organoleptic characteristics of the Madrid region, our laboratories have proposed the use of oenological practices such as skin contact treatment and the use of indigenous yeast strains better adapted to climatic conditions and terroir [11, 12]. As a final step, based on the results of our previous research, this study has focused on the effect of using lees during the aging period in white wines of Malvasia *aromatica* in bottles. This variety was selected for having a great aromatic potential, becoming

Figure 1.
Vitis vinifera *L.cv. Malvasia* aromatica *[13].*

an interesting alternative in the use of new varieties in addition to the traditional ones that could help to improve the value of the local wines. The objective of this research was to optimize the combination of aging on lees techniques and the selection of the adequate native *S. cerevisiae* yeast strain for *Vitis vinifera* L.cv. Malvasia *aromatica* (**Figure 1**) winemaking in PDO "Vinos de Madrid".

2. Material and methods

2.1 Vintage, yeast strains and fermentation procedure

Grapes from the white-berried cv. Malvasia *aromatica* were collected and processed in the Experimental Winery from the IMIDRA Research Institute (Finca El Encín, Alcalá de Henares, Madrid, Spain). The yeast strains of *Saccharomyces cerevisiae* selected for this research were CLI 271 and CLI 889 used in this work were previously isolated from wineries of the PDO "Vinos de Madrid", and selected due to their good fermentative and oenological performance [14–16]. Grape must be macerated at 10°C for 18 h before yeast inoculation at a concentration of 10^6 cells/mL [11]. Fermentations took place in triplicate under a controlled temperature of 16°C and were monitored on a daily basis by measuring density.

2.2 Aging on lees

After the end of the alcoholic fermentation, samples of each wine were racked into 5 L screw-capped glass bottles. Commercial lees (Super Bouquet MN from Agrovin) were used at a concentration of 30 g/hl for a five-month contact period with wine lees (L) at low temperature (8–11°C). Five liters of each wine (without lees) were used as a control. Battonnages in wines aging on lees were carried out twice per week, for 10 minutes per bottle with the aid of a magnetic stirrer. At the end of this treatment, wines were racked to eliminate lees. Wines were also clarified and bottled in the same way as those without lees. Prior to bottling, samples were taken for analysis.

2.3 Wine volatile composition

Wine volatile composition was defined with: major aroma compounds and free varietal minor volatiles. The first were analyzed using gas chromatography coupled with a flame ionization detector (GC-FID). The second, by undertaking a headspace-solid phase microextraction coupled with gas chromatography–mass spectrometry (HS-SPME/GC–MS).

The analysis of major aromatic compounds included the extraction and detection of alcohols, acids, lactones, esters, aldehydes and ketones, according to the method described by Ortega [17], using dichloromethane and a DB-Wax column (60 m × 0.32 mm × 0.5 μm film thickness) from J&W Scientific (Folsom, CA, USA). For sample preparation, conical bottom glass tubes were used, adding 3.9 g of ammonium sulfate, 6.3 mL of milli-Q grade deionized water, 2.7 mL of wine, 20 μL of an internal standard solution (2-Butanol, 4-Methyl-2-pentanol, 4-hydroxy-4-methyl-2-pentanone and 2-Octanol) and 250 μL of dichloromethane. Chromatography conditions included an oven temperature initially programmed at 40°C for 5 min, and then ramped to 200°C. A constant helium flow of 2 mL/min was used. Two mL of aroma extract were injected at 250°C in splitless mode. The total run time was 75 min per sample.

Minor volatiles (terpenoids and C13-norisoprenoids) were determined using headspace-solid phase microextraction coupled with gas chromatography–mass spectrometry (HS-SPME/GC–MS), following the method proposed by Yuan and Qian [18] on an Agilent 6890 gas chromatograph equipped with an Agilent 5973 mass selective detector (Agilent). A volume of two mL of the wine sample was diluted with 8 mL of citric acid (0.5 g/L citric acid, pH 3 saturated with sodium chloride) and 20 μL of 4-octanol (100 μg/L), used as internal standard. The extraction was done by stirring the sample in vials for chromatography (20 mL of volume, Agilent Technologies) tightly capped with a magnetic stir bar for 10 min at 50°C in a thermostatic bath and using SPME fiber (50/30 μm DVB/CAR/PDMS fiber from Supelco Inc., Bellefonte, PA, USA) for 50 min at the same temperature with stirring (1000 rpm) to capture the volatiles. The fiber was manually inserted into the injection port of the GC at 230°C to desorb the analytes. A DB-Wax column from J&W Scientific (Folsom, CA, USA) (60 m × 0.32 mm × 0.5 μm film thickness, Phenomenex, Torrance, CA, USA) was employed to separate the analytes. Carrier gas (helium) was set at a constant flow rate of 1 mL/min. The oven temperature was initially set at 40°C for 2 min, and raised to 230°C at 5°C per min for 15 min. All samples were analyzed in duplicate.

2.4 Incidence of aroma composition

To estimate the contribution of volatile compounds to wine aroma, the odor activity value (OAV) was calculated by estimating the ratio between the concentration of each compound and its perception threshold. A compound was considered to contribute to wine aroma if OAV ≥ 1. The perception threshold used in this work were those found in the literature [19–25].

2.5 Sensorial analysis

Wines were tasted at the tasting room of Experimental Winery from IMIDRA Institute by a sensory panel of eight trained evaluators. Descriptive sensory analyses and triangle tests were performed following the indications of the ISO 4120:2004 to assess the effect of experimental treatments (use of lees) on wine aroma. Sensory descriptive analysis was performed to describe and quantify wine attributes from 1 (low intensity) to 10 (high intensity). The evaluation included a visual phase, an olfactory phase and a gustatory phase, which was interpreted by graphical representation. A hedonic classification was also carried out to determine the order of preference of the wines. A final score for each wine was obtained as the mean with their respective standard deviation.

2.6 Statistical analysis

Statistical analyses were performed using SPSS ver. 20.0 (SPSS, Inc., Chicago, IL, USA). Analysis of variance (ANOVA) was made to examine the differences between treatments in terms of volatile compounds and sensory attributes of the wines. In order to assess the significance ($p < 0.05$) of differences between means, Tukey honestly significant difference (HSD) post-hoc tests were used to establish the significance of differences between means to assess significance ($p < 0.05$).

3. Results and discussion

3.1 Wine volatile composition of Malvasia *aromatica* elaborated wines

An analysis of the aroma profile of Malvasia wines elaborated with the indigenous yeast strains CLI 271 and CLI 889 was made in both conditions, with and without aging on lees, to determine the effects of aging on lees on the volatile composition. **Figure 2** and **Table 1** show the evolution of volatile compounds in wines from the two yeasts after treatment on lees, grouped into chemical families.

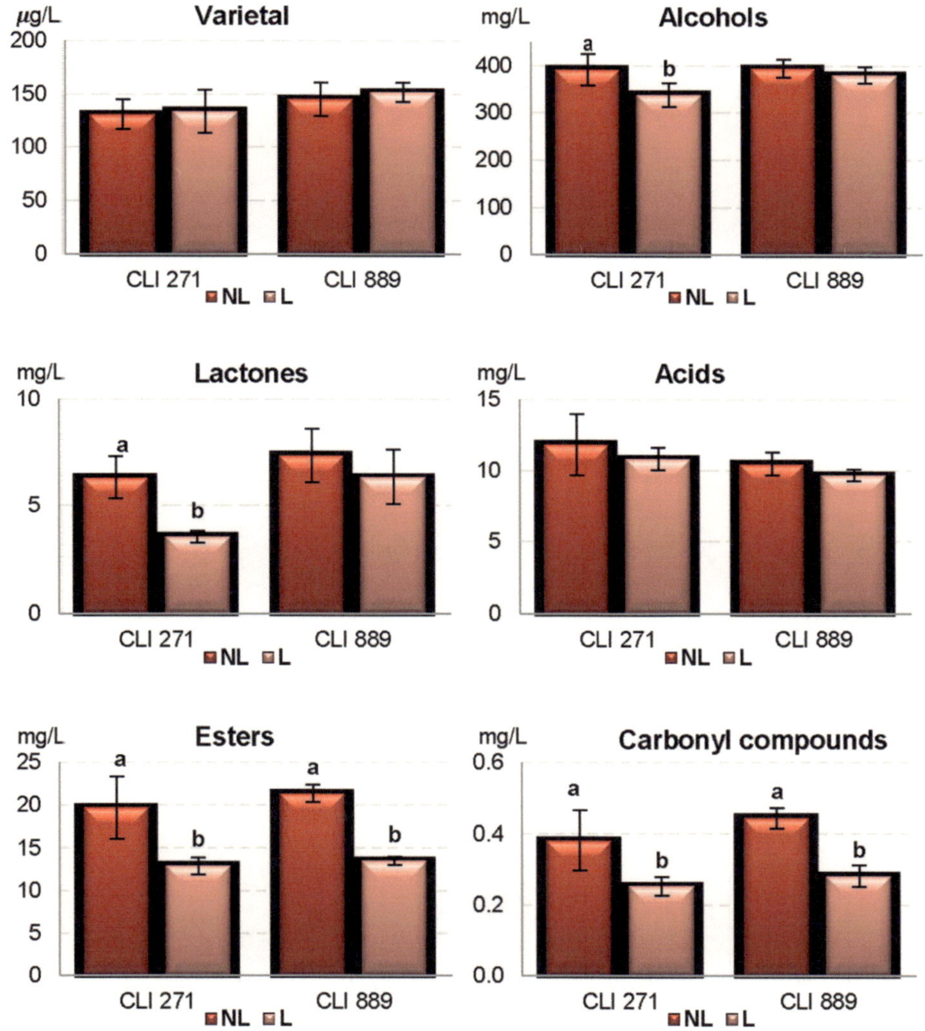

L: Aging on lees
NL: Aging without lees
Values with different letters indicate significant difference between treatments (estimated by ANOVA, followed by Tukey's test, $p<0.05$)

Figure 2.
Effect of aging on lees on the levels of aromatic compounds grouped by families in Malvasia aromatica *wines fermented with* S. cerevisiae *CLI 271 and CLI 889.*

Compounds	OTH	CLI 271			CLI 271 L			Sig[a]	CLI 889			CLI 889 L			Sig.[a]
Terpenols (µg l-1)															
β-Myrcene	—	1.17	±	0.22	**1.55**	±	**0.06**	*	1.86	±	0.06	2.17	±	0.27	Ns
α-Terpinene	—	0.22	±	0.03	0.22	±	0.06	Ns	0.17	±	0.06	0.26	±	0.02	Ns
Limonene	15[b]	0.53	±	0.08	0.56	±	0.08	Ns	0.49	±	0.08	**0.79**	±	**0.08**	*
γ-Terpinene	—	1.23	±	0.12	**1.83**	±	**0.10**	**	1.89	±	0.10	**2.49**	±	**0.27**	*
Linalool	25[c]	69.89	±	5.52	75.14	±	8.29	Ns	78.55	±	8.29	84.17	±	4.79	Ns
α-Terpineol	250[c]	25.21	±	4.12	25.26	±	2.06	Ns	23.39	±	2.06	23.86	±	1.31	Ns
β-Citronellol	100[d]	12.07	±	1.12	11.43	±	2.05	Ns	17.31	±	2.05	16.99	±	0.96	Ns
Geraniol	30[b]	19.17	±	2.78	16.65	±	3.01	Ns	20.74	±	3.01	19.69	±	1.53	Ns
Total		129.48	±	13.97	132.63	±	15.71		144.40	±	15.71	150.41	±	9.22	
C13-norisoprenoids (µg l-1)															
β-Damascenone	0.05[d]	1.24	±	0.06	1.16	±	0.14	Ns	1.11	±	0.02	**1.33**	±	**0.04**	**
Total		1.24	±	0.06	1.16	±	0.14		1.11	±	0.02	1.33	±	0.04	
Alcohols (mg l-1)															
Isobutanol	40[c]	35.58	±	1.74	**28.19**	±	**1.08**	**	42.30	±	0.44	**37.70**	±	**0.17**	***
1-Butanol	150[b]	1.03	±	0.05	**0.90**	±	**0.05**	*	0.40	±	0.02	0.35	±	0.03	Ns
Isoamyl alcohol	30[c]	296.05	±	27.63	262.31	±	22.26	Ns	315.63	±	16.13	309.26	±	16.92	Ns
1-Hexanol	8[c]	1.61	±	0.01	**1.26**	±	**0.01**	***	1.58	±	0.03	**1.49**	±	**0.04**	*
Methionol	1[c]	1.80	±	0.19	**1.36**	±	**0.04**	*	0.85	±	0.07	**0.68**	±	**0.00**	*
β-Phenylethanol	14[c]	56.33	±	3.21	**44.91**	±	**1.86**	*	34.30	±	2.20	31.11	±	0.53	Ns
Total		392.41	±	32.82	338.95	±	25.31		395.09	±	18.90	380.62	±	17.70	

Compounds	OTH	CLI 271			CLI 271 L			Sig[a]	CLI 889			CLI 889 L			Sig.[a]
Lactones (mg l-1)															
γ-Butyrolactone	35[e]	6.33	±	0.98	**3.53**	±	**0.28**	**	7.35	±	1.25	6.34	±	1.27	Ns
Total		6.33	±	0.98	3.53	±	0.28		7.35	±	1.25	6.34	±	1.27	
Fatty acids (mg l-1)															
Isobutyric acid	2.30[f]	1.39	±	0.21	**1.00**	±	**0.11**	*	1.72	±	0.11	**1.39**	±	**0.10**	*
Butyric acid	0.17[c]	0.64	±	0.08	**0.41**	±	**0.02**	**	0.73	±	0.11	0.65	±	0.11	Ns
Isovaleric acid	0.03[c]	2.43	±	0.30	**1.79**	±	**0.18**	*	2.09	±	0.13	**1.67**	±	**0.07**	**
Hexanoic acid	0.42[c]	2.92	±	0.62	3.03	±	0.22	Ns	2.31	±	0.17	2.44	±	0.11	Ns
Octanoic acid	0.50[c]	4.01	±	0.81	4.22	±	0.22	Ns	3.27	±	0.26	3.23	±	0.02	Ns
Decanoic acid	1[c]	0.41	±	0.12	0.37	±	0.00	Ns	0.34	±	0.02	**0.30**	±	**0.01**	*
Total		11.80	±	2.14	10.81	±	0.76		10.47	±	0.80	9.67	±	0.42	
Esters (mg l-1)															
Ethyl butyrate	0.02[c]	0.27	±	0.01	**0.20**	±	**0.01**	**	0.31	±	0.02	0.28	±	0.01	Ns
Ethyl isovalerate	0.003[c]	0.26	±	0.04	**0.17**	±	**0.01**	*	0.26	±	0.01	0.22	±	0.03	Ns
Isoamyl acetate	0.03[c]	1.68	±	0.19	1.40	±	0.01	Ns	1.11	±	0.03	**0.97**	±	**0.04**	*
Ethyl hexanoate	0.01[c]	0.54	±	0.12	0.38	±	0.01	Ns	0.52	±	0.03	**0.36**	±	**0.00**	**
Hexyl acetate	1[g]	0.10	±	0.01	0.10	±	0.00	Ns	0.10	±	0.00	**0.04**	±	**0.00**	***
Ethyl lactate	154[b]	15.48	±	2.86	**9.72**	±	**0.91**	*	17.16	±	0.84	**10.67**	±	**0.40**	***
Ethyl octanoate	0.58[b]	0.52	±	0.08	**0.34**	±	**0.00**	*	0.51	±	0.03	**0.26**	±	**0.01**	***
Ethyl 3-Hydroxy-butyrate	20[e]	0.03	±	0.06	0.04	±	0.04	Ns	0.11	±	0.00	**0.09**	±	**0.00**	***
Diethyl succinate	1.20[c]	0.56	±	0.22	0.24	±	0.01	Ns	1.17	±	0.00	**0.51**	±	**0.01**	**
2-Phenylethyl acetate	0.25[c]	0.26	±	0.01	**0.34**	±	**0.01**	**	0.12	±	0.00	**0.11**	±	**0.00**	*
Total		19.71	±	3.62	12.93	±	1.00		21.38	±	0.98	13.51	±	0.49	

Compounds	OTH	CLI 271			CLI 271 L			Sig[a]	CLI 889			CLI 889 L			Sig.[a]
Carbonyl comp. (mg l-1)															
Diacetyl	0.10[d]	0.16	±	0.05	0.11	±	0.01	Ns	0.23	±	0.02	**0.11**	±	**0.02**	**
Acetoin	150[h]	0.04	±	0.00	**Tr**			Ns	0.30	±	0.07	**0.03**	±	**0.03**	**
Benzaldehyde	5[c]	0.18	±	0.04	0.14	±	0.02	Ns	0.21	±	0.01	0.17	±	0.01	**
Total		0.38	±	0.08	0.25	±	0.03		0.44	±	0.03	0.28	±	0.03	
Total ($mg\ l^{-1}$)		430.77	±	39.6	366.62	±	27.41		434.88	±	21.97	410.57	±	19.93	

[a]*Significance at which means differ as shown by analysis of variance:* * $p < 0.05$; ** $p < 0.01$; *** $p < 0.001$.
[b]*[18]*
[c]*[19]*
[d]*[20]*
[e]*[21]*
[f]*[22]*
[g]*[23]*
[h]*[24].*
OTH: Odor threshold value.*
Ns: not significant.
Tr: traces.
In bold the compounds that seem to have a clear connection with aging on lees treatment.

Table 1.
Effect of aging on lees (L) on the aroma compound levels of Malvasia aromatica *wines fermented with* S. cerevisiae *CLI 271 and CLI 889.*

Thirty-five volatiles were quantified, and distributed in varietal aromas (terpenols and C13-norisoprenoids), alcohols, acids, lactones, esters, aldehydes and ketones; their total content is given in μg/L for varietal aromas and mg/L for the rest of the compounds. **Table 1** also shows the results of the analysis of variance (ANOVA) for each fermentation to determine the compounds that show significant individual differences due to the effect of aging on lees.

The effect of lees treatment was significant in 57% of the compounds identified in wines fermented with CLI 889 strain (20 of 35 compounds identified) and 45% (16 of 35 compounds) in wines from CLI 271 strain (**Table 1**).

Aging on lees caused a significant decrease in the total content of esters, aldehydes and carbonyl compounds in the two winemaking processes (**Figure 2**). The same behavior for the ester family was reported by Bueno et al. [26] and Del Barrio-Galán et al. [27] in Macabeo and Verdejo varieties respectively after a period of aging on lees.

The ANOVA analysis in **Table 1** shows major differences in the CLI 889 wines, 8 of the 10 esters identified decreased significantly. Some studies have shown how the interactions between macromolecules released by the lees and aromatic compounds present in the wine could modify the volatility and aromatic intensity of these compounds; thus, the lower content of some esters compared to the control wine (without lees) could be explained by their association with mannoproteins and polysaccharides released, reducing volatility. These results are in agreement with other authors on model wines [28, 29]. On the other hand, in the esters where there was a notable decrease in terms of concentration, it was not decisive in terms of aroma contribution. Despite the loss, they continued to exceed their perception threshold (isoamyl acetate and ethyl hexanoate) or did not reach the threshold before lees treatment (hexyl acetate, ethyl lactate, ethyl octanoate, diethyl succinate and ethyl 3-hydroxy butyrate). Some of these compounds also showed lower concentrations in Tempranillo wines aged on lees [30]. The authors explained that the lower content of these volatiles can be due to an interaction between esters, acetates and mannoproteins from yeast autolysis and/or the sorption phenomenon onto yeast cell walls. This behavior was also denoted in model wines [28, 29, 31] and in white wines [26]. It could be happened in Malvasia wines from the present work. Wines fermented with CLI 271 showed a significant decrease in 4 of the 10 esters quantified. This decrease was not a disadvantage in terms of aroma contribution because ethyl butyrate and ethyl isovalerate still exceeded their perception threshold despite the loss of concentration and ethyl lactate and ethyl octanoate did not reach the threshold before lees treatment. It is important to note the role of 2-phenylethyl acetate in CLI 271 wines. Its concentration increases significantly with lees treatment. This compound is related to fruity attributes.

Regarding carbonyl compounds, the significant decrease of acetoin (creamy and buttery aromas) in CLI 889 L wines is noteworthy. Acetoin on its own does not have a significant aromatic impact due to its high detection threshold, although it does contribute to the wine bouquet due to its interrelation in the synthesis of diacetyl [32], a compound that, although it decreased significantly in CLI 889 wines, exceeds its perception threshold, contributing buttery aromas. Benzaldehyde also decreased significantly in CLI 889 wines but in all cases, like acetoin, it was far from its threshold.

The effect of aging on lees was different for the lactone family (only represented by γ-butyrolactone). The treatment resulted in a significant decrease in CLI 271 wines and no variations were observed in CLI 889 wines. In all cases, the concentrations of this compound were far from its perception threshold.

The increase of terpenes and norisoprenoids on lees presence is likely related to the release during autolysis of β-glucosidases. These enzymes are able to break the glycoside bound of these volatile compounds releasing the free aromatic form, which has an influence on final aroma of wine [33]. In wines aged on lees, varietals tended to increase in agreement with other works [26, 33]. Individually (**Table 1**), the varietal compounds β-myrcene, limonene and γ-terpinene varied significantly but were found in low concentrations. As for β-damascenone, a significant increase was only seen in CLI 889 wines with lees treatment always exceeding their olfactory perception threshold.

Regarding to acids content, the decrease in their concentration as a consequence of contact with lees had no effect on the aroma of the wines. The maximum values correspond to octanoic acid, and after that, to hexanoic acid in wines aging on lees in good agreement with other studies in white wines [26, 34].

According to the family of alcohols (**Figure 2**) only the wines made with strain CLI 271 aged on lees showed a notable decrease, 5 of the 6 compounds identified (**Table 1**) decreased significantly. Methionol and β-phenylethanol seem to actively contribute to the aroma of these wines with a concentration above their perception threshold even while decreasing in presence of lees. Isobutanol decreased significantly and was found at levels below its threshold in CLI 889 wines aged on lees. These changes in alcohol content compounds were similar to obtained in Macabeo wines [26].

3.2 Odor activity values of Malvasia *aromatica* wines

In order to estimate the sensory contribution of aromatic compounds to overall aroma of wine, the odor activity value (OAV) was calculated for all aroma compounds in study (**Table 2**). An aromatic compound contributes actively in wine when its concentration is higher than its olfactory perception threshold (OTH). The concentration/threshold ratio, known as odor activity value (OAV), shows the contribution of a specific compound to wine aroma properties. Those compounds with an OAV ≥ 1 can be considered active odorants [21].

Sixteen of thirty-five identified compounds presented OAV > 1, contributing to the aroma of the wines studied (**Table 2**). The varietal compounds linalool and β-damascenone had an active contribution to the aroma of all resulting wines providing floral aromas with citrus and lilac attributes respectively. The OAV was always higher than 1. β-damascenone also presented concentrations higher than OTH in all wines showing interesting correlations between the increase in OAVs of wines fermented with CLI 889 yeast strain aged on lees. A study realized by Gallardo-Chacón et al. [35] in Cava aged in contact with lees, they detected some norisoprenoids as β-damascenone more retained in lees surface and thus its presence in the volatile profile of Cava was significant.

The alcohols, characterized by vegetal and bitter aromatic descriptors with the exception of β-phenylethanol (roses), did not show significant differences according to aroma contribution. OAVs were higher than 1 in all wines except isobutanol in CLI 889 L wines.

The wine elaborated with *S. cerevisiae* CLI 271 strain obtained higher concentrations of 2-phenylethyl acetate above its OTH and presented OAV > 1 in traditionally preserved wines (non lees) increasing significantly after lees treatment. This compound is related to fruity attributes. Previous studies have demonstrated the capacity of this locally selected yeast strain CLI 271 to influence in the volatile profile of wines, increasing the concentration of fresh-fruity aroma compounds [15]. In general, esters decreased in both wines treated with lees but their OAVs remained above unity and actively contributed to aroma.

			OAVa			
	Sensory descriptor	**OTH***	**CLI 271**	**CLI 271 L**	**CLI 889**	**CLI 889 L**
Linalool	Floral, citric	25	4.66	5.01	5.24	5.61
β-damascenone	Floral, lilac	0.05	24.80	23.27	22.13	**26.60**
Isoamyl alcohol	Vegetal/herbaceous	30	9.87	8.74	10.52	10.31
Isobutanol	Alcohol	40	<1	<1	1.06	**0.94**
Methionol	Cooked vegetable	1	1.80	1.36	<1	<1
β-Phenylethanol	Roses	14	4.02	3.21	2.45	2.22
Ethyl butyrate	Acid fruit, apple	0.02	13.68	**9.97**	15.33	**13.76**
Ethyl isovalerate	Sweet fruit, orange, blackberry	0.003	87.17	**57.09**	87.52	**74.92**
Isoamyl acetate	Banana	0.03	56.14	**46.65**	37.07	**32.48**
Ethyl hexanoate	Fruit, green apple	0.01	53.80	**38.11**	51.90	**35.60**
2-Phenylethyl acetate	Green apple	0.25	1.04	**1.36**	<1	<1
Butyric acid	Cheese	0.17	3.79	2.38	4.30	3.79
Isovaleric acid	Blue cheese	0.03	81.04	59.82	69.70	55.59
Hexanoic acid	Cheese	0.42	6.95	7.22	5.51	5.80
Octanoic acid	Butter, sour	0.50	8.01	8.43	6.55	6.46
Diacetyl	Butter	0.10	1.62	**1.14**	2.31	**1.14**

**OTH: Odor threshold values*
[a]OAV: Odor activity values calculated by dividing concentration by odor threshold value of the compound.
*OTH and OAV are given in mg l^{-1} except linalool and β-damascenone which are in μg l^{-1}. Sensory descriptor and OTH values according to the references included in **Table 1**.*
In bold the compounds that seem to have a clear connection with aging on lees treatment.

Table 2.
Odor threshold values and odor activity values of the volatile compounds with the greatest influence on the aroma of Malvasia wines fermented with S. cerevisiae *CLI 271 and CLI 889 with lees treatment (L).*

A significant decrease is observed in CLI 889 wines regarding diacetyl during aging on lees (**Table 1**). This compound is related to buttery aroma descriptors although their OAV is still higher than 1 (**Table 2**).

Finally, butyric, isovaleric, hexanoic and octanoic acids, characterized by their unpleasant odors (cheese and rancid attributes) influenced all the wines equally, all wines had OAV > 1. They cannot be considered determinants of the differences in the overall aromatic quality of the wines.

3.3 Sensory characteristics of Malvasia *aromatica* wines

A descriptive tasting of wines was performed at visual, olfactory and gustative levels in order to determine the main organoleptic differences among wines after aging on lees (**Figure 3A–C**). The sensory characterization of the Malvasia *aromatica* wines was obtained from the attributes defined by the tasters. The results were statistically treated and differences between wine samples with a significance level of 0.1% ($p < 0.001$) were considered very significant, of 1% ($p < 0.01$) considered significant and of 5% ($p < 0.05$) considered low significant.

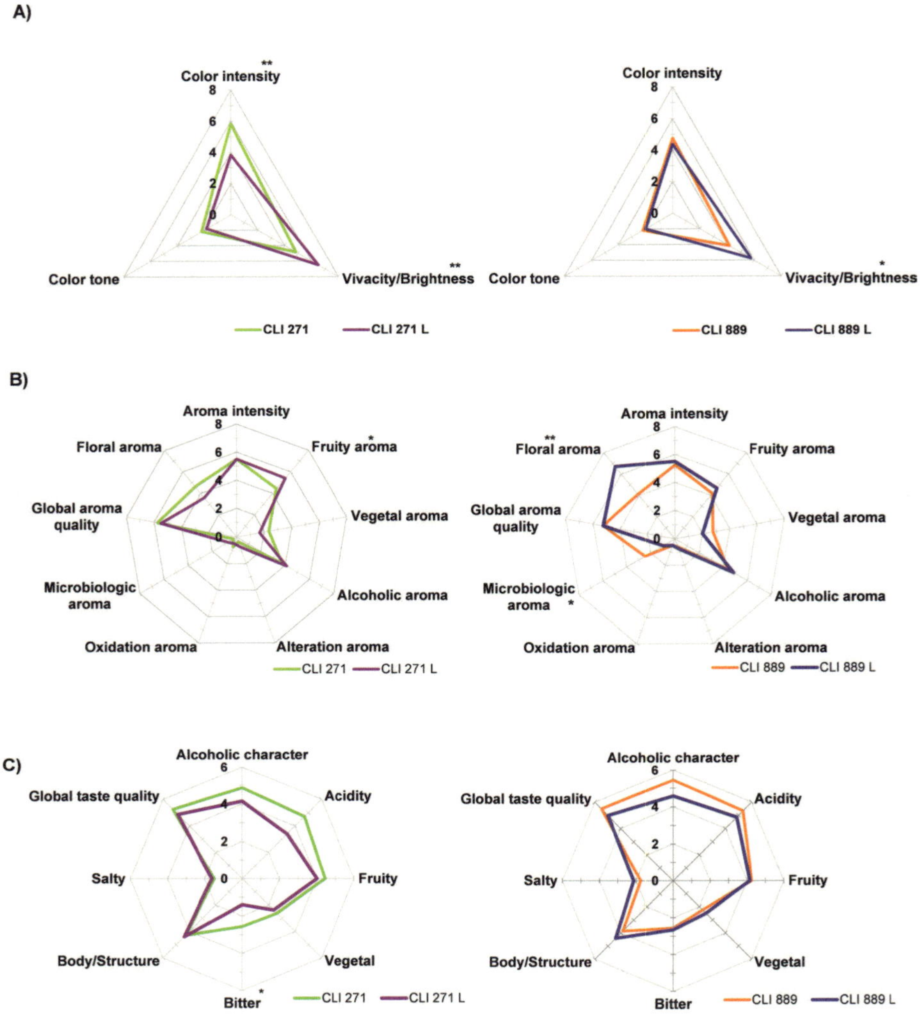

Figure 3.
Descriptive analysis of Malvasia wines elaborated with S. cerevisiae *CLI 271 and CLI 889 with lees treatment (L). A) Visual phase; B) Olfactory phase; C) Gustative phase.*

In the visual phase (**Figure 3A**), aging on lees (L) resulted in Malvasia *aromatica* wines with significantly higher scores in vivacity/brightness descriptor. This could be related to the improved stability of white wines due to the action of some mannoproteins released by yeasts that help to reduce the turbidity produced by the presence of unstable proteins after fermentation [36, 37]. On the other hand, CLI 271 control wines were scored with higher color intensity with statistical significance in difference that wine treated with lees (CLI 271L). The influence of lees in wine color has been pointed out by several authors in artificial wines [38, 39], in white wines [40, 41] and red wines [42]. This fact may be due to capacity of yeast compounds such as mannoproteins and glucans, to interact with different wine phenolic molecules as total polyphenols, flavonols and tannins [43]. In accordance with [27] the use of lees can give rise to a reduction in the color intensity of wines after treatment, so they can be used as agents for reducing browning in white wines.

In the olfactory phase (**Figure 3B**), CLI 889L wines obtained significantly higher scores on floral aroma attribute. As shown in **Table 1**, wines elaborated with CLI 889 strain presented higher concentrations of varietals compounds and lees treatment could have contributed to enhance the floral characteristics of this family of volatiles. According to the results in **Table 2**, there was a noticeable increment of linalool and β-damascenone compounds. Regarding to CLI 271L wines were considered significantly fruitier than CLI 271 control (non lees). These results are in line with those obtained by Del Barrio-Galán et al. [27] in the sensory analysis of the Verdejo variety where the treatment with lees would have displayed stronger fruity and floral aromas and higher olfactory intensity than the control wines which could indicate that these initially retained aromatic compounds are released over time, increasing aroma intensities. Based on the analysis of **Table 2**, 2-phenylethyl acetate with fruit descriptor increased significantly after lees aging. Aging on lees decreased the vegetal character of Malvasia *aromatica* wines from both winemaking but without significant differences. The same behavior for this attribute was reported by other authors in white wines aged on lees [34]. Some tasters noted the presence of microbiological aromas in CLI 889 wines without lees, but this character was not detected in wines aged on lees. There were no significant differences between treatments in the rest of the parameters.

Finally, in the gustatory phase (**Figure 3C**), there are almost no differences between treated wines and their controls. Wines not aged on lees were evaluated as alcoholic and scored higher in global taste quality without significant differences. Aging on lees apparently reduced bitterness in Malvasia wines, but the decrease was only significant in CLI 271. Additionally, there was a tendency for the lees-treated wines to have greater structure and less acidity, although this was not statistically significant. This trend toward a bitterness and acidity reduction and an increment of mouthfeel and balance in wines treated with lees has been described in previous publications [27, 44].

With the aim of completing the sensory study, triangular tests were done to determine whether descriptive analysis was determinant to differentiating the samples. Discriminant triangular tastings were carried out using dark glasses and two series of tastings were performed by each type of wine: CLI 889 vs. CLI 889 L, CLI 271 vs. CLI 271 L. The panel clearly differentiated CLI 271 and CLI 889 wines aged on lees from their respective controls with statistical significance levels of 1% and 0.1% in the two tests performed for CLI 271 wines, and 5% and 1% in the two tests performed for CLI 889 wines. In addition, tasters were asked to assess their hedonic preference for wines in the tests presented, taking into account the preferences of the taster who answered correctly. The tasting panel showed preference by CLI 271 wines in 57% of cases in the first test and 62% in the second. Preferences for the CLI 889 wines were equally divided. In the first test, 50% of the correct judges chose CLI 889 L samples. In the second test, 57% of the correct tasters preferred CLI 889 L.

4. Conclusions

According to the results, we can affirm that lees treatment has had an effect on the result of aroma and sensory composition of Malvasia *aromatica* wines. Some volatiles showed significant variations after lees treatment. Esters were the most affected compounds along with carbonyl compounds in wines fermented with CLI 889, decreasing in concentration after aging on lees. Wines from strain CLI 271 suffered the greatest losses in the alcohol family. According to the yeast strains used for winemaking, we can conclude that aging on lees has caused greater changes in the volatile composition of CLI 889 strain wines. The results of the sensory analysis indicate that Malvasia wines aged on lees have more vivacity and brightness than those aged conventionally, also could reduce acidity, bitter and vegetal character of the wines. The treatment with lees enhanced the floral character of the Malvasia *aromatica* wines fermented with the native yeast strain CLI 889 and the fruity character of the CLI 271. Aging on lees seems to have a positive effect on wines from the native yeast strains CLI 889, since they were preferred over those conventionally aged. In the case of CLI 271 wines, the similar development of wine with and without lees suggests that factors determining aroma and flavor development are influenced by *S. cerevisiae* CLI 271 fermentations. Probably, it could be interesting to test other aging on lees times in CLI 271 wines to find out more differences after lees treatment. Therefore, the employment of aging on lees in wines fermented with locally-selected *S. cerevisiae* can be used to obtain quality wines from the alternative variety Malvasia *aromatica* on PDO "Vinos de Madrid".

Acknowledgements

JMC acknowledges INIA for a PhD grant (FPI-INIA 2012 call). The authors acknowledge IMIDRA colleagues at the El Socorro experimental farm and the Microbiology lab for providing material of Malvasia *aromatica* and yeast strains respectively used in this work and IMIDRA for its financial support.

Conflict of interest

The authors declare no conflict of interest.

Acronyms and abbreviations

PDO	Protected Designation of Origen
L	lees
DVB	divinylbenzene
CAR	carboxen
PDMS	polydimethylsiloxane
GC-FID	gas chromatography with flame-ionization detection
GC/MS	gas chromatography–mass spectrometry
HS-SPME	headspace-solid phase microextraction
C13	C13-norisoprenoids
ANOVA	analysis of variance
OTH	olfactory perception threshold
OAV	odor activity value

Author details

Julia Crespo*, Margarita García, Teresa Arroyo, Valeria Romero and Juan M. Cabellos
Agrifood Research Department, Madrid Institute for Rural, Agriculture and Food Research and Development (IMIDRA), Alcalá de Henares, Madrid, Spain

*Address all correspondence to: julia.crespo.garcia@madrid.org

References

[1] Fornairon-Bonnefond C, Camarasa C, Moutounet M, Salmon JM. New trends on yeast autolysis and wine aging on lees: A bibliographic review. Journal International des Sciences de la Vigne et du Vin. 2002;**36**:49-69. DOI: 10.20870/oeno-one.2002.36.2.974

[2] Charpentier C, Dos Santos AM, Feuillat M. Release of macromolecules by *Saccharomyces cerevisiae* during aging of French flor sherry wine "Vin jaune". International Journal of Food Microbiology. 2004;**96**:253-262. DOI: 10.1016/j.ijfoodmicro.2004.03.019

[3] Pérez-Bibbins B, Torrado-Agrasar A, Salgado JM, Oliveira RP, Domínguez JM. Potential of lees from wine, beer and cider manufacturing as a source of economic nutrients: An overview. Waste Management. 2015;**40**:72-81. DOI: 10.1016/j.wasman.2015.03.009

[4] Lavigne V, Pons A, Dubourdieu D. Assay of glutathione in must and wines using capillary electrophoresis and laser-induced fluorescence detection. Changes in concentration in dry white wines during alcoholic fermentation and aging. Journal of Chromatography A. 2007;**1139**:130-135. DOI: 10.1016/j.chroma.2006.10.083

[5] Chattonnet P. La contamination des vins par *Brettanomyces* au cours de la vinification et de l'élevage Incidence, détection et moyens de lutte. Revue des OEnologues. 2000;**2000**:23-26

[6] Doco T, Vuchot P, Cheynier V, Moutounet M. Structural modification of wine arabinogalactans during aging on lees. American Journal of Enology and Viticulture. 2003;**53**:150-157

[7] Bureau SM, Razungles AJ, Baumes RL. The aroma of Muscat of Frontignan grapes: Effect of the light environment of vine or bunch on volatiles and glycoconjugates. Journal of the Science of Food and Agriculture. 2000;**80**:2012-2020. DOI: 10.1002/1097-0010(200011)80:14<2012: AID-JSFA738>3.0.CO; 2-X

[8] Compés R, Sotés V. El sector vitivinícola frente al desafío del cambio climático. Estrategias públicas y privadas de mitigación y adaptación en el Mediterráneo. Almería, Spain: Cajamar Caja Rural; 2018. p. 372

[9] Ribéreau-Gayon P. New developments in wine microbiology. American Journal of Enology and Viticulture. 1985;**36**:1-10. DOI: 10.5344/ajev.1985.36.1.1

[10] Heard GM, Fleet GH. Growth of natural yeast flora during the fermentation of inoculated wines. Applied and Environmental Microbiology. 1985;**50**:727-728. DOI: 10.1128/aem.50.3.727-728.1985

[11] Crespo J, Romero V, García M, Arroyo T, Cabellos JM. Influence of skin-contact treatment on aroma profile of Malvasia *aromatica* wines in D.O. "Vinos de Madrid". In: Morata A, editor. Grapes Wine. London, UK: IntechOpen; 2021. pp. 137-144. DOI: 10.5772/intechopen.99216

[12] Crespo J, García M, Arroyo T, Romero V, Cabellos JM. Influence of native *Saccharomyces cerevisiae* strains on Malvasia *aromatica* wines. Frontiers in Bioscience-Elite. 2023;**15**:18. DOI: 10.31083/j.fbe1503018

[13] Cabello F, Ortiz J, Muñoz G, Rodríguez I, Benito A. Variedades de vid en España. 1st ed. Madrid, España: Agrícola Española; 2012. p. 504

[14] Arroyo T. Estudio de la influencia de diferentes tratamientos enológicos en la evolución de la microbiota y en la calidad de los vinos elaborados con la variedad "Airén", en la D.O. "Vinos de Madrid." Alcalá de Henares: University of Alcalá; 2000

[15] Balboa-Lagunero T, Arroyo T, Cabellos JM, Aznar M. Yeast selection as a tool for reducing key oxidation notes in organic wines. Food Research International. 2013;**53**:252-259. DOI: 10.1016/j.foodres.2013.04.006

[16] Cordero-Bueso G, Esteve-Zarzoso B, Gil-Díaz M, García M, Cabellos J, Arroyo T. Improvement of Malvar wine quality by use of locally-selected *Saccharomyces cerevisiae* strains. Fermentation. 2016;**2**:7. DOI: 10.3390/fermentation2010007

[17] Ortega C, López R, Cacho J, Ferreira V. Fast analysis of important wine volatile compounds-development and validation of a new method based on gas chromatographic-flame ionisation detection analysis of dichloromethane microextracts. Journal of Chromatography A. 2001;**923**:205-214. DOI: 10.1016/S0021-9673(01)00972-4

[18] Yuan F, Qian MC. Development of C13-norisoprenoids, carotenoids and other volatile compounds in *Vitis vinifera* L. Cv. Pinot noir grapes. Food Chemistry. 2016;**192**:633-641. DOI: 10.1016/j.foodchem.2015.07.050

[19] Etiévant XP. Wine. In: Maarse H, editor. Volatile Compounds in Foods and Beverages. New York: Marcel Dekker; 1991. pp. 483-546

[20] Ferreira V, López R, Cacho JF. Quantitative determination of the odorants of young red wines from different grape varieties. Journal of the Science of Food and Agriculture. 2000;**80**:1659-1667. DOI: 10.1002/1097-0010(20000901)80:11<1659: AID-JSFA693>3.0.CO; 2-6

[21] Guth H. Quantitation and sensory studies of character impact odorants of different white wine varieties. Journal of Agricultural and Food Chemistry. 1997;**45**:3027-3032. DOI: 10.1021/jf9608433

[22] Aznar M, López R, Cacho J, Ferreira V. Prediction of aged red wine aroma properties from aroma chemical composition. Partial least squares regression models. Journal of Agricultural and Food Chemistry. 2003;**51**:2700-2707. DOI: 10.1021/jf026115z

[23] Van Gemert LJ, Nettenbreijer AH. Compilation of Odour Threshold Values in Air and Water. The Netherlands; 1977

[24] Gil M, Cabellos JM, Arroyo T, Prodanov M. Characterization of the volatile fraction of young wines from the denomination of origin "Vinos de Madrid" (Spain). Analytica Chimica Acta. 2006;**563**:145-153. DOI: 10.1016/j.aca.2005.11.060

[25] Ferreira V, Ortín N, Escudero A, López R, Cacho J. Chemical characterization of the aroma of Grenache rosé wines: Aroma extract dilution analysis, quantitative determination, and sensory reconstitution studies. Journal of Agricultural and Food Chemistry. 2002;**50**:4048-4054. DOI: 10.1021/jf0115645

[26] Bueno JE, Peinado RA, Medina M, Moreno J. Effect of a short contact time with lees on volatile composition of Airen and Macabeo wines. Biotechnology Letters. 2006;**28**:1007-1011. DOI: 10.1007/s10529-006-9038-2

[27] Del Barrio-Galán R, Pérez-Magariño S, Ortega-Heras M,

Williams P, Doco T. Effect of aging on lees and of three different dry yeast derivative products on Verdejo white wine composition and sensorial characteristics. Journal of Agricultural and Food Chemistry. 2011;**59**:12433-12442. DOI: 10.1021/jf204055u

[28] Lubbers S, Voilley A, Feuillat M, Charpentier C. Influence of mannaproteins from yeast on the aroma intensity of a model wine. LWT – Food Science and Technology. 1994;**27**:108-114. DOI: 10.1006/fstl.1994.1025

[29] Chalier P, Angot B, Delteil D, Doco T, Gunata Z. Interactions between aroma compounds and whole mannoprotein isolated from *Saccharomyces cerevisiae* strains. Food Chemistry. 2007;**100**:22-30. DOI: 10.1016/j.foodchem.2005.09.004

[30] Rodríguez-Bencomo JJ, Ortega-Heras M, Pérez-Magariño S. Effect of alternative techniques to aging on lees and use of non-toasted oak chips in alcoholic fermentation on the aromatic composition of red wine. European Food Research and Technology. 2010;**230**:485-496. DOI: 10.1007/s00217-009-1189-7

[31] Voilley A, Beghin V, Charpentier C, Peyron D. Interaction between aroma substances and macromolecules in a model wine. LWT – Food Science and Technology. 1991;**24**:469-472

[32] Romano P, Suzzi G. Origin and production of acetoin during wine yeast fermentation. Applied and Environmental Microbiology. 1996;**62**:309-315. DOI: 10.1128/aem.62.2.309-315.1996

[33] Liberatore MT, Pati S, Del NMA, La NE. Aroma quality improvement of chardonnay white wine by fermentation and aging in barrique on lees. Food Research International. 2010;**43**:996-1002. DOI: 10.1016/j.foodres.2010.01.007

[34] Bautista R, Fernández E, Falqué E. Effect of the contact with fermentation-lees or commercial-lees on the volatile composition of white wines. European Food Research and Technology. 2007;**224**:405-413. DOI: 10.1007/s00217-006-0336-7

[35] Gallardo-Chacón J, Vichi S, López-Tamames E, Buxaderas S. Analysis of sparkling wine lees surface volatiles by optimized headspace solid-phase microextraction. Journal of Agricultural and Food Chemistry. 2009;**57**:3279-3285. DOI: 10.1021/jf803493s

[36] Waters EJ, Pellerin P, Brillouet JM. A *Saccharomyces* mannoprotein that protects wine from protein haze. Carbohydrate Polymers. 1994;**23**:185-191. DOI: 10.1016/0144-8617(94)90101-5

[37] Waters E. A review of current knowledge on polysaccharides which "protect" against protein haze en white wine. Australian Grapegrower Winemaker. 2000;**438**:13-17

[38] Mazauric JP, Salmon JM. Interactions between yeast lees and wine polyphenols during simulation of wine aging. Journal of Agricultural and Food Chemistry. 2006;**54**:3876-3881. DOI: 10.1021/jf060037o

[39] Márquez T, Millán C, Souquet JM, Salmon JM. Effect of different yeast strains and their culture conditions on the prevention of wine model solution browning by yeast lees. Journal of Agricultural and Food Chemistry. 2009;**57**:3771-3779. DOI: 10.1021/jf803839s

[40] Razmkhab S, Lopez-Toledano A, Ortega JM, Mayen M, Merida J, Medina M. Adsorption of phenolic compounds and browning products in white wines by yeasts and their cell walls. Journal of Agricultural and

Food Chemistry. 2002;**50**:7432-7437. DOI: 10.1021/jf025733c

[41] Lopez-Toledano A, Mayen M, Merida J, Medina M. Yeasts used to delay browning in white wines. Food Chemistry. 2006;**97**:498-504. DOI: 10.1016/j.foodchem.2005.05.030

[42] Guadalupe Z, Palacios A, Ayestarán B. Maceration enzymes and mannoproteins: A possible strategy to increase colloidal stability and color extraction in red wines. Journal of Agricultural and Food Chemistry. 2007;**55**:4854-4862. DOI: 10.1021/jf063585a

[43] Poncet-Legrand C, Doco T, Williams P, Vernhet A. Inhibition of grape seed tannin aggregation by wine mannoproteins: Effect of polysaccharide molecular weight. American Journal of Enology and Viticulture. 2007;**58**:87-91

[44] Liu L, Loira I, Morata A, Suárez-Lepe JA, González MC, Rauhut D. Shortening the aging on lees process in wines by using ultrasound and microwave treatments both combined with stirring and abrasion techniques. European Food Research and Technology. 2016;**242**:559-569. DOI: 10.1007/s00217-015-2566-z

Chapter 7

Bioprotection in Winemaking

Hervé Alexandre, Maëlys Puyo and Raphaëlle Tourdot-Maréchal

Abstract

Bioprotection in the wine sector is a strategy for protecting grape musts that have been used for a few years now. Bioprotection is intended to be a partial or total alternative to the use of sulfites. The principle of bioprotection consists in providing, from the harvest, on the grapes or on the grape must, yeast biomass, which, by its action, will limit the development of the native microbial flora and consequently avoid microbiological alterations at the early stages of the winemaking process. Most often, the biomasses studied are selected strains of non-*Saccharomyces* such as *Torulaspora delbrueckii* or *Metschnikowia pulcherrima*, but the *Saccharomyces cerevisiae* species can also be used. We propose to present the results of bioprotection used in white and red wine processes obtained in recent years and to underline the limits of this technique. Finally, a section will be devoted to describing proven or potential mechanisms that may explain how the biomass provided limits the development of native flora. Finally, the perspectives on the use of bioprotection in must and wine will be discussed.

Keywords: bioprotection, must, wine, sulfite alternatives, *Brettanomyces bruxellensis*, *Torulaspora delbrueckii*, *Metschnikowia pulcherrima*

1. Introduction

Bioprotection in food refers to the use of natural or controlled microorganisms, enzymes, or antimicrobial compounds to prevent or inhibit the growth of harmful bacteria, fungi, and other microorganisms in food products [1]. The primary goal of bioprotection is to enhance the safety and extend the shelf life of food without relying on synthetic chemical preservatives.

There are several methods of bioprotection in food. In the context of food, certain probiotic strains can inhibit the growth of pathogenic bacteria and spoilage microorganisms, thus contributing to food safety [1]. Bacteriocins are antimicrobial peptides or proteins produced by certain bacteria that can inhibit the growth of closely related bacteria or other pathogens [2]. Some of these natural compounds can be used as food preservatives to prevent the growth of spoilage and pathogenic bacteria. Certain strains of lactic acid bacteria and other microorganisms are used as protective cultures in food products like meat, fish, and dairy to outcompete harmful bacteria and create an acidic or competitive environment that prevents spoilage [2]. Enzymes can play a role in food preservation by breaking down certain components that promote microbial growth or spoilage. For example, enzymes like lysozyme can inhibit the growth of bacteria by disrupting their cell walls [3].

Bioprotection offers several advantages over traditional chemical preservatives. It involves the use of natural and generally recognized as safe (GRAS) microorganisms or compounds, making it more acceptable to consumers seeking clean label products. Additionally, it can have positive effects on the sensory and nutritional properties of food.

Bioprotection in wine refers to the use of natural, non-chemical methods to protect wine from undesirable spoilage microorganisms, primarily spoilage yeasts and bacteria. The aim of bioprotection is to preserve wine quality, prevent off-flavors, and reduce the need for the addition of chemical preservatives. The main reason why winemakers adopt bioprotection is that it reduces or eliminates the need for sulfites in the pre-fermentation phase.

The main factor that might contribute to the increased success of wine bioprotection is the consumer demand. With the growing awareness of wine safety and the desire for natural and clean-label products, consumers are seeking wine products with fewer synthetic chemical additives. Wine bioprotection offers a natural alternative to traditional chemical preservatives, making it a more appealing option for health-conscious consumers.

The main strategy for bioprotection is the use of selected non-*Saccharomyces* yeasts (NS yeasts). Non-*Saccharomyces* yeasts, such as *Torulaspora delbrueckii*, *Metschnikowia pulcherrima* and *Lachancea thermotolerans*, can be added before or alongside the primary fermentation with *Saccharomyces cerevisiae* (traditional yeast). These non-*Saccharomyces* yeasts contribute to flavor complexity and can inhibit spoilage microorganisms through competition for nutrients and space.

It is important to note that while bioprotection can be effective in managing microbial populations and preserving wine quality, they also require careful monitoring and control. The success of bioprotection methods may vary depending on factors such as grape variety, winemaking techniques and environmental conditions. Winemakers need to strike a balance between promoting natural fermentation and ensuring stability and consistency in the final product.

In this chapter, we will make a synthesis of the experiments carried out in bioprotection on white, rosé and red grape musts. In the second part, we will be interested in the mechanisms that can explain the protective effect, and finally, we will take stock of the future prospects.

2. Synthesis of bioprotection experiments

The process of bioprotection generally consists of adding one or more microorganisms in sufficient quantity on harvest or at the vatting. Thus, the high proportion of these microorganisms allows them to colonize quickly the medium preventing the development of spoilage yeasts and bacteria present in smaller quantities. Non-*Saccharomyces* yeasts are the most documented microorganisms today about their potential bioprotection properties.

The use of bioprotection in the wine industry dates back some 10 years, during which time experiments were carried out on an empirical basis. Indeed, the choice of yeast species, the dosage and the timing of addition are all parameters that have been defined without any scientific basis. Optimal bioprotection conditions then emerged from various experiments.

2.1 White must

One of the first experiments carried out and published in a popular journal reported the use of the *Torulaspora delbrueckii* species in the pre-fermentation phase on white grape juice. The *Torulaspora delbrueckii* species, naturally present but not in the majority in must, presents interesting characteristics such as low acetic acid production and cryotolerance. This species can therefore establish itself in musts placed at low temperatures (>5°C) without suffering significant population loss. On the other hand, temperatures >15°C are required to trigger alcoholic fermentation. For example, with inoculation of juice after pressing, without triggering alcoholic fermentation on grape solid [4] and using 5 g/hl of *Torulaspora delbrueckii* in white grape juice just after pressing at 5°C (**Table 1**), the authors reported a similar effect compared with the sulfite modality with a maximum *Saccharomyces* population of 1.4 log/ml, while on the nonsulfited modality, this population reached almost 3 log. Another study [5] demonstrated that *Torulaspora delbrueckii* was successful in limiting the development of spoilage microorganisms (*Brettanomyces*, acetic acid bacteria and lactic acid bacteria) in Aligoté must. In this experiment, *Torulaspora delbrueckii* was added at a dose of 5 g/hl after pressing, and microbiological counts were carried out after settling on sulfited or bioprotected batches. It is noteworthy that no significant difference in populations was observed between either modality.

The two previous studies were carried out with the species *Torulaspora delbrueckii*, but bioprotection can be carried out with the species *Metschnikowia pulcherrima*. Some *Metschnikowia pulcherrima* strains have very low fermentation capacity, good implantation at low temperatures, genetic stability, and a high level of resistance. The use of *Metschnikowia* as a bioprotectant agent on Verdicchio during must clarification during two vintages have shown that without the use of *M. pulcherrima,* there was an increase of about one log CFU/mL, while the inoculated trial showed a containment of the wild yeast population, which remained almost constant [11]. During another vintage, a significant reduction of indigenous yeast population was shown in both trials with and without the inoculation of *M. pulcherrima*. The presence of *Metschnikowia pulcherrima* during cold maceration did not affect significantly the enological parameters. On the other hand, the inoculation of *M. pulcherrima* at the start of cold clarification generally led to wine with a different volatile profile. Indeed, a significant increase in ethyl hexanoate and monoterpene with relevant high Odor Activity Values in comparison with the control was observed [11]. The sensory analysis highlighted the influence of the use of *M. pulcherrima* in cold clarification on the sensory profile of aroma wine.

In another study, Gerbaux et al. [6] used this species in a bioprotection trial on Chardonnay; the microbial flora was not determined, the aim of the study is to look at the protection of wines against oxidation, as it was shown that *Metschnikowia pulcherima* consumed the oxygen in musts and thus prevented oxidation of musts in the absence of oxygen [6]. The sensorial analysis concluded a higher fruity intensity for the bioprotection modality despite the lack of statistical analysis.

The problem of must oxidation when bioprotection is used is just as important as limiting the development of spoilage flora. As we have seen, in white musts, under certain conditions, bioprotection protects against both microbial development and oxidation [5, 6]. A study carried out at a lab scale confirmed the oxygen consumption capacity of *Metschnikowia pulcherrima* [10]. In this study, it was shown in Muscat

Yeast species	Addition step	Dose	Temperature	Presence of a control	Grape variety	Sensorial analysis	Scale	Reference
Torulaspora delbrueckii	After pressing	5 g/hl (10^6 cells/ml)	5 °C	Yes	Sauvignon	No	Indust.	[4]
Torulaspora delbrueckii	After pressing	5 g/hl (10^6 cells/ml)	15°C	No	Aligoté	No	Indust.	[5]
Metschnikowia pulcherrima	Not precised	10 g/hl	15°C	Yes	Chardonnay	Yes	Lab. and pilot scale	[6]
Metschnikowia pulcherrima	Before skin maceration	20 g/hl	8°C	Yes	Sauvignon	No	Indust	[7]
Metschnikowia pulcherrima	half in the giraffe leading to the press and half in the tray under the press	5 g/hl	From 7 to 18°C	No	Chardonnay	Yes	Pilot scale	[8]
Metschnikowia pulcherrima	Harvest containers	10 g/hl	Non-specified	No	Pinot noir	No	Indust	[9]
Metschnikowia pulcherrima		20 g/hl	22°C	Yes	Muscat grape must + buffer	No	Lab. scale	[10]
Metschnikowia pulcherrima		10 g/l	10°C	Yes	Verdicchio	Yes	Indust	[11]

Table 1.
Synthesis of reported bioprotection experiments in white and rosé musts.

juice that the addition of *Metschnikowia pulcherrima* as a bioprotectant increases the total oxygen consumption capacity.

This problem also exists in rosé musts. In an experiment on rosé must, it was shown by carrying out colorimetric assays associated with chemical analyses of anthocyanins and phenolic compounds that the use of bioprotection alone did not protect the wine from oxidation [9]. An addition of oenological tannins on musts stabilized the color of bioprotected rosé wine in a similar way that SO_2 addition did. Quebracho tannins appeared more efficient than gall nut tannins.

Based on these different experiments, some discrepancies in the results regarding must oxidation could be observed. These differences could be linked to differences in the experimental procedure (temperature of must settling, bioprotectant dose, protection of must from oxygen, etc.). Indeed, controlling these parameters is of utmost importance for successful bioprotection.

Comparing six different maceration routes before alcoholic fermentation (AF) of a bioprotected must, varying the duration and temperature parameters, Simonin et al. [8] have shown that a temperature value ≤12°C was the main factor independent of the duration which allowed good implantation of the bioprotectant. An increase of the maceration duration at 12°C led to browning of the must, without significant effect on the final color of the wine, which was felt as more "floral," with more length in the mouth [8].

Across all the trials, it seems that sulfiting remains the safest practice for significantly reducing the pre-fermentative microbial population. Bioprotection can contain the development of indigenous populations to a non-detrimental level, but on condition of guaranteeing colonization of the medium by the yeast selected for more than 90% of the total population [12]. Below this proportion, even if the selected yeast is mainly established, it does not prevent the development of an indigenous population at the same level as the sulfite-free modality. The use of non-*Saccharomyces* fermentative yeast or *Saccharomyces cerevisiae* is not recommended for white and Rosé vinification. The risk of triggering alcoholic fermentation during settling is too high. It is therefore advisable to use non-fermentative non-*Saccharomyces* yeasts such as *Metschnikowia pulcherrima* or *Metschnikowia fructicola* [12].

All of the tests published and summarized in **Table 1** have made it possible to provide recommendations to winemakers wishing to use bioprotection as an alternative to sulfites and to show certain limits of bioprotection. However, many parameters need to be specified. Indeed, the impact of bioprotection on bacterial populations has been little studied, as has the sensory impact. Other species that could have interesting properties in bioprotection should also be tested. For example, *Pichia kluyveri* has been successfully tested in red must [13], what about white must?

The bioprotection strategy in white or rosé and red is quite different. Indeed, in white or rosé, bioprotection is used to limit the development of indigenous yeasts throughout the phase before settling. Since the settling is then carried out at low temperature, it helps to limit the development of native flora, including spoilage yeasts and bacteria. In red, the bioprotection intervenes early and as there is no settling, the bioprotectant remains in the tank throughout the fermentation and probably contributes more to the sensory profile of the wines than in white.

2.2 Red must

In red winemaking, cold pre-fermentative maceration is a widely used technique throughout the world. The choice of temperature and the dose of sulfite added strongly influence the microbial flora present. The temperature range generally used is 10-15°C. This preferential maceration is often used to improve the color and the fruity aromatic intensity. This increase in fruity character would be linked to the microbial flora present. However, this microbial flora consists of yeasts belonging to the species *Hanseniaspora uvarum*, which is known for its ability to produce ethyl acetate and acetic acid. Under these conditions, an alternative to sulfites to limit the development of spoilage yeasts and promote aromatic expression is the use of non-fermentative yeasts such as *Metschnikowia pulcherrima* [14].

The use of *Metschnikowia pulcherrima* as a bioprotectant agent at vatting on pinot noir at a dose of 5 g/hl (5.10^5 cell/ml) has been shown to limit the development of spoilage microbiota. A slight increase of *B. bruxellensis* species was observed in all the wineries and the evolution of acetic bacteria depended on the experimental site.

It is important to note that bioprotection is not always as effective as sulfites in limiting the development of indigenous yeasts and that this depends, among other things, on the initial population level. Trials conducted on Merlot in 2017 and 2018 showed that bioprotection did not limit the growth of *Hanseniaspora* species compared to sulfites [15]. On the other hand, these authors demonstrate that the use of bioprotective non-*Saccharomyces* yeast limited the abundance of filamentous fungi that are systematically associated with a decline in grape must quality.

2.2.1 Effect of bioprotection on phenolic compounds, volatile composition and sensoriality

In this study, it was hypothesized that the replacement of sulfites by a bioprotection strain affects proanthocyanidin and anthocyanin levels and leads to a higher degree of polymerization. But bioprotection had no influence on the phenolic compounds protecting musts and wine from oxidation. Wines produced from bioprotected or sulfited musts had different metabolic signatures, probably reflecting the production of specific metabolites by *M. pulcherrima* or the presence of chemical adducts due to sulfites. The metabolomics approach carried out by FT-ICR-MS analyses revealed statistical discriminations contrary to analyses of volatile compounds and conventional oenological analyses [16]. Bioprotection with *Metschnikowia pulcherrima* has no impact on the volatile compounds of red wines, although and specific sensory differences were perceived according to the winery. This was probably due to the fact that the low-temperature value of grape must during pre-fermentative maceration (12°C) limits the growth of the *Metschnikowia pulcherrima* yeast added. The bioprotectant population was 100-fold lower than the *Saccharomyces cerevisiae* population present in both modalities. The impact of the bioprotection strain was therefore limited on the production of volatile compounds during alcoholic fermentation and, therefore, on the sensory analyses [16].

In a popular science magazine, a synthesis of different experimental assays is presented. These include experiments with *Saccharomyces cerevisiae* and *Torulaspora delbrueckii*. Of the six trials conducted on Syrah, Cabernet-Sauvigon, Grenache and Mourvèdre, half showed that the indigenous flora was present in quantities comparable to those of the sulfited modality and in lesser quantities than the non-sulfited control [17]. The same authors reported trials at an industrial scale. The use of *Saccharomyces cerevisiae* in bioprotection at 30 g of yeast/100 kg of grapes did not reduce the indigenous yeast population compared with the control. However, the sensory profile is more qualitative [18].

It should be underlined, that while Simonin et al. [16] reported the absence of differences in aromatic profile between the bioprotection modality and the control, Windholtz et al. [19] observed that the use of non-*Saccharomyces* yeasts as a bioprotection has a significant impact on the aromatic profile of wines. In a bioprotection trial with a *Lachancea thermotolerans* and *Lactobacillus plantarum* pairing on Tempranillo grapes, it was shown that anthocyanin composition was barely altered by the use of bioprotectors. From a sensory point of view, the wines were better evaluated, with fruitier, stewed, peppery, lactic and fresher aromas, bitter and astringent than the control [20].

The experiments summarized in **Table 2** show contradictory results in terms of bioprotectant implantation, inhibition of indigenous flora development and sensory profile. However, recommendations can be made to encourage bioprotectant implantation and benefit from bioprotection from both a microbiological and sensory point of view. Bioprotectors should be added as early as possible, temperatures should be controlled (particularly during pre-fermentation), and the bioprotection yeast ratio should be as high as possible.

One study focused specifically on the sensory aspect of sulfite-free wines, including wines produced using bioprotection (**Table 2**) [21]. Sensory evaluation was carried out after 2 years of bottling. The sensory profiles of Merlot wines without sulfites, with and without bioprotection, were very similar to and significantly different from those of wines with sulfites. Four descriptors ("Fresh blackcurrant", "Cooked black cherries", "Mint" and "Coolness") were significantly more intense in wine without sulfites and/or wines with bioprotection treatment.

3. Wine bioprotection during aging

Wines can undergo microbial alterations during aging. The most studied alteration concerns the *Brettanomyces bruxellensis* yeast responsible for wine depreciation. To prevent the development of this spoilage yeast, we recommend sulfiting wines after malolactic fermentation (MLF). However, the resistance of these yeasts to sulfites is well-known [22]. It has been shown that the use of selected lactic acid bacteria to conduct malolactic fermentation could act as a bioprotector against *Brettanomyces* [23]. In the absence of stabilization after MLF, the lactic acid bacteria here play a protective role against *Brettanomyces* contamination. Sensitivity appears to be universal: when exposed to a given biomass of lactic acid bacteria, genetically distinct strains of *Brettanomyces,* which may exhibit distinct characteristics such as sulfite resistance, are inhibited in a comparable manner. One condition for this protection to be effective is a high level of *Oenococcus oeni*, which prevents contamination by *Brettanomyces*. Indeed, without this bioprotection, the yeast can develop within a few weeks and produce volatile phenols. This bioprotection effect is specific to lactic acid bacteria, at least to *O. oeni*. No protection is obtained by maintaining a living *Saccharomyces cerevisiae* flora after AF. Maintaining the bacterial flora after MLF is therefore a good way of combating *Brettanomyces*. This bioprotection phase is favored by good hygiene and fermentation control, which promote low initial volatile acidity [23].

4. How bioprotection works?

It remains difficult to answer this question today, as the fact that oenological conditions imply complex microbial ecosystems including yeasts, bacteria and mold.

Yeast species	Addition step	Dose	Temperature	Volatile composition effect/ sensorial effect	Grape variety	Scale	Reference
Metschnikowia pulcherrima	Vatting	5 g/hl	12°C	Yes/No	Pinot noir	Industrial	[16]
Torulaspora delbrueckii /Metschnikowia pulcherrima	On grapes at harvest and vatting	5 to 30 g/hl	10°C	No/Yes	Merlot	Industrial	[19]
Torulaspora delbrueckii/Metschnikowia pulcherrima	On grapes at harvest	5 g/hl	10°C	No/No	Merlot	Pilot	[15]
L. plantarum/L. thermotolerans	Non-specified	Non-specified	Non-specified	Yes/Yes	Tempranillo	Pilot	[20]
Torulaspora delbrueckii/Metschnikowia pulcherrima	On grapes at vating	5 g/hl	10°C	Yes/Yes	Merlot	Pilot	[21]

Table 2.
Synthesis of reported bioprotection experiments in red must.

Hence, many interactions could be set up within native flora and also with microorganisms inoculated during the winemaking process. The efficacy of bioprotection is based on the hypothesis that the added bioprotector induces a negative interaction with the indigenous microbiota, although no mechanism has been clearly identified yet in field trials. The expected effects are of two kinds: either the metabolism of the bioprotector significantly slows down the growth of indigenous microorganisms, allowing them to predominate during pre-fermentation phases, or its metabolism induces the death of these microorganisms. The implied mechanisms could be indirect such as nutrient competition or production of antimicrobial compounds or direct such as cell-cell contact [24, 25], but the main mechanisms put forward to date to explain bioprotective effects appear to be indirect interactions.

4.1 Oxygen competition

At the beginning of the fermentation process, the dissolved oxygen in must is about 8 mg/L (at 20°C) and some practices, as crushing, pressing or pumping can add significative quantities of oxygen into the grape must [26, 27]. Taking as the reference the alcoholic fermentation yeast *S. cerevisiae*, Visser et al. had demonstrated a higher consumption of oxygen by NS yeasts [28]. Among the species recommended for bioprotection, *M. pulcherrima* remains one that has the most important oxygen needs [29]. Recent work by Windholtz et al. determined the rates of dissolved oxygen consumption (DOC) on grape juice in 47 strains belonging to six different genera and species [30]. The highest OCR values were obtained for *M. pulcherrima* strains, compared to the values obtained for strains of *Saccharomyces* genus, confirming the data of Quirós et al. [29]. In contrast, *T. delbrueckii* and *L. thermotolerans* strains showed consumption classified as intermediate, with a high heterogeneity of the values obtained within both genera and species. These measurements, which were also carried out on eight strains of *Hanseniaspora uvarum*, the species most commonly found on grape must, showed interestingly oxygen consumption rates comparable to those obtained in *L. thermotolerans* or *T. delbrueckii*. Under oenological conditions, the difference in oxygen consumption and fermentative capacity can be partly explained by the Crabtree effect. In the presence of a high concentration of sugars (as found in must), the respiratory metabolism of Crabtree-positive yeasts is inhibited from redirecting the central carbonaceous metabolism to a fermentative metabolism [31, 32]. The impact of this catabolic repression depends on the yeast species, which will lead to more or less important oxygen consumption. *T. delbrueckii* as *L. thermotolerans* are Crabtree-positive yeasts [33], while *M. pulcherrima* and *H. uvarum* are mainly reported as Crabtree-negative yeasts, with high respiration capacities [34, 35]. These data suggest that there may be strong competition for dissolved oxygen during the grape must bioprotection, particularly through the application of *M. pulcherrima* strains as bioprotectors. However, the results obtained by Windholtz et al. [30] and earlier by Visser et al. [28], demonstrating that strains of *Hanseniaspora* strains (sp. *uvarum* and *valbyensis*) were able to grow in anaerobic conditions, suggest that competition for oxygen to explain the effectiveness of bioprotection appears to highly depend on the biodiversity of the *Hanseniaspora* strains present on the grape must.

4.2 Nitrogen compounds competition

It is well known that the quantity and quality of nitrogen resources are essential for yeast anabolism including the production of compounds that are inextricably linked to wine quality, such as higher alcohols and their esters. In the case of non-*Saccharomyces* yeasts, literature data remain still scarce, and few data are available not only on their minimum nitrogen requirements for growth but also on the nitrogen resources preferentially consumed. Preferential amino acid uptakes are found to be highly variable between yeast species. But, bibliographic data comparison appears complex due to the wide disparities of experimental conditions including nitrogen composition of media and growth parameters (temperature, oxygenation, etc.). The work of Kemsawasd et al. established the specific N-source influence on growth (in synthetic medium; 25°C) among five wine-related yeast species [36]. The N-sources that more particularly improved performance parameters were arginine, asparagine, glutamine and isoleucine for *T. delbrueckii*, with mainly same N-source influence profiles as *S. cerevisiae*. These N-sources were alanine and asparagine or serine, for *M. pulcherrima* and *L. thermotolerans*, respectively. *H. uvarum* was characterized by the most different N-sources influence pattern, with alanine as a booster for its growth.

In comparison with the previous data and focusing on *M. pulcherrima*, the study of Gobert et al. realized that grape juice showed preferential assimilation of isoleucine, leucine, lysine, methionine, glutamine, and cysteine at 28°C and alanine, cysteine, glutamine, histidine, lysin and tryptophan at 20°C [37]. The preferential assimilation of lysine has been confirmed on a wider panel of *M. pulcherrima* strains, as the good assimilation of branched-chain amino acids, including leucine and isoleucine [38–40], hardly suggesting that this assimilation pattern is characteristic of this species.

Concerning *H. uvarum*, Roca-Mesa reported that leucine and isoleucine were among the highest nitrogen sources consumed in synthetic must, with an early depletion of those amino acids (less than 48 h at 22°C) [41].

The highlighting of common N-sources consumed preferentially by both *Hanseniaspora* strains and bioprotectors could result from competition phenomena to the predominance of the bioprotective strain over the indigenous microbiota of a grape must. Indeed, a comparison of assimilation kinetics on preferential amino acids would help to confirm this hypothesis. Much broader studies conducted specifically within the framework of possible interactions between genera and species generated specifically by the practice of bioprotection (taking into account nitrogen composition of the grape must, the initial indigenous population level with the biodiversity of the genera and species, the temperature applied and the duration of pre-fermentative stage) will implement the data needed to refine the knowledge of the mechanics behind these nutrients competitions. In addition, the regulatory pathways implied in nitrogen consumption are still unknown among NS yeasts, unlike *S. cerevisiae* where, depending on the concentration of assimilable nitrogen in the medium, two main systems are involved: Nitrogen Catabolic System (NCR) and regulation involving a specific sensor of the plasma membrane [42].

4.3 Antimicrobial compounds production

4.3.1 Killer toxins

The production of these antimicrobial proteins was described in *T. delbruecki* by Villalba et al. [43]. The toxin Tdkt, presenting a molecular mass of around 30 kDa, exhibits a broad spectrum against wine spoilage yeasts, more particularly on yeasts belonging to *Brettanomyces* and *Hanseniaspora* genera [44]. Tdkt possesses glucanase and chitinase activities, and its putative mode of action includes binding to β 1-6 glucan and chitin with sensitive cells, with a potential degradation of these polysaccharides, cell wall disruption and finally, cell death initially by necrosis, then by apoptosis [43]. Most recently, another smaller killer toxin (TK) with a molecular mass of 15 kDa was purified in *T. delbrueckii* [45].

Among the yeasts used for bioprotection, *T. delbrueckii* is not the only one with the potential to produce killer toxins. The work of Büyüksırıt-Bedir & Kuleaşan [46] have demonstrated the production of a killer toxin by one strain of *M. pulcherrima* strain isolated from grapevine. This toxin is a peptide with a molecular weight of 10.3 kDa. The amino-sequence analysis showed that a part of the sequence was similar to the KHR killer toxin of *S. cerevisiae* previously characterized by Goto et al. [47]. Antimicrobial peptides production (molecular mass around 10 kDa) has also been demonstrated in *M. pulcherrima* by the work of Hicks et al. with strains showing killer phenotypes against pathogenic bacteria [48].

If, as in the case of the Tdkt toxin where the mode of action involves specific enzymatic activities targeting the degradation of parietal compounds of undesirable yeasts [43], it cannot be ruled out that enzymatic activities highly expressed in *M. pulcherrima* may also be involved. In fact, this species is characterized by a production of a very diverse range of enzymes, more particularly lipase, chitinase and β-1,3-glucanase [49–54], which may also be involved in killer phenotypes.

The known modes of action of killer toxins produced by yeast are summarized in **Figure 1**.

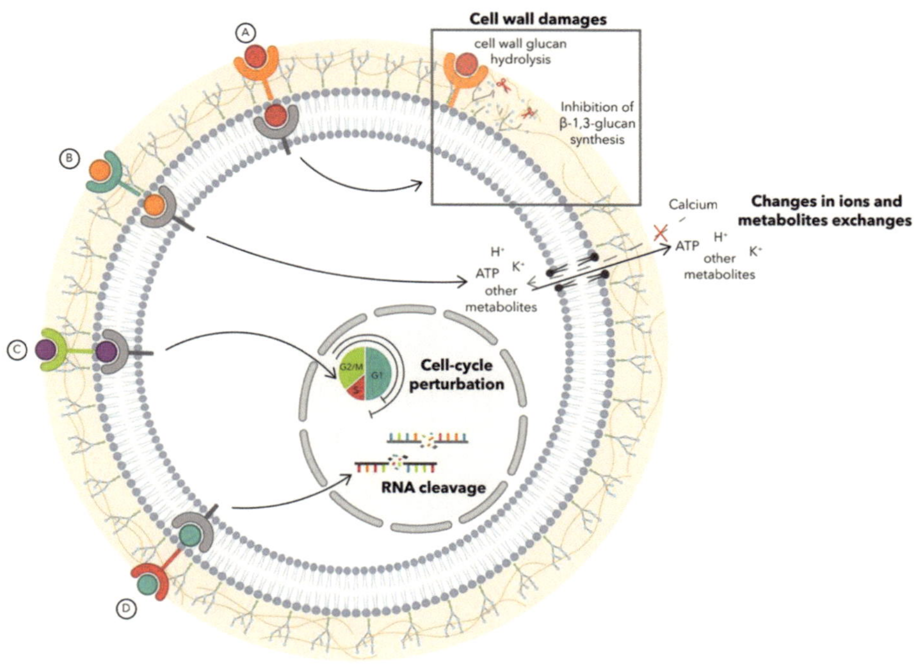

Figure 1.
Known toxin-killer mechanisms in yeasts. A: Cell wall damage by glucan hydrolysis through enzymatic activity of killer toxin or by synthesis inhibition of cell wall component. B: Plasmic membrane perturbations leading to iron and other metabolites leakage or inhibition of calcium uptakes. C: Cell-cycle perturbation by locking up cell between G1/S phase or inhibiting the completion of G1 phase. D: RNA cleavage especially for 18S and 25S RNA and tRNA (adapted from Mannazzu et al. [44]).

4.3.2 Production of pulcherriminic acid and pulcherrimin

Firstly characterized in the bacteria *Bacillus* [55–57], the pulcherriminic acid biosynthesis pathway implies two leucyl-tRNA that are cyclized in cyclo-(L-Leu-L-Leu). This intermediate compound is oxidized in pulcherriminic acid inside the cell and then excreted to the extracellular medium. Pulcherriminic acid chelates iron ion (Fe^{3+}) by a non-enzymatic reaction to form the red pigment named pulcherrimin [58, 59] (**Figure 2**). In *Metschnikowia pulcherrima*, as well as in other pulcherrimin-producers *Metschnikowia* species [60], pulcherriminic acid appears to be the main antimicrobial compound. Its action is not direct as it is due to the chelation of iron (Fe^{3+}), which becomes unavailable for yeasts whose growth requires this ionic form of iron.

This mechanism could explain the inhibitory effect of *M. pulcherrima* observed on agar plate and in grape juice medium on spoilage yeasts such as apiculate yeast and *Brettanomyces bruxellensis* [61]. However, pulcherriminic acid production is dependent of the initial iron concentration in the medium [58]. Studies conducted *in vitro* on agar plates showed that the intensity of the pigmented halo around *Metschnikowia*, due to pulcherrimin production, increased with the iron concentration in the medium in contrast to the zone of inhibition [50, 62–64]. These results suggest that the chelation of iron by pulcherriminic acid, as an antagonistic mechanism toward target yeasts such as *B. bruxellensis*, can only be effective in grape musts containing a very limited concentration of ferric iron.

4.3.3 Antimicrobial effect of L-lactic acid?

This question may be asked with regard to the negative effects on *B. bruxellensis* observed during malolactic fermentation by Gerbaux et al. [23]. The study of

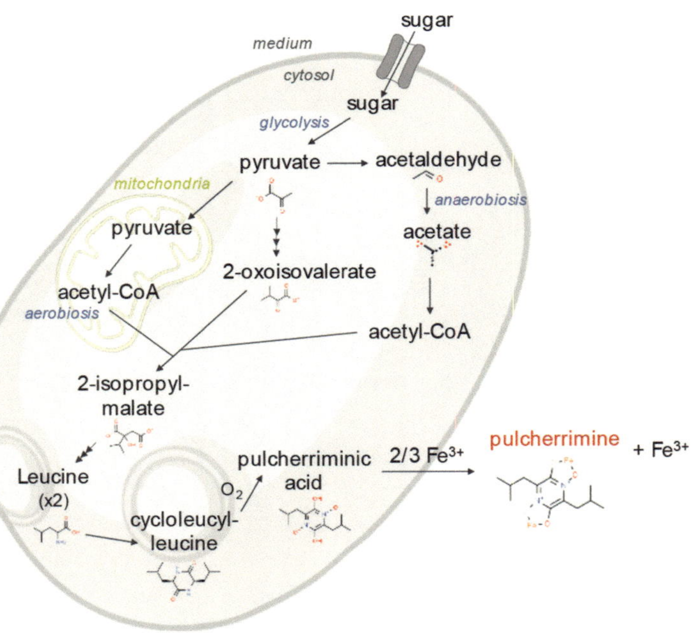

Figure 2.
Metabolic pathway of pulcherrimin production in M. pulcherrima (from [56, 57]; metabolic pathways, sigma-Aldrich).

intraspecific diversity among the species *B. bruxellensis* [64, 65] has enabled to clearly identify different genetic groups, which may explain the difference in sensitivity to sulfites observed within this species. The absence of L-lactate dehydrogenase in some strains of *B. bruxellensis* (genetic group 1 – data not published) could explain the inhibitory effect of L-lactate production during MLF on their growth. In the same vein, when applying *L. thermotolerans* for grape must acidification thanks to the production of L-lactate from sugars [66, 67], it cannot be ruled out that this yeast plays simultaneously a bioprotective role against *B. bruxellensis*.

5. Future perspectives

Numerous field trials have already demonstrated the effectiveness of bioprotection in oenological conditions [15, 19, 68]. However, there is still little information on mechanisms that bioprotection uses to limit the development of indigenous flora. In literature, numerous interactions have been identified in the oenological context [24, 25], but none have been studied in the bioprotection context, leaving many possibilities unexplored.

The main hypothesis on the bioprotection implantation and efficiency are based on nutrient competition between indigenous microorganisms and bioprotection, especially nitrogen resources. It has been shown that nitrogen is an essential resource for yeast's growth. In the winemaking context, nitrogen deficit could lead to stuck or sluggish fermentation [69]. Nitrogen needs and mechanisms regulating nitrogen consumption are well-known for *Saccharomyces cerevisiae* [42], but still poorly understood for non-*Saccharomyces* (NS) yeasts.

In the last decades, more studies have started to investigate NS nitrogen requirement [36–39, 41], but those studies investigate mainly yeast genera such as *Lachancea*, *Torulaspora,* or *Metschnikowia* corresponding mainly to NS genera used in winemaking including in bioprotection strategy. Those results give important information about the nitrogen requirement of important yeasts used in winemaking, but the temperature and medium used in those studies were not always close to pre-fermentative conditions. Nitrogen consumption of yeast is highly impacted by the initial nitrogen concentration and the temperature, in order to obtain a better characterization of the nitrogen needs of bioprotective yeasts, it is important to study their consumption on conditions closer to those of winemaking steps where bioprotection is used. However, the nitrogen requirements on NS yeasts associated with main microbial alteration (such as *Hanseniaspora* yeasts genera) are very limited [36, 41], and to determine if a possible competition could be involved in inhibitory mechanisms, it is also important to characterize the nitrogen needs of spoilage flora. It would also be important to monitor nitrogen consumption during the growth of bioprotective yeasts in coculture with target native flora. If some nitrogen resources appear to be preferentially consumed by bioprotective yeasts and spoilage microorganism, a supplementation in these resources could help to determine the involvement of this competition in the inhibitory effect.

The same reasoning could be applied to other nutrient found in must, such as vitamins and lipids. These latter are not produced by yeasts under anaerobic conditions, so yeasts are dependent on lipids naturally present on grapes and musts for their metabolism and membrane integrity. Some vitamin deficiencies have also been found to be highly deleterious to yeast growth [70].

At the beginning of the winemaking process, oxygen was found in the mud in significative concentration. Information about the respiro-fermentative mechanisms of

NS yeast in enology is still scarce. Studying these mechanisms, as well as the oxygen requirement of *M. pulcherrima* under oenological conditions, could allow a better understanding of the conditions supporting its establishment, as well as the interactions between *M. pulcherrima* and the other yeasts. Indeed, it was previously shown that the presence of *M. pulcherrima* has a strong influence on the respiro-fermentative metabolism of *S. cerevisiae*. After 3 h of coculture, the presence of *M. pulcherrima* induced a strong repression of genes involved in the respiratory metabolism of *S. cerevisiae*, as well as an over expression of genes involved in the fermentative pathway [35]. Sadoudi et al. have also shown that the presence of *M. pulcherrima* in coculture with *S. cerevisiae* impacts the metabolism and expression level of genes implied in PDH-pathway in *S. cerevisiae* [71]. In order to better understand interactions set up by this species during its use in bioprotection, it could be interesting to apply this strategy for a fine analyze of the effects of *M. pulcherrima* on the metabolism of other NS yeasts, more particularly *H. uvarum* or *B. bruxellensis*.

Strong competition for some resources essential to the growth of indigenous flora could explain the observed bioprotective effect, but the low temperatures used in winemaking condition induce a limited nutrient competition. The implication of nutrient competition in bioprotective efficiency remains to be proven.

Other, less passive, interactions have also been reported in the literature, such as the production of antimicrobial compounds. Among them, we can found the pulcherriminic acid production by yeasts of the *Metschnikowia* genera, and in particular by *M. pulcherrima* species. Many experiments have already shown the antimicrobial efficiency of these compounds by its capability to chelate iron in the medium [48, 61–63, 72, 73]. However, these experiments were generally carried out on petri dishes or *in vivo* on fruit (apples, grapes and oranges). To date, there are only few data available on the effect of pulcherriminic acid in oenological conditions [61]. Teams have set up protocols to quantify pulcherriminic acid [58], making it possible to follow pulcherriminic acid production during *M. pulcherrima* growth, and possibly to see whether the pulcherriminic acid concentration of different *M. pulcherrima* strains correlates with cell death of sensitive flora in liquid medium. In addition, other teams have created mutant strains which exhibited a lower or higher pulcherriminic acid production, but their antimicrobial efficiency was investigated in fruits or on agar plate [62, 63]. Studying the antimicrobial efficiency of these mutants under oenological conditions (synthetic must or real grape must) would also be a promising way to determine the impact of these compounds on the bioprotective effect of this yeast.

Pulcherriminic acid, produced by the *Metschnikowia* genera, is not the only toxic compound produced by NS species used in bioprotection. Numerous NS yeasts have been reported in the literature to produce killer toxins, including *Torulaspora delbrueckii* and *Metschnikowia pulcherrima*, which are the main yeasts genera commercialized. Most of the studies were carried out on synthetic liquid medium or in agar plate, and few of them were conducted in must/grape juice or grapes [74–77]. Furthermore, a lot of studies investigate the "killer phenotype" on plate but did not correlate the inhibition area to the secretion of peptidic-type compounds. It would be interesting to investigate the killer toxin production in marketed bioprotectors and to characterize those toxins as well as to obtain their sequences in order to improve protein databases.

In addition, Kemsawasd et al. [78] have shown that the inhibition found between *L. thermotolerans* and *S. cerevisiae* involves the production of a killer toxin, but that this efficiency is also due to direct contact between cells. Indeed, when growing in a membrane-separated fermenter, *L. thermotolerans* mortality is lower than when growing in coculture. Cell-cell interactions are still poorly understood in NS yeasts and need more investigation. A study has already referred to *M. pulcherrima* ability to form biofilms [54], thus demonstrating the potential for cells to aggregate with others. The ability to form cell adhesion would not necessarily mean that yeasts will be capable to form cell-cell adhesion with other yeast species. In order to study the direct interaction between species, it could be possible to create genetically modified fluorescent mutants of commercialized bioprotection strains in order to follow by fluorescent microscopy their possible aggregation with spoilage microorganisms in mixed culture. Impact of cell-cell contact interaction could also be studied by liquid culture in double-compartment fermenters with and without spatial separation [78]. Furthermore, some genes responsible for cell adhesion are already known in *S. cerevisiae*, it could be interesting to sequence more bioprotective strains and search for homologous genes to those involved in attachment in *S. cerevisiae*. Those genes could thereafter be studied by mutagenesis (targeted or un-targeted) in order to confirm their implication in cell adhesion.

Indirect interactions also include quorum sensing (QS), which is a way of communication mediated by the production of signal molecules that lead to phenotypic change at the population, not cell, level. This mode of communication has been extensively studied in bacteria, and signal molecules and phenotype changes are well known. However, the yeast *Candida albicans* has been found in the literature to have a quorum-sensing system capable of mediating certain reactions such as hyphae morphology by higher alcohol production [79, 80]. Quorum sensing molecules involved in QS of *C. albicans* have also been found produced by *S. cerevisiae* [81–83], but their role in a QS mechanisms in *S. cerevisiae* remain controversial [84, 85]. NS yeasts were also reported in literature as able to produce some of the higher alcohol involved in QS in *C. albicans* [38, 86]. However, the involvement of these compounds in QS still needs to be demonstrated. Identification and quantification of *ARO* genes, which are involved in the production of higher alcohols, have to be investigated in bioprotective yeasts by molecular biology. Furthermore, it is still important to correlate their production to possible inhibition against spoilage flora in must. In a larger scale, metabolomic analyses are increasingly used in order to explore yeasts interaction in must and wine [87–91]. Implementation and improvement of databases could be helpful in characterizing compounds involved in interactions between bioprotection and spoilage microorganisms.

Bioprotection was firstly investigated in order to reduce the SO_2 addition during pre-fermentative steps of winemaking by using NS yeasts to protect most against microbial spoilage. But another positive effect of the use of some bioprotective yeasts could also be pointed. For example, the use of *L. thermotolerans* leads to the acidification of the medium by sugar transformation into lactic acid. On preliminary results, this lactic acid production was reported in a French review for wine industry professionals as responsible for the potential inhibition of *B. bruxellensis* growth and so limits its possible production of off-flavors, but more investigations are needed [92]. Furthermore, only a fraction of the total SO_2 is active against microbial spoilage, and the proportion of each SO_2 form is influenced by must and wine pH [93]. The acidification of the medium led to a higher proportion of active SO_2 [93], and so to a lower use of SO_2 to obtain the necessary part of active SO_2. The use of *L. thermotolerans* could help to protect most against *B. bruxellensis* by lactic acid production as well as enhancing the aromatic profile of wines and reducing SO_2 addition. Yeasts could also influence wine color by metabolite production such as pyruvic acid or acetaldehyde which are able to interact with anthocyanins to produce more stable pigments

[94–98]. More investigations are needed in order to determine the capability of marketed bioprotective strains to modify and stabilize wine color. Selection of yeast strains that are able to limit indigenous yeast proliferation as well as protect wine color, especially for sensitive matrix such as rosé wine, could help to replace SO_2 and limit its use during winemaking.

Bioprotection was used at the early steps of winemaking (harvest or after pressing), but microbial alteration could also occur after alcoholic fermentation, especially by yeasts *B. bruxellensis*, which could alter wine organoleptic properties after malolactic fermentation and wine aging. In order to avoid this problem, SO_2 was traditionally added, but another strategy could be investigated. Some lactic acid bacteria have been reported in literature as bacteriocins producers able to inhibit fungi proliferation [2]. Some lactic acid bacteria isolated from wines have also proven their ability to produce toxic compounds *in vitro*, but their efficiency have not yet been proven in oenological context [99]. The use of lactic acids bacteria strain producing bacteriocin could be helpful to achieve malolactic fermentation and protect wine against potential microbial spoilage. Furthermore, the lactic acid bacterium *O. oeni*, have demonstrated some capabilities to inhibit *B. bruxellensis* growth after malolactic fermentation as bioprotector. It is a newly developed concept in enology, and more investigations are needed in order to understand and develop the use of *O. oeni* in the bioprotection context.

Even if we already have some evidence and clues of the ability of microorganisms to protect must against spoilage microorganisms and to influence wine color at the end of the fermentations, there are still a lack of knowledge about the medium- and long-term effect of the use of bioprotection to reduce SO_2 during wine aging, as well as diverse interactions and mechanisms set up by bioprotection.

Moreover, the cost of bioprotection is still high today, compared with the cost of sulfiting. In fact, bioprotection yeast preparations cost more than active dry yeasts (ADYs), starters of alcoholic fermentation. The development of production and stabilization strategies better adapted for these NS yeasts should make it possible to reduce the cost of the bioprotection strategy in the future.

Author details

Hervé Alexandre*, Maëlys Puyo and Raphaëlle Tourdot-Maréchal
Laboratoire Valmis, UMR PAM, IUVV, Université de Bourgogne, Dijon, France

*Address all correspondence to: rvalex@u-bourgogne.fr

References

[1] Gaggia F, Di Gioia D, Baffoni L, Biavati B. The role of protective and probiotic cultures in food and feed and their impact in food safety. Trends in Food Science & Technology. 2011;**22**:S58-S66

[2] Siedler S, Balti R, Neves AR. Bioprotective mechanisms of lactic acid bacteria against fungal spoilage of food. Current Opinion in Biotechnology. 2019;**56**:138-146

[3] Ulpathakumbura CP, Ranadheera CS, Senavirathne ND, Jayawardene LPINP, Prasanna PHP, Vidanarachchi JK. Effect of biopreservatives on microbial, physico-chemical and sensory properties of Cheddar cheese. Food. Bioscience. 2016;**13**:21-25

[4] Pillet O, Davaux F, Gabillot P, Peyrot S, Sylvano A, Robillard B. Stratégies de limitations des sulfites dans les vins: Quelles alternatives? Partie1/3: l'axe microbiologique, bioprotection et étapes préfermentaires. Revue des Œnologues. 2016;**160**:21-24

[5] Simonin S, Alexandre H, Nikolantonaki M, Coelho C, Tourdot-Maréchal R. Inoculation of *Torulaspora delbrueckii* as a bio-protection agent in winemaking. Food Research International. 2018;**107**:451-461

[6] Gerbaux V, Daventure I, Julien-Ortiz A, Bastien M, Silvano A. Développement d'une levure de bioprotection des moûts blanc en phase préfermentaire. Partie 1/2: intérêt du concept. Revue des Œnologues. 2021;**48**(179):45-47

[7] Bastien M, Silvano A, Ortiz-Julien A, Gerbaud V, Daventure I. Développement d'une levure de bioprotection des moûts blancs, en phase préfermentaire. Partie 2/2: mécanismes d'action et validation en cave. Revue des Œnologues. 2021;**180**:39-41

[8] Simonin S, Honoré-Chedozeau C, Monnin L, David-Vaizant V, Bach B, Alexandre H, et al. Bioprotection on chardonnay grape: Limits and impacts of settling parameters. Australian Journal of Grape and Wine Research. 2022;**2022**:e1489094

[9] Puyo M, Simonin S, Klein G, David-Vaizant V, Quijada-Morín N, Alexandre H, et al. Use of oenological tannins to protect the colour of Rosé wine in a bioprotection strategy with *Metschnikowia pulcherrima*. Food. 2023;**12**(4):735

[10] Giménez P, Just-Borras A, Pons P, Gombau J, Heras JM, Sieczkowski N, et al. Biotechnological tools for reducing the use of sulfur dioxide in white grape must and preventing enzymatic browning: Glutathione; inactivated dry yeasts rich in glutathione; and bioprotection with *Metschnikowia pulcherrima*. European Food Research and Technology. 2023;**249**(6):1491-1501

[11] Agarbati A, Canonico L, Ciani M, Comitini F. Metschnikowia pulcherrima in cold clarification: Biocontrol activity and aroma enhancement in Verdicchio wine. Fermentation. 2023;**9**(3):302

[12] Pladeau V, Philippe C, Pic L, Richard N. Comment s'affranchir du SO_2 en vinification en bio. Partie 1/2: Gestion des premières étapes de vinification. Revue des Œnologues. 2020;**47**(176):56-59

[13] Englezos V, Gianvito PD, Peyer L, Giacosa S, Segade SR, Edwards N, et al.

Bioprotective effect of *Pichia kluyveri* and *Lactiplantibacillus plantarum* in Winemaking Conditions. American Journal of Enology and Viticulture. Oct 2022;**73**:(4):294-307. DOI: 10.5344/ajev.2022.22008

[14] Gerbaux V, Davanure I, Guilloteau A, Ortiz-Julien A, Raginel F, Silvano A. Macération préfermentaire à froid des vins rouges. Revue des Œnologues. 2015;**42**(155):29-33

[15] Windholtz S, Vinsonneau E, Farris L, Thibon C, Masneuf-Pomarède I. Yeast and filamentous fungi microbial communities in organic red grape juice: Effect of vintage, maturity stage, SO_2, and bioprotection. Frontiers in Microbiology. 2021;**12**:1-13. [Internet]. Disponible sur: https://www.frontiersin.org/articles/10.3389/fmicb.2021.748416

[16] Simonin S, Roullier-Gall C, Ballester J, Schmitt-Kopplin P, Quintanilla-Casas B, Vichi S, et al. Bio-protection as an alternative to sulphites: Impact on chemical and microbial characteristics of red wines. Frontiers in Microbiology. 2020;**11**:1-14. [Internet]. Disponible sur: https://www.frontiersin.org/articles/10.3389/fmicb.2020.01308

[17] Pic L, Philippe C, Granès P, Laurens N, Pladeau V, Richard N. Bioprotection des vins biologiques. Partie 1/2: Résultats des essais en conditions expérimentales. Revue des Œnologues. 2019;**46**(171):46-49

[18] Pic L, Philippe C, Granès P, Laurens N, Pladeau V, Richard N. Bioprotection des Vins Biologique. Quel impact en vinification en rouge traditionnel? Partie 2/2: Résultats des essais à l'échelle de la production. Revue des Œnologues. 2019;**46**(172):31-33

[19] Windholtz S, Redon P, Lacampagne S, Farris L, Lytra G, Cameleyre M, et al. Non-*saccharomyces* yeasts as bioprotection in the composition of red wine and in the reduction of sulfur dioxide. LWT. 2021;**149**:111781

[20] Rubio-Bretón P, Gonzalo-Diago A, Iribarren M, Garde-Cerdán T, Pérez-Álvarez EP. Bioprotection as a tool to free additives winemaking: Effect on sensorial, anthocyanic and aromatic profile of young red wines. LWT. 2018;**98**:458-464

[21] Pelonnier-Magimel E, Mangiorou P, Philippe D, de Revel G, Jourdes M, Marchal A, et al. Sensory characterisation of Bordeaux red wines produced without added sulfites. OENO one. 2020;**54**(4):733-743

[22] Avramova M, Vallet-Courbin A, Maupeu J, Masneuf-Pomarède I, Albertin W. Molecular diagnosis of *Brettanomyces bruxellensis'* Sulfur dioxide sensitivity through genotype specific method. Frontiers in Microbiology. 2018;**9**:1260

[23] Gerbaux V, Thomas J, Briffox C, Matéo A. The advantage of using lactic acid bacteria for the biopreservation of wines against *Brettanomyces*. Revue Française d'Œnologie. 2020;**301**:28-35

[24] Bordet F, Joran A, Klein G, Roullier-Gall C, Alexandre H. Yeast-yeast interactions: Mechanisms, methodologies and impact on composition. Microorganisms. 2020;**8**(4):600

[25] Zilelidou EA, Nisiotou A. Understanding Wine through Yeast Interactions. Microorganisms. 2021;**9**(8):1620

[26] Du Toit WJ, Marais J, Pretorius IS, du Toit M. Oxygen in must and wine: A review. South African Journal of Enology and Viticulture. 2006;**27**(1): 57-67. [Internet]. Disponible sur:

http://www.journals.ac.za/index.php/sajev/article/view/1610

[27] Moenne MI, Saa P, Laurie VF, Pérez-Correa JR, Agosin E. Oxygen incorporation and dissolution during industrial-scale red wine fermentations. Food and Bioprocess Technology. 2014;**7**(9):2627-2636

[28] Visser W, Scheffers WA, Batenburg-van der Vegte WH, van Dijken JP. Oxygen requirements of yeasts. Applied and Environmental Microbiology. 1990;**56**(12):3785-3792

[29] Quirós M, Rojas V, Gonzalez R, Morales P. Selection of non-*saccharomyces* yeast strains for reducing alcohol levels in wine by sugar respiration. International Journal of Food Microbiology. 2014;**181**:85-91

[30] Windholtz S, Nioi C, Coulon J, Masneuf-Pomarede I. Bioprotection by non-*saccharomyces* yeasts in oenology: Evaluation of O_2 consumption and impact on acetic acid bacteria. International Journal of Food Microbiology. 2023;**405**:110338

[31] Crabtree HG. The carbohydrate metabolism of certain pathological overgrowths. Biochemical Journal. 1928;**22**(5):1289-1298. DOI: 10.1042/bj0221289

[32] Dashko S, Zhou N, Compagno C, Piškur J. Why, when, and how did yeast evolve alcoholic fermentation? FEMS Yeast Research. 2014;**14**(6):826-832

[33] Gonzalez R, Quirós M, Morales P. Yeast respiration of sugars by non-*saccharomyces* yeast species: A promising and barely explored approach to lowering alcohol content of wines. Trends in Food Science & Technology. 2013;**29**(1):55-61

[34] Venturin C, Boze H, Fahrasmane L, Moulin G, Galzy P. Influence de la concentration en glucose et en oxygène sur la capacité fermentaire de la souche *Hanseniaspora uvarum* K5 (Niehaus). Sciences des aliments. 1994;**14**(3):321-333

[35] Mencher A, Morales P, Curiel JA, Gonzalez R, Tronchoni J. *Metschnikowia pulcherrima* represses aerobic respiration in *Saccharomyces cerevisiae* suggesting a direct response to co-cultivation. Food Microbiology. 2021;**94**:103670

[36] Kemsawasd V, Viana T, Ardö Y, Arneborg N. Influence of nitrogen sources on growth and fermentation performance of different wine yeast species during alcoholic fermentation. Applied Microbiology and Biotechnology. 2015;**99**(23):10191-10207

[37] Gobert A, Tourdot-Maréchal R, Morge C, Sparrow C, Liu Y, Quintanilla-Casas B, et al. Non-*saccharomyces* yeasts nitrogen source preferences: Impact on sequential fermentation and wine volatile compounds profile. Frontiers in Microbiology. 2017;**8**:2175

[38] Prior KJ, Bauer FF, Divol B. The utilisation of nitrogenous compounds by commercial non-*saccharomyces* yeasts associated with wine. Food Microbiology. 2019;**79**:75-84

[39] Su Y, Seguinot P, Sanchez I, Ortiz-Julien A, Heras JM, Querol A, et al. Nitrogen sources preferences of non-*saccharomyces* yeasts to sustain growth and fermentation under winemaking conditions. Food Microbiology. 2020;**85**:103287

[40] Seguinot P, Ortiz-Julien A, Camarasa C. Impact of nutrient availability on the fermentation and

production of aroma compounds under sequential inoculation with *M. Pulcherrima* and *S. Cerevisiae*. Frontiers in Microbiology. 2020;**11**:305

[41] Roca-Mesa H, Sendra S, Mas A, Beltran G, Torija MJ. Nitrogen preferences during alcoholic fermentation of different non-*saccharomyces* yeasts of oenological interest. Microorganisms. 2020;**8**(2):157

[42] Crépin L, Nidelet T, Sanchez I, Dequin S, Camarasa C. Sequential use of nitrogen compounds by *Saccharomyces cerevisiae* during wine fermentation: A model based on kinetic and regulation characteristics of nitrogen Permeases. Applied and Environmental Microbiology. 2012;**78**(22):8102-8111

[43] Villalba ML, Susana Sáez J, del Monaco S, Lopes CA, Sangorrín MP. TdKT, a new killer toxin produced by *Torulaspora delbrueckii* effective against wine spoilage yeasts. International Journal of Food Microbiology. 2016;**217**:94-100

[44] Mannazzu I, Domizio P, Carboni G, Zara S, Zara G, Comitini F, et al. Yeast killer toxins: From ecological significance to application. Critical Reviews in Biotechnology. 2019;**39**(5):603-617

[45] Abu-Mejdad NMJA, Al-Badran AI, Al-Saadoon AH. Purification and characterization of two killer toxins originated from *Torulaspora delbrueckii* (Lindner) and *Wickerhamomyces anomalus* (E.C.Hansen) Kurtzman, Robnett, and Basehoar-powers. Bulletin of the National Research Centre. 2020;**44**(1):48

[46] Büyüksırıt-Bedir T, Kuleaşan H. Purification and characterization of a *Metschnikowia pulcherrima* killer toxin with antagonistic activity against pathogenic microorganisms. Archives of Microbiology. 2022;**204**(6):337

[47] Goto K, Iwatuki Y, Kitano K, Obata T, Hara S. Cloning and nucleotide sequence of the KHR killer gene of *Saccharomyces cerevisiae*. Agricultural and Biological Chemistry. 1990;**4**(54):979-984

[48] Hicks RH, Moreno-Beltrán M, Gore-Lloyd D, Chuck CJ, Henk DA. The oleaginous yeast *Metschnikowia pulcherrima* displays killer activity against avian-derived pathogenic bacteria. Biology. 2021;**10**(12):1227

[49] Saravanakumar D, Ciavorella A, Spadaro D, Garibaldi A, Gullino ML. *Metschnikowia pulcherrima* strain MACH1 outcompetes *Botrytis cinerea*, *Alternaria alternata* and *Penicillium expansum* in apples through iron depletion. Postharvest Biology and Technology. 2008;**49**(1):121-128

[50] Banani H, Spadaro D, Zhang D, Matic S, Garibaldi A, Gullino ML. Postharvest application of a novel chitinase cloned from *Metschnikowia fructicola* and overexpressed in *Pichia pastoris* to control brown rot of peaches. International Journal of Food Microbiology. 2015;**199**:54-61

[51] Pretscher J, Fischkal T, Branscheidt S, Jäger L, Kahl S, Schlander M, et al. Yeasts from different habitats and their potential as biocontrol agents. Fermentation. 2018;**4**(2):31

[52] Freimoser FM, Rueda-Mejia MP, Tilocca B, Migheli Q. Biocontrol yeasts: Mechanisms and applications. World Journal of Microbiology and Biotechnology. 2019;**35**(10):154

[53] Morata A, Loira I, Escott C, del Fresno JM, Bañuelos MA, Suárez-Lepe JA.

Applications of *Metschnikowia pulcherrima* in wine biotechnology. Fermentation. 2019;**5**(3):63

[54] Oztekin S, Karbancioglu-Guler F. Bioprospection of *Metschnikowia* sp. isolates as biocontrol agents against postharvest fungal decays on lemons with their potential modes of action. Postharvest Biology and Technology. 2021;**181**:111634

[55] Kluyver AJ, van der Walt JP, van Triet AJ. Pulcherrimin, the pigment of *Candida Pulcherrima*. Proceedings of the National Academy of Sciences of the United States of America. 1953;**39**(7):583-593

[56] MacDonald J. Biosynthesis of pulcherriminic acid. Biochemical Journal. 1965;**96**(2):533-538

[57] RobertL U, Canale-Parola E. Synthesis of Pulcherriminic acid by *bacillus subtilis*. Journal of Bacteriology. 1972;**111**(1):86-93

[58] Wang D, Zhan Y, Cai D, Li X, Wang Q, Chen S. Regulation of the synthesis and secretion of the iron Chelator Cyclodipeptide Pulcherriminic acid in *bacillus licheniformis*. Applied and Environmental Microbiology. 2018;**84**(13):e00262-18. DOI: 10.1128/AEM.00262-18

[59] Yuan S, Yong X, Zhao T, Li Y, Liu J. Research Progress of the biosynthesis of natural bio-antibacterial agent Pulcherriminic acid in *bacillus*. Molecules. 2020;**25**(23):5611

[60] Piombo E, Sela N, Wisniewski M, Hoffmann M, Gullino ML, Allard MW, et al. Genome sequence, assembly and characterization of two *Metschnikowia fructicola* strains used as biocontrol agents of postharvest diseases. Frontiers in Microbiology. 2018;**9**(593):1-17. [Internet]. Disponible sur: http://journal.frontiersin.org/article/10.3389/fmicb.2018.00593/full

[61] Oro L, Ciani M, Comitini F. Antimicrobial activity of *Metschnikowia pulcherrima* on wine yeasts. Journal of Applied Microbiology. 2014;**116**(5):1209-1217

[62] Wang S, Zhang H, Ruan C, Yi L, Deng L, Zeng K. *Metschnikowia citriensis* FL01 antagonize Geotrichum citri-aurantii in citrus fruit through key action of iron depletion. International Journal of Food Microbiology. 2021;**357**:109384

[63] Sipiczki M. *Metschnikowia* strains isolated from Botrytized grapes antagonize fungal and bacterial growth by iron depletion. Applied and Environmental Microbiology. 2006;**72**(10):6716-6724

[64] Lebleux M, Denimal E, De Oliveira D, Marin A, Desroche N, Alexandre H, et al. Prediction of genetic groups within *Brettanomyces bruxellensis* through cell morphology using a deep learning tool. Journal of Fungi. 2021;**7**(8):581

[65] Avramova M, Vallet-Courbin A, Maupeu J, Masneuf-Pomarède I, Albertin W. Molecular diagnosis of *Brettanomyces bruxellensis*' Sulfur dioxide sensitivity through genotype specific method. Frontiers in Microbiology. 2018;**9**:1260

[66] Morata A, Loira I, Tesfaye W, Bañuelos M, González C, Suárez LJ. *Lachancea thermotolerans* applications in wine technology. Fermentation. 2018;**4**(3):53

[67] Gobbi M, Comitini F, Domizio P, Romani C, Lencioni L, Mannazzu I, et al. *Lachancea thermotolerans* and *Saccharomyces cerevisiae* in simultaneous

and sequential co-fermentation: A strategy to enhance acidity and improve the overall quality of wine. Food Microbiology. 2013;**33**(2):271-281

[68] Windholtz S, Dutilh L, Lucas M, Maupeu J, Vallet-Courbin A, Farris L, et al. Population dynamics and yeast diversity in early winemaking stages without Sulfites revealed by three complementary approaches. Applied Sciences. 2021;**11**(6):2494

[69] Gobert A, Tourdot-Maréchal R, Sparrow C, Morge C, Alexandre H. Influence of nitrogen status in wine alcoholic fermentation. Food Microbiology. 2019;**83**:71-85

[70] Evers MS, Roullier-Gall C, Morge C, Sparrow C, Gobert A, Alexandre H. Vitamins in wine: Which, what for, and how much? Comprehensive Reviews in Food Science and Food Safety. 2021;**20**(3):2991-3035

[71] Sadoudi M, Rousseaux S, David V, Alexandre H, Tourdot-Maréchal R. *Metschnikowia pulcherrima* influences the expression of genes involved in PDH bypass and Glyceropyruvic fermentation in *Saccharomyces cerevisiae*. Frontiers in Microbiology. 2017;**8**:1137

[72] Türkel S, Ener B. Isolation and characterization of new *Metschnikowia pulcherrima* strains as producers of the antimicrobial pigment Pulcherrimin. Zeitschrift für Naturforschung C. 2009;**64**(5-6):405-410

[73] Kregiel D, Nowacka M, Rygala A, Vadkertiová R. Biological activity of Pulcherrimin from the *Meschnikowia pulcherrima* clade. Molecules. 2022;**27**(6):1855

[74] Mehlomakulu NN, Setati ME, Divol B. Characterization of novel killer toxins secreted by wine-related non-*saccharomyces* yeasts and their action on *Brettanomyces* spp. International Journal of Food Microbiology. 2014;**188**:83-91

[75] Mehlomakulu NN, Prior KJ, Setati ME, Divol B. *Candida pyralidae* killer toxin disrupts the cell wall of *Brettanomyces bruxellensis* in red grape juice. Journal of Applied Microbiology. 2017;**122**(3):747-758

[76] Velázquez R, Zamora E, Álvarez ML, Hernández LM, Ramírez M. Effects of new *Torulaspora delbrueckii* killer yeasts on the must fermentation kinetics and aroma compounds of white table wine. Frontiers in Microbiology. 2015;**6**:1222

[77] Liu Z, Du S, Ren Y, Liu Y. Biocontrol ability of killer yeasts (*Saccharomyces cerevisiae*) isolated from wine against *Colletotrichum gloeosporioides* on grape. Journal of Basic Microbiology. 2017;**58**(1):1-8

[78] Kemsawasd V, Branco P, Almeida MG, Caldeira J, Albergaria H, Arneborg N. Cell-to-cell contact and antimicrobial peptides play a combined role in the death of *Lachanchea thermotolerans* during mixed-culture alcoholic fermentation with *Saccharomyces cerevisiae*. FEMS Microbiology Letters. 2015;**362**(14):fnv103

[79] Hogan DA. Talking to themselves: Autoregulation and quorum sensing in fungi. Eukaryot Cell. 2006;**5**(4):613-619

[80] Jagtap SS, Bedekar AA, Rao CV. Quorum sensing in yeast. In: Dhiman SS, éditeur. ACS Symposium Series Washington, DC: American Chemical Society; 2020. pp. 235-250. [Internet]. Disponible sur: https://pubs.acs.org/doi/abs/10.1021/bk-2020-1374.ch013

[81] Sprague GF Jr, Winans SC. Eukaryotes learn how to count: Quorum sensing by yeast. Genes & Development. 2006;**20**:1045-1049

[82] Avbelj M, Zupan J, Kranjc L, Raspor P. Quorum-sensing kinetics in *Saccharomyces cerevisiae*: A symphony of *ARO* genes and aromatic alcohols. Journal of Agricultural and Food Chemistry. 2015;**63**(38):8544-8550

[83] Avbelj M, Zupan J, Raspor P. Quorum-sensing in yeast and its potential in wine making. Applied Microbiology and Biotechnology. 2016;**100**(18):7841-7852

[84] Winters M, Arneborg N, Appels R, Howell K. Can community-based signalling behaviour in *Saccharomyces cerevisiae* be called quorum sensing? A critical review of the literature. FEMS Yeast Research. 2019;**19**(5):foz046

[85] Winters M, Aru V, Howell K, Arneborg N. *Saccharomyces cerevisiae* does not undergo a quorum sensing-dependent switch of budding pattern. Scientific Reports. 2022;**12**(1):8738

[86] González B, Vázquez J, Morcillo-Parra MÁ, Mas A, Torija MJ, Beltran G. The production of aromatic alcohols in non-*saccharomyces* wine yeast is modulated by nutrient availability. Food Microbiology. 2018;**74**:64-74

[87] Petitgonnet C, Klein GL, Roullier-Gall C, Schmitt-Kopplin P, Quintanilla-Casas B, Vichi S, et al. Influence of cell-cell contact between *L. thermotolerans* and *S. Cerevisiae* on yeast interactions and the exo-metabolome. Food Microbiology. 2019;**83**:122-133

[88] Roullier-Gall C, David V, Hemmler D, Schmitt-Kopplin P, Alexandre H. Exploring yeast interactions through metabolic profiling. Scientific Reports. 2020;**10**(1):6073

[89] Roullier-Gall C, Bordet F, David V, Schmitt-Kopplin P, Alexandre H. Yeast interaction on chardonnay wine composition: Impact of strain and inoculation time. Food Chemistry. 2022;**374**:131732

[90] Bordet F, Romanet R, Bahut F, Ballester J, Eicher C, Peña C, et al. Expanding the diversity of chardonnay aroma through the metabolic interactions of *Saccharomyces cerevisiae* cocultures. Frontiers in Microbiology. 2023;**13**:1032842

[91] Lebleux M, Alexandre H, Romanet R, Ballester J, David-Vaizant V, Adrian M, et al. Must protection, sulfites versus bioprotection: A metabolomic study. Food Research International. 2023;**173**:113383

[92] Hranilovic A, Jiranek V, Coulon J, Masneuf-Pomarede I, Albertin W, Bely M. Bioacidification des vins par la levure *Lachancea thermotolerans*. Revue des Oenologues. 2022;**183**:39-42

[93] Boulton RB, Singleton VL, Bisson LF, Kunkee RE. The role of sulfur dioxide in wine. In: Principles and Practices of Winemaking. Boston, MA: Springer; 1999

[94] Medina K, Boido E, Dellacassa E, Carrau F. Yeast interactions with Anthocyanins during red wine fermentation. American Journal of Enology and Viticulture. 2005;**56**(2):104-109

[95] Martin V, Valera M, Medina K, Boido E, Carrau F. Oenological impact of the *Hanseniaspora/Kloeckera* yeast genus on wines—A review. Fermentation. 2018;**4**(3):76

[96] Echeverrigaray S, Menegotto M, Delamare APL. A simple and reliable

method for the quantitative evaluation of anthocyanin adsorption by wine yeasts. Journal of Microbiological Methods. 2019;**157**:88-92

[97] Echeverrigaray S, Scariot FJ, Menegotto M, Delamare APL. Anthocyanin adsorption by *Saccharomyces cerevisiae* during wine fermentation is associated to the loss of yeast cell wall/membrane integrity. International Journal of Food Microbiology. 2020;**314**:108383

[98] Morata A, Loira I, González C, Escott C. Non-*saccharomyces* as biotools to control the production of off-Flavors in wines. Molecules. 2021;**26**(15):4571

[99] Knoll C, Divol B, Dutoit M. Genetic screening of lactic acid bacteria of oenological origin for bacteriocin-encoding genes. Food Microbiology. 2008;**25**(8):983-991

Chapter 8

Characteristic Metabolites Drive the Self-Assembly of Microeukaryotic Communities during Spontaneous Fermentation of Icewine

Kai Chen

Abstract

Wine can be regarded as a nutritional source for the human diet. It contains many nutrients such as vitamins and minerals, organic acids, flavonoids, and terpenoids. The varietal aroma of wines originated from the symbionts of the grapes and epiphytic microbiota, which combinedly grew from the sexual reproduction of the plant through seeds or during clonal reproduction. Nowadays, more and more studies focus on the assembly process of epiphytic microbiota and design a synthetic microbial community based on regional characteristics to improve wine quality and biofunctions. This review synthesizes the current concepts on the construction of synthetic microbiota, analyzes the advantages and difficulties in designing a functional yeast community, and lists the practical tools for data processing and model construction. A well-designed yeast community will possess high robustness against environment interference, higher efficiency of fermentation, and higher yield of targeted bioproducts.

Keywords: non-*Saccharomyces*, top-down strategy, systematic biology, untargeted metabolomics, icewine, spontaneous fermentation

1. Introduction

The application of commercial yeast in an incompatible fermentation scenario potentially causes the risk of stuck and sluggish fermentation and product homogenization. It even has a detrimental effect on the composition of regional microbiota [1], which play important roles in characterizing aroma profile and wine quality [2]. Currently, more and more researchers and wineries have realized the limitations of commercial yeasts and are gradually turning their attention to the study of non-*Saccharomyces* strains and the interactions between local microbiota and environmental factors [3–5]. Spontaneous fermentation has returned to the fore as a natural and regionally characterized winemaking mode [6]. Due to microbial succession, spontaneous fermentation usually presents higher sensorial complexity [4].

Although spontaneous fermentation can largely strengthen the '*terroir*' characteristics of wines, the contamination by undesired microbes is the first issue and easily cause unpleasant flavors, such as a musty taste and mousiness [7, 8]. Besides, the high dimension of ecology of spontaneous fermentation results in many difficulties in control of fermentation and stable quality of wine production. With the development of high-throughput sequencing technology, researchers can conduct parallel analyses of the epiphytic microorganisms from vineyards and wines. It is effective to acquire an overview of microbial structure and position the pathogenic factors [2, 9, 10]. Therefore, the prerequisite for a well-controlled spontaneous fermentation is to identify the characteristic microbial resources in the production area, reduce the ecological dimension of the microeukaryotes community using self-assembled yeast communities, and establish a proper fermentation condition according to regional characteristics. In our previous study, two main environmental factors have been determined, namely, initial abiotic factors around vineyards and initial microbial community on the grape skin. Through the 'top-down' approach, we specially designed a self-assembled non-*Saccharomyces* yeasts group, which was composed of six selected yeast strains and highly expresses carbohydrate-active enzymes [2, 5]. The yeast group was first used to obtain a predictable and controllable fermentation mode during the spontaneous fermentation of icewine.

Icewine is a type of dessert wine produced from grape juice extracted from frozen grapes [11]. It is normally characterized by a fruity and flowery aroma and smooth taste. Icewine making is heavily dependent on the local environmental abiotic factors during late harvest. Fully ripe grapes hang on the vines for several months to suffer natural freeze-thaw cycles and be desiccated for the concentration of sugar (>35°Brix) [12]. This period has been proven to be effective for the accumulation of varietal aroma in grapes [2, 5]. For instance, high-odor-active compounds, such as terpenes and phenylalanine-derivates, can be largely developed during late harvest through a series of interactions between epiphytic microeukaryotes and grapes [2, 5]. Our previous studies have revealed the freeze-thaw cycles are the inevitable abiotic factors that cause cellular degradation and compartmentation of grape skin [2] and improve the aroma complexity of grapes during late harvest [5]. Therefore, selecting an icewine region is the essential prerequisite for producing high-quality icewine, which must present the regional characteristics [13]. Huanren area (Liaoning Province, Northeast China) is a representative icewine-producing area in China. It is characterized by a year-round cold climate, fewer problems of pests and disease [2]. Unique ecology shapes distinctive and stress-tolerant microeukaryotic communities, which remain active in pressed grape juice and become determinant to icewine fermentation [2, 13]. However, the mechanism and driving factors of the microeukaryotic assembly process are still unclear during icewine fermentation.

Metabolomics mainly studies the variety, quantity, and change rule of the metabolites with molecular weight less than 1500 Da caused by external stimulation, pathophysiological changes, and gene mutation. It is an extension of transcriptomics and proteomics, which accurately reflect the physiological state of organisms. Therefore, it effectively reveals the biological processes of biomarkers, the mechanism of the biological activities, and the regulatory pathways [14]. In terms of detection modes, metabolomics is mainly divided into untargeted analysis and targeted analysis. The untargeted metabolomics is usually based on a high-resolution mass spectrometer (triple TOF or QE) [15]. It can perform unbiased, large-scale, and systematic detection of various metabolites in experimental samples, providing an 'aerial photography' perspective to reflect the metabolic disturbance in the plants and microorganisms [16].

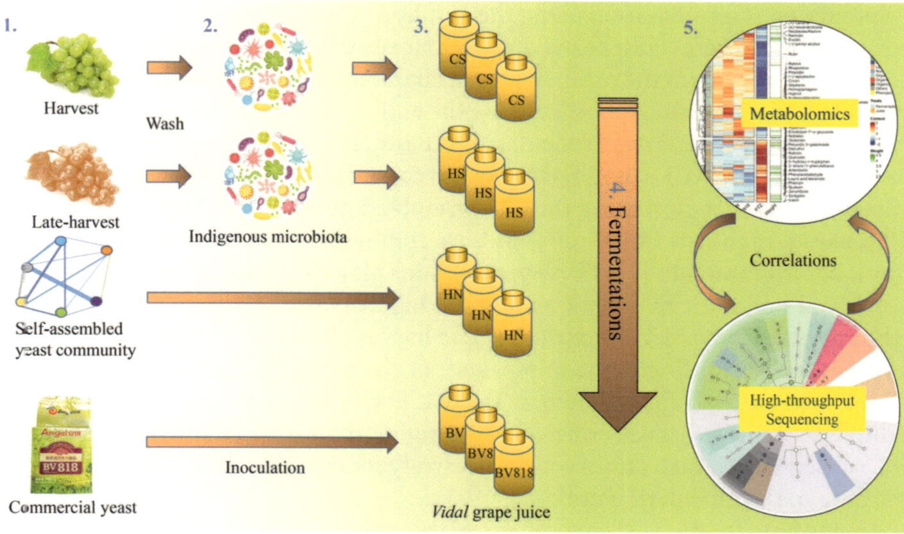

Figure 1.
Four inoculation methods were carried out for icewine fermentation. CS and HS were inoculated using the epiphytic microorganisms from the fully ripe grape berries and the grape berries in late harvest, respectively. HN was inoculated using the six core epiphytic yeast strains, which were important microeukaryotes in the grape skin. Control was conducted in icewine fermentation using a commercial yeast strain (BV818). Untargeted metabolomics and high-throughput sequencing technology were jointly used to analyze the self-assembly process of micro eukaryotic communities and the driving factors (important metabolites) during icewine fermentation.

This study jointly used untargeted metabolomics and high-throughput sequencing technology to verify the effects of the self-assembled non-*Saccharomyces* yeasts group on the development of metabolites during AF of icewine. Meanwhile, we also studied two types of spontaneous fermentation using the extracted microbiota from the ripe *Vidal blanc* grape and the frozen grape after late harvest, respectively (see **Figure 1**). Our work first elucidated the mechanism and driving factors of microeukaryotic assembly in the different fermentation scenarios. It provides a reference to artificially design functional microeukaryotic communities and to improve the stability of spontaneous fermentation.

2. Materials and methods

2.1 Vinification of *V. blanc* icewine

Spontaneous fermentation using the epiphytic microorganisms from the fully ripe grape berries (CS): The ripe *V. blanc* grape berries were manually collected in October 2021 and preserved at −20°C. To collect epidermal microorganisms, 200 g of frozen grape berries were soaked in 500 mL sterilized 10× PBS solution for 20 min and vortexed to elute epiphytic microorganisms [2]. The 50 mL suspension was centrifuged (10,000 rpm, 10 min) to remove the supernatant. Sediment was collected and first inoculated into 100 mL filtered grape juice for cultivation (25°C, 24 h). Subsequently, the cultivation medium was inoculated into 2 L filtered grape juice to carry out AF. The fermentation was supplied with 0.4 g/L nutrients (Nutrient-Vit, Lallemand, Montreal, PQ, Canada) and maintained the temperature at 18°C. At the beginning of AF, pumpover was carried out twice a day to facilitate yeast growth [4].

Spontaneous fermentation using the epiphytic microorganisms from the grape berries in late harvest (HS): based on previous research, *V. blanc* grape at the end of December can perform the optimal aroma characteristics [5]. Herein, the *V. blanc* grape berries were manually collected at the end of December 2021, and the epidermal microorganisms were inoculated into the filtrated grape juice. The specific procedures of inoculation and fermentation can refer to that of CS.

Icewine fermentation using the self-assembled non-*Saccharomyces* yeasts group (HN): based on the previous study, six core epiphytic yeast strains were separated from *V. blanc* grape skin [2]. These yeast strains play an important role in the self-assembly of epiphytic microeukaryotes and highly contribute to the development of the varietal aroma of *V. blanc* during late harvest. The percentage of each strain was calculated according to the importance of the random forest. The inoculation radio of each yeast strain is shown in **Table 1**. Each strain was cultured using a YPD medium to ensure yeast cell concentration higher than 10^8 CFU/mL before inoculation. Subsequently, each strain was respectively inoculated into 2 L filtered grape juice according to the inoculation radio.

Icewine fermentation using commercial yeast as control (BV818): a commercial yeast BV818 (AngelYeast Co., Ltd., Yichang) was inoculated into the filtrated grape juice for starting AF as control. This yeast belongs to *Saccharomyces bayanus*, which is stress-tolerant and highly active in a hyperosmotic juice. The fermentation was supplied with 0.4 g/L nutrients (Nutrient-Vit, Lallemand, Montreal, PQ, Canada) and maintained the temperature at 18°C. No sulfur dioxide or other antimicrobial agent was used in these four fermentation scenarios. All fermentations were conducted in triplicate.

2.2 Viticulture and grape juice filtration

V. blanc grapes were manually collected in the core icewine-producing area of Northeast China at the end of December 2021. The grape vines were planted at a commercial vineyard in Sidaolingzi village adjacent to Huanlong Lake, Huanren County, Benxi City, Liaoning Province, China (latitude 41°30′N, elevation 125°23′E) on a 25° slope. A vertical trellis system was installed in the vineyard with east-west oriented rows. The vines were 9 years old with a plant density of 2.5 m × 1 m. The vines were pruned to five to six shoots, and the overall crop yield was maintained at approximately 10,000 kg/ha [11]. Grape berries were destemmed and pressed to collect grape juice. Subsequently, the grape juice was carried through centrifugation (Kc > 6000) and cross-flow filtration (0.2 μm) for the removal of particles and indigenous microorganisms.

No.	Yeast species	Importance	Initial ratio (%)	Inoculation (mL)
1	*Pichia kudriavzevii*	88.05	21.7	8.7
2	*Kluyveromyces lactis*	80.59	19.8	7.9
3	*Lachancea thermotolerans*	77.00	18.9	7.6
4	*Issatchenkia orientalis*	63.69	15.7	6.3
5	*Torulaspora delbrueckii*	57.59	14.2	5.7
6	*Candida dubliniensis*	39.62	9.7	3.9

Table 1.
*The percentage of each non-*Saccharomyces *yeast in artificial self-assembled yeast community.*

2.3 Determination of physicochemical parameters of icewine

Basic wine physicochemical parameters, such as hue, color intensity, Brix°, total sugar, total acid, glucose/fructose, lactic acid, malic acid, acetic acid, and glycerol, were determined using Y15 enzymatic autoanalyzer (Biosystems S.A., Barcelona, Spain). These analyses were performed using the appropriate enzymatic reaction kits purchased from Biosystems. Before detection, the Y-15 was calibrated with external standards that were technically supported by the Biosystems enterprise (www.biosystems.es). The pH value was measured using a PB-10 pH meter (Sartorius, Göttingen, Germany). The alcoholic degree of icewine was determined based on the National Standards of P. R. China (GB/T 15038-2006) [2, 4].

2.4 Untargeted metabolomics analysis

2.4.1 Sample collection and preparation

Icewine samples were centrifuged (4°C, 10,000 rpm, 10 min). The supernatant was added into precooled methanol/acetonitrile/water solution (2:2:1, v/v). The mixture was carried out vortex, ultrasonic homogenization (4°C, 30 min), stewing (−20°C, 10 min), and centrifugation (4°C, 12,000 rpm, 20 min). Then, the supernatant was conducted vacuum freeze-drying. Before metabolomics analysis, 100 μL acetonitrile aqueous solution (acetonitrile:water = 1:1, v/v) was added to redissolve the dry sample. Subsequently, the solution was vortexed and centrifuged (4°C, 12,000 rpm, 15 min) for UHPLC-MS analysis. Twenty microliters of each sample were mixed as a QC sample [17].

2.4.2 UHPLC conditions

Analyses were performed using a UHPLC (1290 Infinity LC, Agilent Technologies) coupled to a quadrupole time-of-flight (AB Sciex Triple TOF 6600). For HILIC separation, samples were analyzed using a 2.1 mm × 100 mm ACQUIY UPLC BEH 1.7 μm column (waters, Ireland). In both ESI positive and negative modes, the mobile phase contained A = 25 mM ammonium acetate and 25 mM ammonium hydroxide in water, and B = acetonitrile. The gradient was 85% B for 1 min, linearly reduced to 65% in 11 min, then reduced to 40% in 0.1 min, kept for 4 min, and then increased to 85% in 0.1 min, with a 5 min re-equilibration period employed.

2.4.3 ESI-Q-TOF MS/MS conditions

The ESI source conditions were set as follows: Ion Source Gas1 (Gas1) as 60, Ion Source Gas2 (Gas2) as 60, curtain gas (CUR) as 30, source temperature: 600°C, IonSpray Voltage Floating (ISVF) ± 5500 V. In MS-only acquisition, the instrument was set to acquire over the *m/z* range 60–1000 Da, and the accumulation time for TOF MS scan was set at 0.20 s/spectra. In auto MS/MS acquisition, the instrument was set to acquire over the *m/z* range 25–1000 Da, and the retention time for product ion scan was set at 0.05 s/spectra. The product ion scan is acquired using information-dependent acquisition (IDA) with a high sensitivity mode selected. The parameters were set as follows: the collision energy (CE) was fixed at 35 V with ±15 eV; declustering potential (DP), 60 V (+) and −60 V (−); exclude isotopes within 4 Da, candidate ions to monitor per cycle: 10.

2.4.4 Data processing

The raw MS data (wiff.scan files) were converted to MzXML files using ProteoWizard MSConvert before importing them into freely available XCMS software. CAMERA (Collection of Algorithms of Metabolite Profile Annotation) was used for the annotation of isotopes and adducts [18]. In the extracted ion features, only the variables having more than 50% of the nonzero measurement values in at least one group were kept. Compound identification of metabolites was performed by comparing the accuracy *m/z* value (<10 ppm) and MS/MS spectra with an in-house database established with available authentic standards [19].

2.5 Metabarcoding of internal transcribed spacer (ITS) sequence

The quantity and quality of extracted DNA were measured using a NanoDrop ND-1000 spectrophotometer (Thermo Fisher Scientific, Waltham, MA, USA) and agarose gel electrophoresis, respectively. To analyze the taxonomic composition of the microeukaryotes communities, ITS1 of microeukaryotes 18S rRNA genes via a two-step amplification procedure using primers ITS1F (5′-CTTGGTCATTTAGAGGAAGTAA-3′), and ITS1R (5′-GCTGCGTTCTTCATCGATGC-3′) were carried out. Specific DNA extraction, PCR, and Illumina MiSeq sequencing (2- by 150-bp reads) were performed using the Illumina MiSeq platform at Shanghai Personal Biotechnology Co., Ltd. (Shanghai, China) [4]. Each sample was extracted for use in three replicates, and each extraction was analyzed twice. The Quantitative Insights into Microbial Ecology (QIIME2) pipeline was employed to process the sequencing data. Amplicon sequence variants were obtained using the procedure of denoise, quality control, splicing, and mosaicking, which were carried out using denoise-paired QIIME DADA2.

2.6 Statistical analysis

Statistics for the characteristic metabolites in *V. blanc* icewine, such as principle component analysis, partial least squares discriminant analysis (PLS-DA), and heatmap cluster with Euclidean distance, were carried out using the MetaboAnalyst 5.0 platform. ANOVA was conducted using Duncan's multiple-range test ($p < 0.05$). Additionally, data visualization of α-diversity and β-diversity, spearman correlation ($p < 0.05$) of characteristic metabolites and microeukaryotes, volcano plots, principal coordinates analysis (PCoA), and random forest algorithm were carried out using relative packages in R version 4.2.2. The icewine metabolomics was mapped to the Kyoto Encyclopedia of Genes and Genomes (KEGG) database to obtain the metabolic pathways in AF.

3. Results

3.1 Identification of the characteristic metabolites

The results of PCA and heatmap cluster of metabolomic illustrated different metabolite profiles between the four fermentation scenarios and *V. blanc* grape juice (**Figure 2A**). The metabolomics result was divided into two clusters. Organic acids and derivatives, benzenoids, hydrocarbon derivatives, and nucleosides were

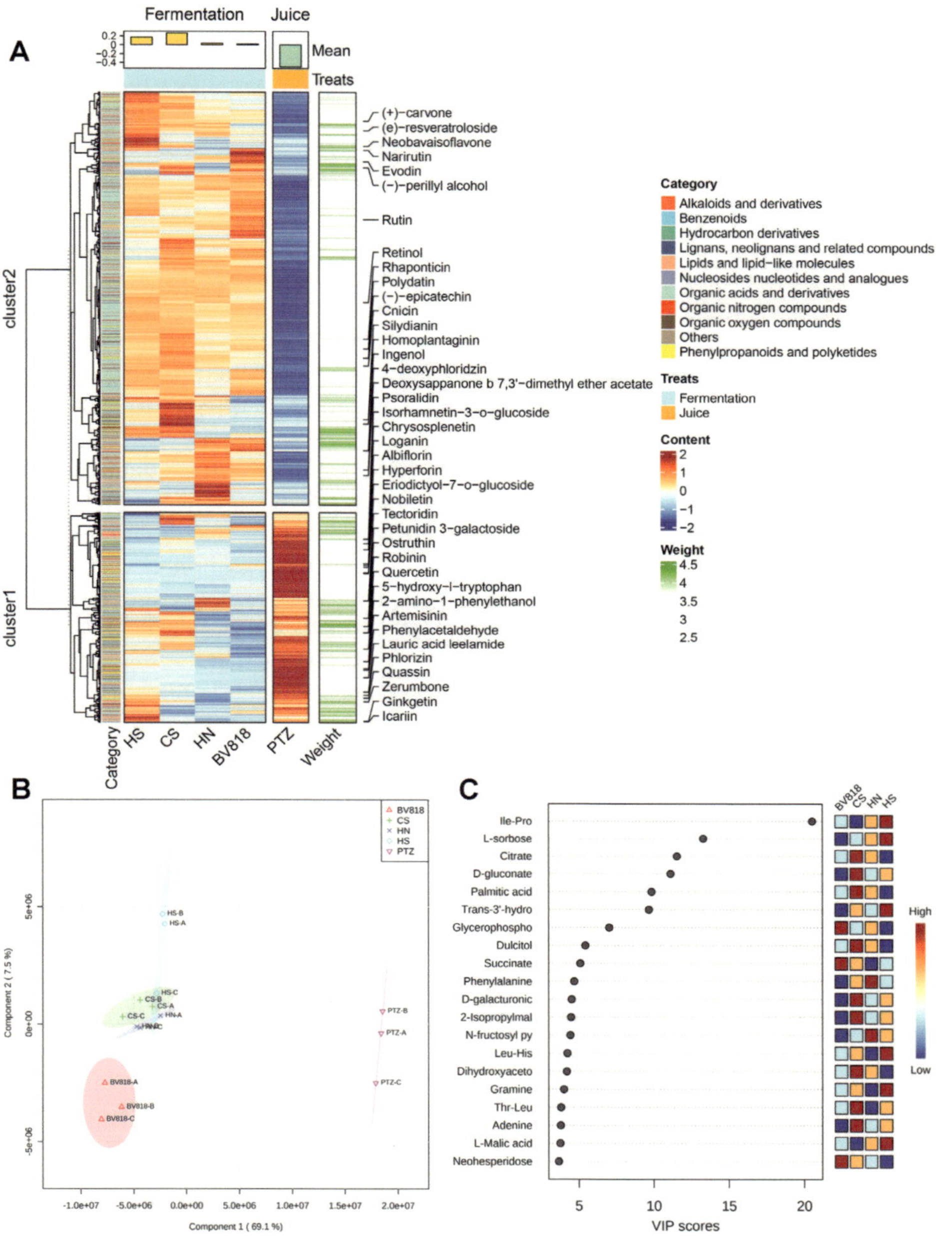

Figure 2.
*Comparison of characteristic metabolites in the icewine fermentation and grape juice. (A) Heatmap cluster of all identified metabolites in the four icewine fermentation and grape juice, especially the high-valued terpenes and flavonoids, were calculated and highlighted on the left side of the heatmap diagram. The importance of each treatment was shown on the top of the heatmap diagram. Herein, the Euclidean method was used for distance measurement, and the clustering algorithm was Ward. (B) Score plot of PLS-DA for analysis of observation of four icewine fermentations. (C) Ranking of characteristic metabolites by weighting the sum of VIP scores in compound 1 of PLS-DA. Colored boxes on the right side indicated the correlations between each metabolite and fermentation. Note: CS: spontaneous fermentation using the epiphytic microorganisms from the fully ripe grape berries; HS: spontaneous fermentation using the epiphytic microorganisms from the grape berries in late harvest; HN: icewine fermentation using the self-assembled non-*Saccharomyces *yeasts group icewine fermentation using a commercial yeast strain, which belongs to* Saccharomyces bayanus.

promoted during AF and mainly grouped in cluster 2. Their concentration increases, resulting from the growth of microorganisms. The mean of normalized scores in spontaneous fermentations (CS and HS) showed the highest positive values. In other words, spontaneous fermentation developed more organic compounds than the external yeast inoculation fermentations (HN and BV818). The weight of each metabolic compound represented its importance in the fermentation [20]. Obviously, fermentations generated abundant key metabolites and showed different metabolite profiles from that of grape juice. The important terpenes and flavonoids were grouped and marked on the right side of the heatmap. Some of them are first identified in icewine and have specific medical functions such as resveratroloside, neobavaisoflavone, narirutin, homoplantaginin, chrysosplenetin, and deoxysappanone b 7,3′-dimethyl ether acetate, etc. Besides, many high-valued compounds enriched in the grape juice were inhibited after AF, such as nobiletin, tectoridin, robinin, quercetin, Icariin, ginkgetin, etc.

Partial least squares discrimination analysis (PLS-DA) was carried out to compare the differences in each fermentation scenario (**Figure 2B**). Component 1 and component 2 contributed 76.6% of PLS-DA in total. It indicated the metabolite profile of each fermentation can be discriminated with the first two components. The four fermentations showed a far distance from the grape juice. Herein, the spontaneous fermentations (HS and CS) were close to HN, while BV818 showed a different metabolite profile from the others. To study the characteristic metabolites and their correlations to each fermentation. The top 20 characteristic metabolites were ranked according to their VIP scores of PLS-DA (**Figure 2C**). Most characteristic metabolites in icewine were positively correlated to the spontaneous fermentation and HN except glycerophosphocholine, succinate, and neohesperidose. In terms of spontaneous fermentation, some characteristic metabolites commonly showed positive correlations to CS and HS, such as D-gluconate, trans-3′-hydroxycotinine o-β-D-glucuronide, D-galacturonic acid, 2-isopropylmalic acid, gramine, adenine, and two dipeptides (Leu-His and Thr-Leu).

3.2 Spatial dispersal of microeukaryotic communities and correlations to the characteristic metabolites during AF

The spatial dispersal and co-occurrence network of each fermentation scenario jointly showed the dynamic changes of microeukaryotes during AF (**Figure 3**). Accordingly, the microeukaryotes in BV818 were highly developed at the end stage of AF, where ASVs of the microeukaryotes showed a higher enrichment but lower diffusivity (**Figure 3A**). The inner interactions between microeukaryotes in each module were positive. The major microeukaryotic modules in BV818 were modules 1, 2, 9, 7, and 14. Herein, the cross interactions between modules 1, 2, 7, and 9 showed higher negative similarity. It indicated the major activities between these modules were dominated by negative cohesion, such as competition, parasitism, or amensalism [21]. Moreover, module 14 showed fewer interactions with other modules. It indicated the related microeukaryotic communities could independently develop and be neutralized to other species during AF. HN showed a contrary spatial dispersal but a similar co-occurrence network to BV818 (**Figure 3C**). The strongest diffusivity of microeukaryotes was observed at the beginning of AF, while the weakest diffusivity was presented at the end of AF. The major microeukaryotic modules of HN were 1, 2, 6, and 12. The cross interactions between modules 1, 2, and 6 were also dominated by negative cohesion;

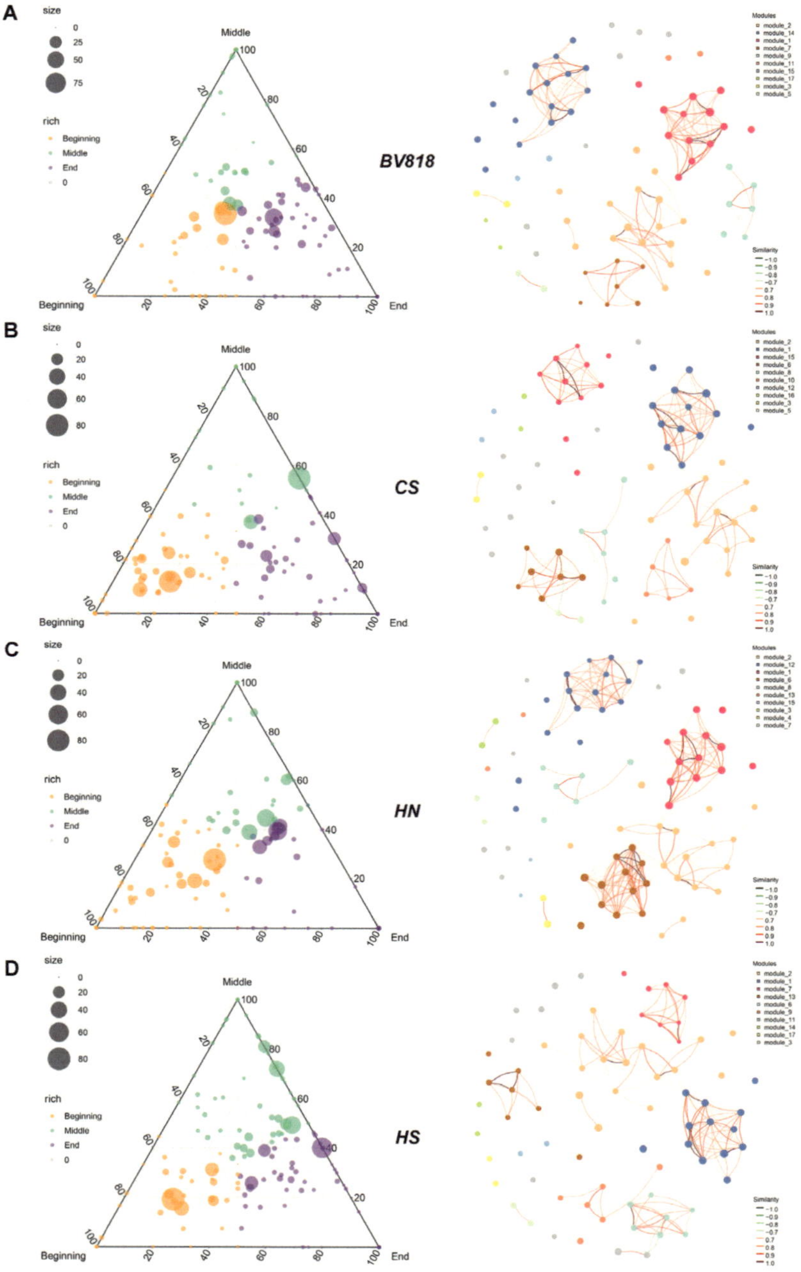

Figure 3.
*(A) Spatial dispersal and co-occurrence networks of fungi genera of BV818 on three stages during icewine fermentation. (B) Spatial dispersal and co-occurrence networks of fungi genera of CS on three stages during icewine fermentation. (C) Spatial dispersal and co-occurrence networks of fungi genera of HN on three stages during icewine fermentation. (D) Spatial dispersal and co-occurrence networks of fungi genera of HS on three stages during icewine fermentation. Each node represents one ASV that is labeled by genus. A node was verified by a robust (Spearman's correction coefficient R > 0.6) and significant (pFDR < 0.05) correlation. The size of each node is relational to the number of connections, while nodes with the same color display the same module. The thickness of each connection between two nodes is relational to the strength of Spearman's correlation coefficient. Note: CS. spontaneous fermentation using the epiphytic microorganisms from the fully ripe grape berries; HS: spontaneous fermentation using the epiphytic microorganisms from the grape berries in late harvest; HN: icewine fermentation using the self-assembled non-*Saccharomyces *yeasts group, icewine fermentation using a commercial yeast strain, which belongs to* Saccharomyces bayanus.

meanwhile, these modules showed fewer interactions with module 12. Both CS and HS belonging to spontaneous fermentation and the same microbial origination from a vineyard resulted in similar spatial dispersals and co-occurrence networks of microeukaryotes during AF (**Figure 3B** and **D**). ASVs of microeukaryotes were evenly distributed in all the stages of AF. Comparatively, the microeukaryotes of CS mainly developed at the beginning and end stages of AF, while the microeukaryotes' development of HS was vigorous at the middle stage of AF. The spontaneous fermentation presented more modules and more complex interactions than BV818 and HN. The major microeukaryotic modules in CS were 1, 2, 6, 8, 10, and 15. Herein, module 1 and module 10 showed positive cross interactions, namely, cooperation and mutualism. However, other microeukaryotic modules were mainly controlled by negative cohesion. The major modules in HS were 1, 2, 6, 7, 9, and 10. Herein, module 1 and module 7 showed positive cross interactions, while other modules were controlled by negative cohesion. In general, the cross-interaction between microeukaryotic phyla during AF was dominated by negative cohesion, and the abundance patterns of microeukaryotic communities depend on different fermentation modes [22].

Mantel test was used to study the correlations between the specific order of microeukaryotes and fermentation mode (**Figure 4A**). As a result, *Saccharomycetes*, *Microbotryomycetes*, and *Tremellomycetes* showed strong positive correlations ($p < 0.05$), which potentially constructed a stable microeukaryotic community during AF and increase the efficiency of alcohol conversion. However, significant negative correlations were found between some orders, such as *Sgaricomycetes*, *Lecanoromycetes*, *Pezizomycetes*, *Cystobasidiomycetes*, *Zoopagomycetes*, *Exobasidiomycetes*, *Sordariomycetes*, *Leotiomycetes*, *Wallemiomycetes*, *Microbotryomycetes*, and *Tremellomycetes*. It probably results in interspecific competition and internal friction of resources during fermentation. Moreover, *Ustilaginomycetes* and *Zoopagomycetes* were highly correlated to BV818 and HN, respectively.

The correlation matrix between the functional compounds and the top 25 active microeukaryotic genera was established using a random forest model and Pearson's correlation coefficients (**Figure 4B**). Accordingly, many microeukaryotes showed positive correlations. They potentially contributed to the development of terpenoids and flavonoids in icewines, such as *Starmerella*, *Mrakia*, *Tausonia*, *Rhodotorula*, *Mrakiella*, *Verticillium*, and *Curvibasidium*. Herein, *Starmerella* played an important role in promoting the metabolism of bioflavonoids and polyflavonoids, flavans, and flavones. Fatty alcohols were strongly correlated to many microeukaryotes, such as *Mrakia*, *Rhodotorula*, *Cystofilobasidium*, *Mrakiella*, *Filobasidium*, *Bettsia*, *Verticillium*, and *Curvibasidium*. Homoisoflavans positively correlated to *Mrakia*, *Rhodotorula*, *Mrakiella*, *Verticillium*, and *Curvibasidium*. Moreover, *Cystofilobasidium* and *Golubevia* contributed to the development of tryptamines and derivatives, while *Pichia* was regarded as an important factor in affecting the development of monoterpenoids and tryptamines.

3.3 Characteristic metabolites of *V. blanc* grape induced stochastic assembly of microeukaryotic communities during AF

One of the reasons why spontaneous fermentation could not be widely applied is that many winemakers treat it as a 'black-box', a complex, multi-strains participated system. With the development of systematic biology, the neutral community model (NCM) will be a useful tool to excavate the deep principle of microbial assembly in

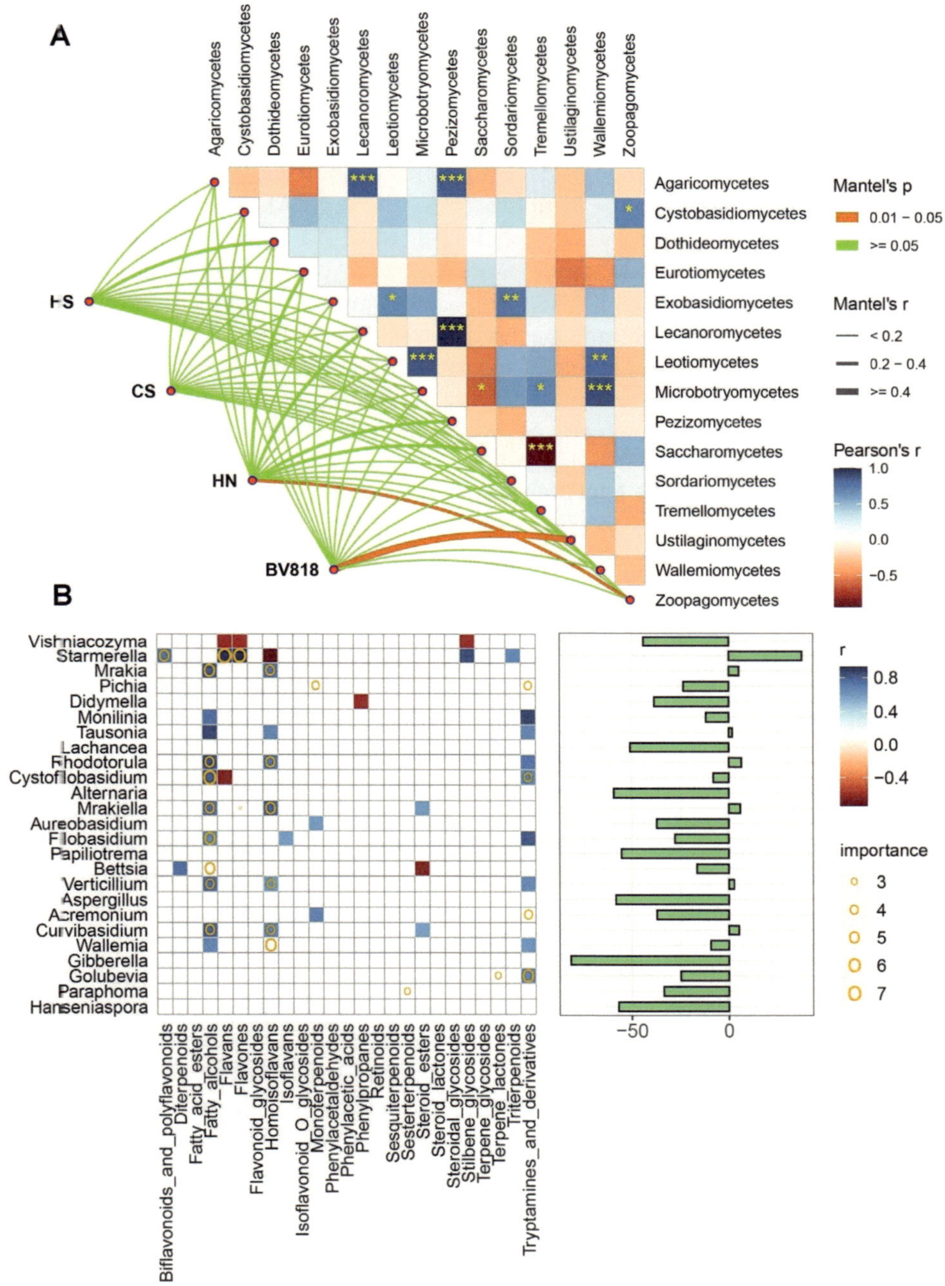

Figure 4.
(A) Correlations of the networked community structures (Bray–Curtis distance) with important fungi orders. The edge width corresponds to Mantel's r value, and the edge color denotes the statistical significance. Significance levels were labeled as "" (p-value < 0.1), "**" (p-value < 0.05), and "***" (p-value < 0.01), respectively. Pairwise correlations of these variables were shown with a color gradient denoting Pearson's correlation coefficients. (B) The correlation matrix for the top 25 high-valued terpenes and flavonoids and the top 25 active fungi genus was established using a random forest algorithm, and the importance of the coefficient was shown with different cycles. Meanwhile, the explained ratio of the top 25 active fungi genera was represented on the right side of the diagram. Note: CS: spontaneous fermentation using the epiphytic microorganisms from the fully ripe grape berries; HS: spontaneous fermentation using the epiphytic microorganisms from the grape berries in late harvest; HN: icewine fermentation using the self-assembled non-*Saccharomyces *yeasts group, icewine fermentation using a commercial yeast strain, which belongs to* Saccharomyces bayanus.

co-fermentation or spontaneous fermentation [23]. In this study, the NCM predicted a large fraction of the total correlation between the occurrence frequency of ASVs and their relative abundance variations (**Figure 5A**), with 83%, 73%, 80%, and 76% of the explained microeukaryotic community variations for the BV818, CS, HN, and HS, respectively. The NCM indicates that microeukaryotic communities comply with stochastic assembly instead of deterministic assembly in different fermentation scenarios. Higher R^2 suggests the sample is closer to the NCM. Compared to spontaneous fermentation, commercial yeast inoculation (BV818) fits the stochastic process. Moreover, a smaller Nm value suggests more restrictive species dispersal. Therefore, the level of species dispersal in BV818 (Nm = 374) was the lowest in all fermentations, while HS had the maximum species dispersal (Nm = 1005). CS and HN showed similar Nm values. This result is also in accord with the spatial dispersal of microeukaryotic communities during AF; namely, the microeukaryotic communities of HS can evenly develop at three stages of AF, while the ASVs of microeukaryotes of BV818 are mainly concentrated at the end of AF.

β-Nearest taxon index (β-NTI) between samples was calculated to describe the phylogenetic turnover of microeukaryotic communities in the different fermentation scenarios. Linear regression curves between |β-NTI| and the relative abundance of the six characteristic metabolites were constructed (**Figure 5B**). Among these six characteristic metabolites, β-sitosterol is a plant steroid, empenthrin, gentiopicroside, zerumbone, and qingyangshengenin belong to terpenoids, and phlorizin is a kind of flavonoids. Empenthrin, β-sitosterol, gentiopicroside, phlorizin, and zerumbone positively correlated to |β-NTI| while gentiopicroside and qingyangshengenin showed negative correlations to |β-NTI|. However, β-NTI of all the fermentations were in the range of −2 to 2 (**Figure 5C**). Although gentiopicroside and qingyangshengenin could affect the stochasticity of microeukaryotic communities, the stochastic process still dominated the phylogenetic turnover of microeukaryotic communities during AF. To assess the relative importance of determinism and stochasticity, modified stochasticity ratio (MST) was carried out to supply the results of NCM (**Figure 5D**). In four fermentation scenarios, MST of CS (36.1%) was much lower than the boundary (50%), which indicated determinism affected the microeukaryotic assembly of CS. The MST of HS (54.7%) was slightly higher than the MST of HN (53.0%) without a significant difference ($p < 0.05$). As expected, BV818 showed the highest MST (63.8%), which indicated stochasticity played an important role in shaping microeukaryotic communities of BV818. This result suggested the assembly process of spontaneous fermentation could be more stable than the inoculation of commercial yeasts as long as the source of grape is consistent and fermentative conditions are standard.

3.4 The specific process of stochastic assembly of microeukaryotic communities and the selection of fermentative specialists

Principal coordinates analysis (PCoA) was carried out to observe the degree of similarity between the microeukaryotic communities of different fermentation scenarios (**Figure 6A**). The total contribution of the first two components of PCoA can explain 66.11% of variations. The confidence intervals of HN and BV818 have been separated from the spontaneous fermentations (CS and HS). Comparatively, the degree of distribution of BV818 was more concentrated than that of the others while HN and HS showed higher degree of distributions than CS and BV818. The result of

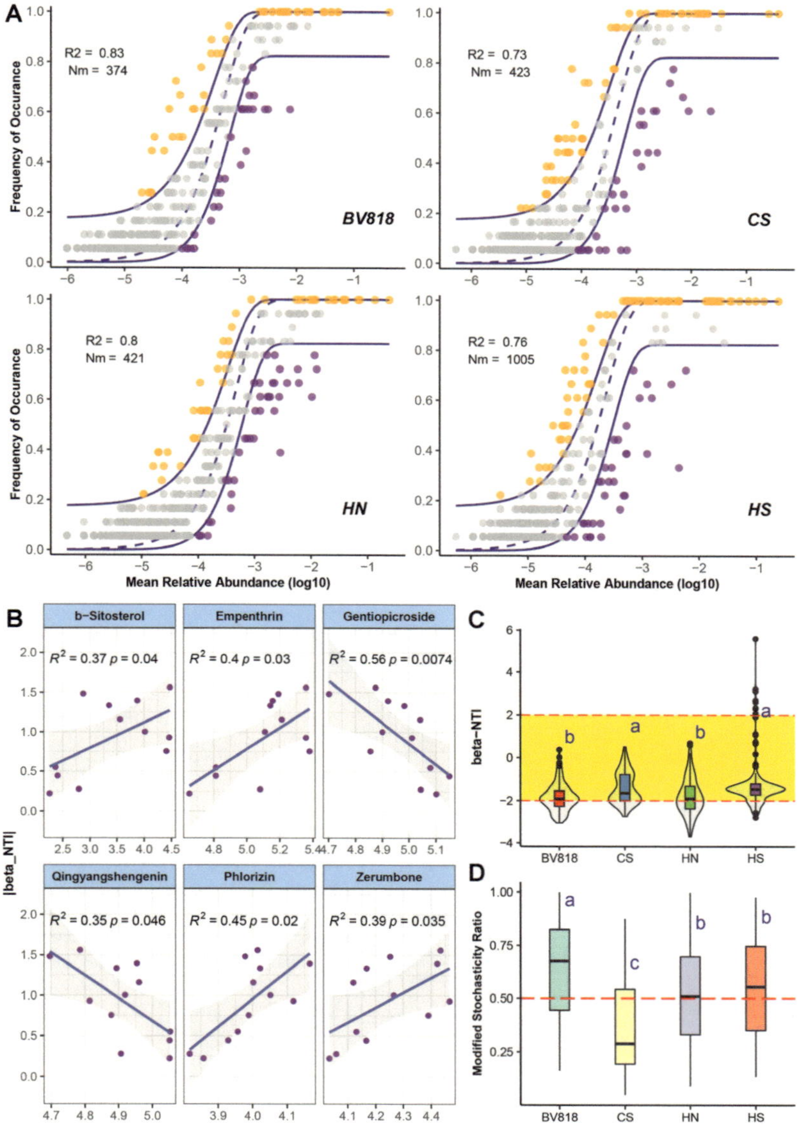

Figure 5.
*(A) Fit of the neutral community models of fungi communities assembly in the four fermentations. The solid blue lines indicate the best fit to the NCM, and the dashed blue lines represent 95% confidence intervals around the model prediction. OTUs that occurred more or less frequently than predicted by the NCM are shown in different colors (gray dots represent the frequency of occurrence within the 95% confidence interval ranging around the model prediction, orange dots represent above prediction, and purple dots represent below prediction). Nm indicates the metacommunity size times immigration, R^2 indicates the fit to this model. Meanwhile, a modified stochasticity ratio was used to assess the relative importance of stochasticity. (B) Coefficients between |β-NTI| and the relative abundance of the six featured metabolites were calculated to evaluate the correlative fitting curves, p-values were evaluated, and significant correlations were determined at $p < 0.05$. (C) Contributions of deterministic and stochastic processes on community assembly within four icewine fermentations. β-NTI calculations of phylogenetic turnover between different icewine fermentations indicate that variable selection has greater effects on fungi dynamics and fermentation progress. (D) A modified stochasticity ratio was carried out to assess the relative importance of stochasticity, which was developed with 50% as the boundary line in each icewine fermentation. Note: CS: spontaneous fermentation using the epiphytic microorganisms from the fully ripe grape berries; HS: spontaneous fermentation using the epiphytic microorganisms from the grape berries in late harvest; HN: icewine fermentation using the self-assembled non-*Saccharomyces *yeasts group, icewine fermentation using a commercial yeast strain, which belongs to* Saccharomyces bayanus.

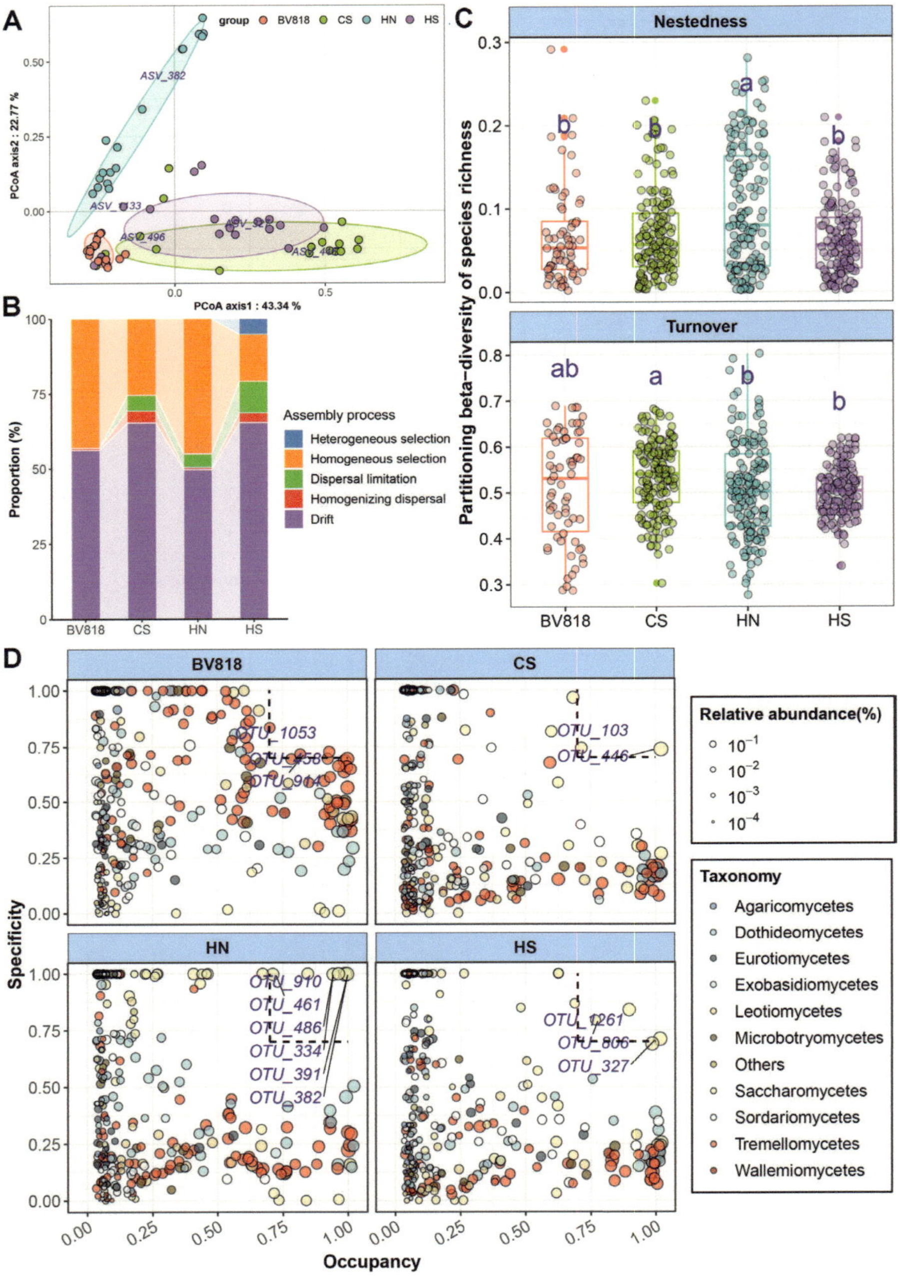

Figure 6.

(A) Principal coordinate analysis (PCoA) based on Sørensen dissimilarly displaying differences in community composition of fungal genera in the four icewine fermentations, the top 10 highest active fungi genera were selected and labeled. (B) The relative influence of each community assembly process among four icewine fermentation was defined by the percentage of site pairs governed by each process. (C) Partitioning the turnover and nestedness components of β-diversity of fungi communities in four icewine fermentations. Duncan's multiple range test was carried out for the significant differences of the model at $p < 0.05$. (D) The SPEC-OCCU plots showed the most abundant OTUs in each icewine fermentation; the x-axis represents occupancy, i.e., how well an OTU is distributed across the specific fermentation scenario; and the y-axis represents specificity, i.e., whether they are also found in other fermentation scenarios. The fermentative specialists of fungi species in each fermentation were selected as both occupancy and specificity higher than 0.7.

PCoA indicated that the microeukaryotic composition of BV818 was different from co-fermentations (HN, CS, and HS), and the concentrated distribution of BV818 also indicated low diversity of microbial participation during AF. Moreover, the top five abundant microeukaryotic genera were selected and located on the PCoA plot. Herein, *Pichia* (ASV_382) was located in the confidence ellipse of HN ($p < 0.05$), while *Vishniacozyma* and *Starmerella* (ASV_1133, ASV_496, ASV_327, and ASV_446) highly related to CS and HS.

Except for the NCM, phylogenetic-bin-based null model analysis was carried out further to infer community assembly mechanisms in the stochastic process (**Figure 6B**) [20]. The important assembly processes of microeukaryotic communities were heterogeneous selection, homogeneous selection, dispersal limitation, homogenizing dispersal, and drift. Specifically, the assembly process was mainly composed of homogeneous selection (43.1%) and drift (56.2%) in BV818. For HN, dispersal limitation (0.05%), homogeneous selection (45.1%), and drift (49.7%) were the important parts of the stochastic assembly. For CS, the specific stochastic assembly included homogeneous selection (25.5%), dispersal limitation (0.05%), homogenizing dispersal (0.04%), and drift (65.4%). The heterogeneous selection was the characteristic assembly process in HS, which possessed 0.05% of the stochastic assembly. Besides, the proportion of homogeneous selection, dispersal limitation, homogenizing dispersal, and drift was 15.7%, 10.5%, 0.03%, and 65.4% of the stochastic assembly in HS, respectively. In general, drift and homogeneous selection were the major assembly processes during AF of icewine, and HS showed a more complex composition of the assembly process than other fermentations. Partitioning β-diversity of richness was carried out to quantify the result of species replacement between different fermentations (turnover) and species gains or losses between different microeukaryotic communities (nestedness) (**Figure 6C**). As a result, species replacement played an important role in shaping microeukaryotic communities. The greatest difference between the contributions of β-diversity components was observed in the four fermentation scenarios, where the turnover was about six times higher than nestedness. Specifically, the range of turnover in each fermentation was from 50% to 55% while the range of nestedness was from 5% to 8%. The nestedness of HN was the highest level among the four fermentations. For turnover, BV818 and CS were higher than HN and HS.

To inspect the distribution of OTUs from each fermentation and the specificity of these microeukaryotes during AF, specificity and occupancy were calculated for each OUT, which was then projected onto a plot (SPEC-OCCU plot, **Figure 6D**). As indicated by the spread of OTUs across occupancy, OTUs from BV818 and HN communities showed highly varied occupancy while the majority of OTUs from CS and HS exhibited more homogenous occupancy, which mainly concentrated at the range of 0–0.25. To find specialist species attributable to each fermentation, we selected OTUs with specificity and occupancy greater or equal to 0.7 (dotted boxes). These microeukaryotes are specific and could be highly fermentative in their fermentation scenarios. The number of these specialist OTUs was BV818 (3 OTUs represent), HN (6 OTUs), CS (2 OTUs), and HS (3 OTUs), respectively. Specifically, *Mrakia frigida* is the fermentative specialist in BV818, the specialists of HN included *Lachancea thermotolerans* and *Pichia kudriavzevii*, while *Starmerella bacillaris* was the fermentative specialist in spontaneous fermentation. Meanwhile, the changes in the relative abundance of microeukaryotic species verified these yeasts are the key factors to influence fermentation process.

3.5 The analysis of metabolic pathways in four icewine fermentation scenarios

A structural equation modeling (SEM) accounting for the yeast strains, characteristic metabolites, and key assembly factors was built to analyze the interactions of variables in four icewine fermentation scenarios (**Figure 7A**). Based on the previous results, five yeast strains (*Vishniacozyma victoriae*, *Lachancea thermotolerans*, *Pichia kudriavzevii*, *Mrakia frigida,* and *Starmerella bacillaris*) were assigned as important components of microeukaryotic community in the SEM. Average variation degree (AVD), alpha diversity, MST, and bNTI were assigned as key factors of the assembly process of microeukaryotic community. Terpenoids, flavonoids, and steroids were assigned as characteristic metabolites of icewine fermentations. As a result, *M. frigida* showed a strong connection to yeast combination, which also significantly related to microeukaryotes assembly process during icewine fermentation and except for *P. kudriavzevii*, other four yeast strains, *V. victoriae*, *L. thermotolerans*, *S. bacillaris*, and *M. frigida* negatively affected the microeukaryotes assembly process. Moreover, the characteristic metabolites showed negative effects on the microeukaryote assembly process. Steroids showed a strong and positive influence on the total metabolites of icewine. AVD showed an indirect effect on the microeukaryotes assembly process, while AD, MST, and bNTI directly and positively related to the microeukaryotes assembly process during alcoholic fermentation.

The top 20 enriched KEGG pathways of four icewine fermentations were shown in **Figure 7B**, where the rich factors of each KEGG pathway were the ratio of differential genes in this pathway, while the p-value indicated the importance of this pathway in the icewine fermentation. Accordingly, protein digestion, absorption, and ABC transporters were the most important pathways of microeukaryotic metabolism in all fermentation treatments. Compared to commercial yeast, CS, HN, and HS showed similar categories of important KEGG pathways, which also contained mineral absorption, biosynthesis of amino acids, and aminoacyl-tRNA biosynthesis. Lysine degradation was the characteristic pathway for the spontaneous fermentations (CS and HS). Starch and sucrose metabolism were relatively important to HS and BV818, while proximal tubule bicarbonate reclamation was the common pathway in BV818 and CS. Interestingly, valine, leucin, and isoleucine biosynthesis and nicotinate and nicotinamide metabolism were the specific pathways in CS. It indicated more potential volatile metabolites could be synthesized by applying the indigenous yeast community during alcoholic fermentation.

4. Discussion

In this study, the top-down approach was first used to design a self-assembled non-*Saccharomyces* yeasts group (HN) [24]. Herein, 'top' referred to the special ecosystem in which the local epiphytic yeasts of the *V. blanc* grape performed a strong ability to hydrolyze the glycosidic bound terpenes and enhanced the varietal aroma of the grape during late harvest [2, 21]. We tested the availability of HN to conduct AF and explored its biological functions of producing high-valued metabolites in icewine. Besides, two types of spontaneous fermentations were carried out to observe the assembly process of indigenous microeukaryotes during AF. The microbial starter of the spontaneous fermentation was extracted from the fully ripe *V. blanc* grape and the grape berries in the late harvest. Therefore, the spontaneous fermentations simultaneously represented the regional microeukaryotic composition and the "*terroir*"

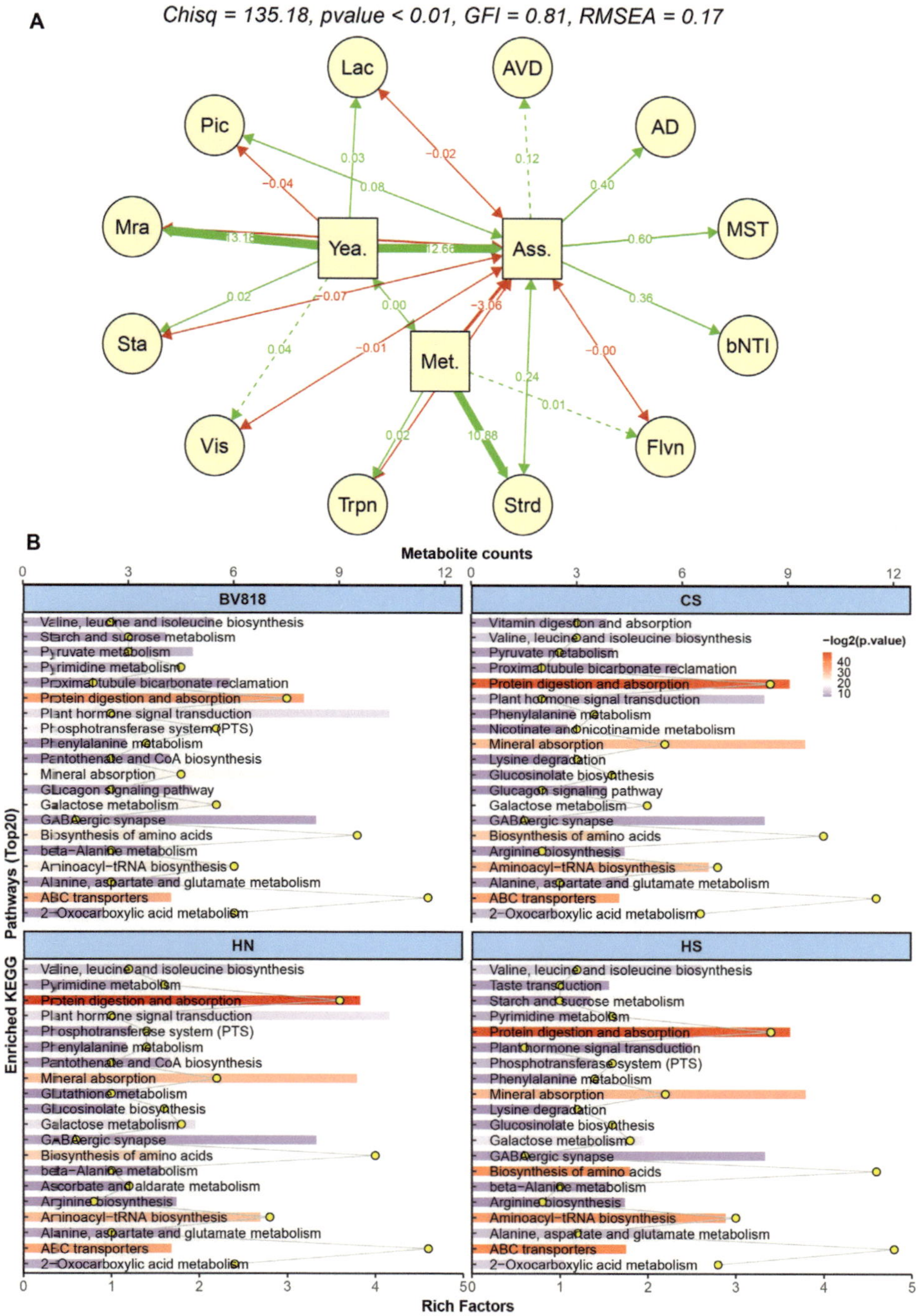

Figure 7.
*(A) Structural equation modeling (SEM) accounting for the yeast strains, characteristics, metabolites, and key assembly factors in four icewine fermentation scenarios. (B) Top 20 enriched KEGG pathways of the fungal community in four icewine fermentation scenarios. Numbers adjacent to measured variables are their coefficients with composite variables. Numbers adjacent to arrows are path coefficients are the directly standardized effect size of the relationship. The thickness of the arrow represents the strength of the relationship. Note: CS: spontaneous fermentation using the epiphytic microorganisms from the fully ripe grape berries; HS: spontaneous fermentation using the epiphytic microorganisms from the grape berries in late harvest; HN: icewine fermentation using the self-assembled non-*Saccharomyces *yeasts group, icewine fermentation using a commercial yeast strain, which belongs to* Saccharomyces bayanus.

characteristics of icewine. As a control, BV818 showed a different metabolite profile from the other fermentations according to PLS-DA. It indicated the application of commercial yeasts may affect the indigenous microbial communities and change wine regional characteristics [25, 26]. Terpenoids and flavonoids were identified as the characteristic metabolites in icewine. Some high-valued metabolites of icewine were first reported in our work. For instance, resveratroloside is a monoglucosylated form of stilbene in red wine. It effectively alleviates postprandial hyperglycemia and cardioprotective effects [27, 28]. Neobavaisoflavone has the potential for anti-human glioma cancer and inhibition of inflammatory mediators. Deoxysappanone b 7,3′-dimethyl ether acetate has been reported to have anti-cryptosporidial activities [29]. Based on PLS-DA, the spontaneous fermentation, and self-assembled non-*Saccharomyces* yeasts group had the advantage of producing more characteristic metabolites.

The spatial dispersal and co-occurrence network of each fermentation scenario performed a dynamic distribution of microeukaryotes and the correlations between microeukaryotic communities during AF. As a result, negative cohesion, such as competition, parasitism, or amensalism, played an important role in the cross-interactions between microeukaryotic communities. The microeukaryotic dispersal of BV818 was mainly concentrated at the end of fermentation, while HN showed the opposite result that the microeukaryotes mainly distributed at the beginning of fermentation. Understandably, most commercial yeasts were *Saccharomyces cerevisiae*, which usually performs killer activity against other fungi at the beginning of AF, then dominates and completes the fermentation. Although the spontaneous fermentation obtained quite a balanced development of microeukaryotic communities at the three stages of AF, fermentation stress and frequent microbial succession jointly resulted in more complex interactions between microeukaryotic communities in the spontaneous fermentations [22, 30]. Moreover, many microeukaryotic genera were positively related to the development of terpenoids and flavonoids in icewine. It is due to these compounds commonly present in glycosidic bound forms, while the epiphytic microeukaryotes of grapes can generate a series of glycoside hydrolases (GHs) and glycosyl transferases (GTs), which efficiently break glycosidic bonds and release the aglycones [2, 31]. For instance, monoterpenyl disaccharide glycosides originated from plants and microbes, such as α-arabinofuranosidase, β-glycosidase, and β-galactosidase, hydrolyze glycosidic bound terpenes into free forms [32]. The result of NCM indicated the microeukaryotic community assembly in icewine was predominately driven by births, deaths, and immigration (stochastic process) instead of external abiotic factors and niche differentiation (deterministic process) [23]. Therefore, the hydrolytic/fermentative microeukaryotes in AF are inclined to spontaneously conduct stochastic assembly through the response or sharing of certain organic compounds [33]. Based on the result of random forest ranking of metabolites and β-NTI of ASVs, β-sitosterol, four terpenoids (empenthrin, gentiopicroside, zerumbone, and qingyangshengenin), and phlorizin were identified as the characteristic metabolites to affect the stochastic assembly of microeukaryotic communities. These metabolites may induce the common response of microeukaryotes to biotic and abiotic stresses and contribute to the assembly process of microeukaryotic communities. For instance, β-sitosterol could be used as an antibiotic to enhance the antibacterial activity of microeukaryotes [34]. Gentiopicroside is significantly correlated with the abundance of dominant endophyte genera among different tissues of *Gentiana officinalis* [35]. Moreover, fungal life history strategies determined homogenous selection and dispersal limitation attributed to the large proportion in structuring microeukaryotic communities, e.g., growth, dispersal, or dormancy methods [36].

People had been drinking natural wines for thousands of years before sulfur dioxide was applied in the wine industry. Current studies have admitted the importance of a certain strain of commercial yeast in the winemaking industry, while indigenous fungi have been misunderstood as microbial contaminators during AF. In our study, the dominator of AF was indigenous non-*Saccharomyces* yeasts instead of commercial yeast. Some fermentative specialists not only complete AF but highly contribute to the stochastic assembly of microeukaryotic communities. In particular, *S. bacillaris* dominated the spontaneous fermentation and has been reported high glycerol production, moderate nitrogen demand, and more active adaptation to extremely high sugar concentration [37]. *L. thermotolerans* and *P. kudriavzevii* are multi-stress-tolerant and widely present in different kinds of fruits [38]. Appropriate use of these indigenous non-*Saccharomyces* yeasts can enhance the aroma complexity of icewine [4]. *S. cerevisiae* was difficult to be the advantage yeast strain since the growth was always inhibited throughout AF of icewine. Inoculation of BV818 induced the development of *M. frigida*, which is highly active in producing β-galactosidase and killer toxins [39]. The result of SEM also confirmed *M. frigida* could be a key defender to restrict the development of the invasive exotic species and ensure the stability of indigenous microeukaryotic communities. Our results elucidated the importance of indigenous non-*Saccharomyces* yeasts in stochastic assembly, while inoculation of commercial yeast will induce reconstruction of microeukaryotic communities and potentially affect the "*terroir*" characteristics of icewine. Nevertheless, steroids, especially β-sitosterol, are not only commonly upregulated in different fermentation scenarios but play an important role in driving the stochastic assembly of microeukaryotic communities during AF. Vitamin digestion, absorption, and lysine degradation are the characteristic enriched pathways in spontaneous fermentation. It mainly relates to the stress-resistant functions of microeukaryotes, such as antioxidation, low-temperature resistance, and salt resistance [40].

5. Conclusion

This study first explored the assembly process of microeukaryotic communities in four typical icewine fermentation scenarios. Herein, top-down design was carried out to construct a non-*Saccharomyces* yeasts group for icewinemaking. Untargeted metabolomics revealed terpenoids and flavonoids were the characteristic metabolites of icewine. Besides, some microeukaryotic families were strongly correlated to the development of the key classes of terpenoids and flavonoids. Through NCM and the β-diversity of microeukaryotes, a stochastic process dominated microeukaryotic community assembly during AF. Moreover, the spontaneous fermentation showed an evenly spatial distribution of microeukaryotes at the three stages of AF. Co-occurrence networks suggested that negative cohesion played a pivotal role in microeukaryotic interactions between different modules. Five terpenoids and a flavonoid showed their high contributions in driving stochastic assembly. Meanwhile, mechanisms of microeukaryotic community assembly are mainly involved in the homogeneous selection, drift, and dispersal limitation during AF of icewine. Indigenous non-*Saccharomyces* yeasts, *M. frigida*,

L. thermotolerans, *P. kudriavzevii*, and *S. bacillaris* were identified as the fermentative specialists during AF of icewine. Furthermore, the enriched KEGG pathways of microeukaryotes related to their stress-resistant functions, which result in higher production of terpenoids, flavonoids, and steroids in the spontaneous fermentation.

Acknowledgements

We acknowledge Huanren Senpatina Icewine Domaine Co., Ltd. for kindly providing the experimental raw materials.

Author details

Kai Chen
College of Food Science and Technology, Henan Agricultural University, Zhengzhou, Henan Province, China

*Address all correspondence to: chenkai96@foxmail.com

References

[1] Llauradó JM, Rozès N, Constantí M, Mas A. Study of some Saccharomyces cerevisiae strains for winemaking after preadaptation at low temperatures. Journal of Agricultural and Food Chemistry. 2005;**53**:1003-1011

[2] Chen K, Zhang L, Qiu S, Wu X, Li J, Ma L. Freeze–thaw cycles characterize varietal aroma of Vidal blanc grape during late harvest by shaping self-assembled microeukaryotic communities. Food Chemistry. 2022;**384**:132553

[3] Chen K, Escott C, Loira I, Fresno JMD, Morata A, Tesfaye W, et al. Use of non-Saccharomyces yeasts and oenological tannin in red winemaking: Influence on colour, aroma and sensorial properties of young wines. Food Microbiology. 2018;**69**:51

[4] Chen K, Liu C, Wang Y, Wang Z, Li F, Ma L, et al. Predominance of indigenous non-saccharomyces yeasts in the traditional fermentation of greengage wine and their significant contribution to the evolution of terpenes and ethyl esters. Food Research International. 2021;**143**:110253

[5] Chen K, Qiu S, Liu C, Zhang L, Wu X, Ma L, et al. Abiotic factors play important roles in complexity and characterization of aroma precursors in *Vidal blanc* grape. Food Research International. 2022;**162**:112015

[6] Qiu S, Chen K, Liu C, Wang Y, Chen T, Yan G, et al. Non-saccharomyces yeasts highly contribute to characterisation of flavour profiles in greengage fermentation. Food Research International. 2022;**157**:111391

[7] Li R, Lin M, Guo S, Yang S, Han X, Ren M, et al. A fundamental landscape of fungal biogeographical patterns across the main Chinese wine-producing regions and the dominating shaping factors. Food Research International. 2021;**150**:110736

[8] Wang QJ, Spence C. Wine complexity: An empirical investigation. Food Quality and Preference. 2018;**68**:238-244

[9] Bokulich NA, Thorngate JH, Richardson PM, Mills DA. Microbial biogeography of wine grapes is conditioned by cultivar, vintage, and climate. Proceedings of the National Academy of Sciences. 2014;**111**:E139-E148

[10] Chen Y, Zhang W, Yi H, Wang B, Xiao J, Zhou X, et al. Microbial community composition and its role in volatile compound formation during the spontaneous fermentation of ice wine made from Vidal grapes. Process Biochemistry. 2020;**92**:365-377

[11] Chen K, Wen J, Ma L, Wen H, Li J. Dynamic changes in norisoprenoids and phenylalanine-derived volatiles in off-vine Vidal blanc grape during late harvest. Food Chemistry. 2019;**289**:645-656

[12] Jing W, Min L, Jixin L, Tengzhen M, Shunyu H, Morata A, et al. Chapter 10—Biotechnology of ice wine production. In: Holban AM, Grumezescu AM, editors. Advances in Biotechnology for Food Industry. Academic Press; 2018. pp. 267-300

[13] Ma Y, Xu Y, Tang K. Aroma of icewine: A review on how environmental, viticultural, and oenological factors affect the aroma of icewine. Journal of Agricultural and Food Chemistry. 2021;**69**:6943-6957

[14] López-Rituerto E, Savorani F, Avenoza A, Busto JH, Peregrina JM, Engelsen SB. Investigations of La Rioja terroir for wine production using 1H NMR metabolomics. Journal of Agricultural and Food Chemistry. 2012;**60**:3452-3461

[15] Schrimpe-Rutledge AC, Codreanu SG, Sherrod SD, McLean JA. Untargeted metabolomics strategies—Challenges and emerging directions. Journal of the American Society for Mass Spectrometry. 2016;**27**:1897-1905

[16] Patti GJ, Yanes O, Siuzdak G. Metabolomics: The apogee of the omics trilogy. Nature Reviews Molecular Cell Biology. 2012;**13**:263-269

[17] Dunn WB, Erban A, Weber RJM, Creek DJ, Brown M, Breitling R, et al. Mass appeal: Metabolite identification in mass spectrometry-focused untargeted metabolomics. Metabolomics. 2013;**9**:44-66

[18] Blaženović I, Kind T, Ji J, Fiehn O. Software tools and approaches for compound identification of LC-MS/MS data in metabolomics. Metabolites. 2018;**8**:31

[19] Chen Y, Wang H, Wang K, Zhu G, Yang Z, Wang M, et al. Serum metabolomic patterns in patients with aldosterone-producing adenoma. Frontiers in Molecular Biosciences. 2022;**9**:816469

[20] Ning D, Yuan M, Wu L, Zhang Y, Guo X, Zhou X, et al. A quantitative framework reveals ecological drivers of grassland microbial community assembly in response to warming. Nature Communications. 2020;**11**:4717

[21] Karkaria BD, Fedorec AJH, Barnes CP. Automated design of synthetic microbial communities. Nature Communications. 2021;**12**:672

[22] Hernandez DJ, David AS, Menges ES, Searcy CA, Afkhami ME. Environmental stress destabilizes microbial networks. The ISME Journal. 2021;**15**:1722-1734

[23] Sloan WT, Lunn M, Woodcock S, Head IM, Nee S, Curtis TP. Quantifying the roles of immigration and chance in shaping prokaryote community structure. Environmental Microbiology. 2006;**8**:732-740

[24] Lawson CE, Harcombe WR, Hatzenpichler R, Lindemann SR, Löffler FE, O'Malley MA, et al. Common principles and best practices for engineering microbiomes. Nature Reviews Microbiology. 2019;**17**:725-741

[25] Lopez-Toledano A, Mayen M, Merida J, Medina M. Yeasts used to delay browning in white wines. Food Chemistry. 2006;**97**:498-504

[26] Martínez C, Cosgaya P, Vásquez C, Gac S, Ganga A. High degree of correlation between molecular polymorphism and geographic origin of wine yeast strains. Journal of Applied Microbiology. 2007;**103**:2185-2195

[27] Naumenko ES, Latysheva VT, Gilinsky AM, Rogachev DA, Komarova IN, Salakhutdinov FN, et al. Cardioprotective effect of resveratrol and resveratroloside. Cardiovascular & Hematological Agents in Medicinal Chemistry. 2013;**11**:207-210

[28] Zhao X, Tao J, Zhang T, Jiang S, Wei W, Han H, et al. Resveratroloside alleviates postprandial hyperglycemia in diabetic mice by competitively inhibiting α-glucosidase. Journal of Agricultural and Food Chemistry. 2019;**67**:2886-2893

[29] Jin Z, Ma J, Zhu G, Zhang H. Discovery of novel anti-cryptosporidial activities from natural products by in vitro high-throughput phenotypic screening. Frontiers in Microbiology. 2019;**10**:1999

[30] Zhang H, Wang L, Tan Y, Wang H, Yang F, Chen L, et al. Effect of Pichia on shaping the fermentation microbial community of sauce-flavor baijiu. International Journal of Food Microbiology. 2021;**336**:108898

[31] Hollá V, Karkeszová K, Antošová M, Polakovič M. Transglycosylation properties of a Kluyveromyces lactis enzyme preparation: Production of tyrosol β-fructoside using free and immobilized enzyme. Process Biochemistry. 2021;**110**:168-175

[32] Barata A, Malfeito-Ferreira M, Loureiro V. The microbial ecology of wine grape berries. International Journal of Food Microbiology. 2012;**153**:243-259

[33] Zhang L, Guo K, Wang L, Xu R, Lu D, Zhou Y. Effect of sludge retention time on microbial succession and assembly in thermal hydrolysis pretreated sludge digesters: Deterministic versus stochastic processes. Water Research. 2022;**209**:117900

[34] Raimundo e Silva JP, Policarpo IdS, Chaves TP, Coutinho HDM, Alves HdS. A glycosylated β-sitosterol, isolated from Tacinga inamoena (Cactaceae), enhances the antibacterial activity of conventional antibiotics. South African Journal of Botany. 2020;**133**:193-200

[35] Hou QZ, Chen DW, Wang YP, Ehmet N, Ma J, Sun K. Analysis of endophyte diversity of Gentiana officinalis among different tissue types and ages and their association with four medicinal secondary metabolites. PeerJ. 2022;**10**:13949

[36] Osburn ED, Aylward FO, Barrett JE. Historical land use has long-term effects on microbial community assembly processes in forest soils. ISME Communications. 2021;**1**:48

[37] Horvath BO, Sardy DN, Kellner N, Magyar I. Effects of high sugar content on fermentation dynamics and some metabolites of wine-related yeast species *Saccharomyces cerevisiae*, *S. uvarum* and *Starmerella bacillaris*. Food Technology and Biotechnology. 2020;**58**:76-83

[38] Li CS, Xu Y, Li LH, Yang XQ, Wang YQ. Acid stress induces cross-protection for cadmium tolerance of multi-stress-tolerant *Pichia kudriavzevii* by regulating cadmium transport and antioxidant defense system. Journal of Hazardous Materials. 2019;**366**:151-159

[39] Liu GL, Wang K, Hua MX, Buzdar MA, Chi ZM. Purification and characterization of the cold-active killer toxin from the psychrotolerant yeast *Mrakia frigida* isolated from sea sediments in Antarctica. Process Biochemistry. 2012;**47**:822-827

[40] Yang CY, Hao RJ, Du XD, Wang QH, Deng YW, Sun RJ, et al. GC-TOF/MS-based metabolomics studies on the effect of protein sources in formulated diet for pearl oyster *Pinctada fucata martensii*. Aquaculture. 2018;**486**:139-147

Chapter 9

Update on the Role of *Saccharomyces cerevisiae* in Sherry Wines

Gustavo Cordero-Bueso, Marina Ruiz-Muñoz, Antonio Florido-Barba and Jesús Manuel Cantoral Fernández

Abstract

Sherry wines undergo a complex, two-stage production process. Initially, the *Palomino Fino* grape must undergo alcoholic fermentation, resulting in the base wine. This wine is fortified and enters the dynamic biological aging system known as "criaderas y soleras." Despite the wide variety of wine yeasts available, there's growing interest in developing new yeast strains with specific traits to enhance wine quality, safety, and consumer acceptance. Rising temperatures are expected to impact alcoholic fermentation stability and flor yeast film development during biological aging, potentially reducing wine quality. This chapter explores oenological advancements, such as reducing hydrogen sulfide and ethyl carbamate concentrations in Jerez's base wines. Non-genetic modification techniques that enhance sensory complexity in industrial-scale winemaking are discussed. Additionally, a diverse range of yeasts, including *Saccharomyces cerevisiae* species with novel phenotypic traits, is found during biological aging, offering potential value in winemaking and biotechnology. The presence of mycoviruses in flor yeasts of the *Saccharomyces* genus, providing evolutionary advantages in dominance and establishment in "Fino" and "Manzanilla" wines, is examined. The chapter also delves into how these yeasts affect flor yeast film stability under varying temperatures and ethanol conditions, and alternative methods for veil of regeneration using amino acids as nitrogen sources or inert supports are explored.

Keywords: veil of flor, sherry base wine, evolved yeasts, biotechnology, sensory, mycoviruses, fortification

1. Introduction

The practice of biological aging of wines is a traditional method observed in various regions worldwide. Examples of such regions include Jura in France, Szamorodni and Aszú in Tokaj-Hegyalja, Hungary, as well as Vernaccia di Oristano in Sardinia, Italy. However, the most renowned biologically aged wines are notably crafted in the Jerez-Xèrés-Sherry and Manzanilla de Sanlúcar de Barrameda Designation of Origin (D.O.) located in southern Spain [1, 2].

Jerez-Xèrés-Sherry wines undoubtedly represent one of the most acclaimed and globally recognized Spanish national products. This recognition is attributed not only to the unique geographical region but also to the winemaking practices associated with their production, preserved for generations and esteemed to the extent of being considered an international oenological reference. Thus, the wines of the Jerez region encompass a wide range of wines with unquestionable personality and authenticity [3].

In October 2022, the Consejo Regulador of the Denominations of Origin, a public law corporation, implemented changes and modifications. The recent agreement specifies that Fino can only be produced in Jerez de la Frontera or El Puerto de Santa Maria. Sherry wines made in Sanlúcar can exclusively be labeled as Manzanilla. Another significant change in the regulations is that sherry wines can no longer be categorized as fortified wines if they reach a minimum alcohol content of 15% (v/v ethanol) without the addition of alcohol. These modifications also involve the utilization of several outdated local pre-phylloxera grape varieties for sherry production, notably the ancient varieties *Mantúo Castellano*, *Mantúo de Pilas*, *Vejeriego*, *Perruno*, *Cañocazo*, and *Beba*. These adjustments and considerations offer winemakers the opportunity to prepare for the future while innovating within the longstanding tradition of sherry production [4].

The production of Sherry and "Manzanilla" wines involves a two-step process. Initially, they must undergo alcoholic fermentation by commercial or naturally occurring yeasts on the surface of *Vitis vinifera* var. Palomino Fino grapes, resulting in the production of a "young" wine. Young base wines are crafted from grapes cultivated in nine different regions, including Jerez de la Frontera, Sanlúcar de Barrameda, and El Puerto de Santa Maria. Additionally, wines are produced in the wine cellars of Trebujena, Chipiona, Chiclana, Rota, Puerto Real, and Lebrija, which are located in Seville [1, 2, 4].

Subsequently, this base wine undergoes biological aging, facilitated by other specific yeasts known as flor yeasts. Biological aging comprises two distinct phases. The first phase, known as 'añadas' or static aging, involves storing the wine in a barrel (known as "bota") for several years. This static phase is followed by a dynamic phase called "criaderas-solera" or "soleraje," which consists of a series of oak barrels containing sherry in various stages of maturation (**Figure 1**).

The Jerez barrels, painted in matte black for easy identification of leaks and with a capacity of 600 liters (equivalent to 36 arrobas), are the most common in the wineries of the Sherry region, although there are also different sizes such as the small butt (500 liters), the half butt (250 liters), or the hogshead (700 liters), among others. In the past, Sherry wineries used different types of wood, including cherry, chestnut, and oak. However, it was oak that turned out to be the best choice for making, storing, and transporting wine. This preference for oak was partly due to the thriving trade with the New World, where oak was readily available. Ships that carried various goods, including wine in barrels (which was essential for trade and as a food source for the crews), often brought back oak wood from the Americas on their return voyages [5]. Today, Jerez wines are primarily aged in American white oak barrels, although some are also aged in barrels made from Spanish or French oak (**Figure 2**).

The initial stage, referred to as "sobretablas," occurs when the wine is young and contains approximately 15.5% (v/v) ethanol. The intermediate stages are called

Figure 1.
The solera and criadera system operates in a way that provides several advantages to the oldest wines. These wines benefit from regular refreshments (known as "Saca-Rocío") with younger wines. Additionally, they acquire unique characteristics over years of aging. (photo of Manuel Fernández Barcell).

"criaderas," while the final stage, known as "solera," contains the oldest wine. From the solera, the finished wine is withdrawn. However, the 600 L barrels are only partially emptied during each withdrawal, never exceeding one-third of their contents. The transfer of wine from one oak barrel to the next is termed "rocío." This process typically occurs twice a year. A common characteristic of these barrels, in the case of Fino and Manzanilla wines, is the presence of the veil of flor, approximately 1–3 cm thick, formed by yeasts and other microorganisms that grow on the free surface of the wine (**Figure 3**).

Figure 2.
Material and parts of a barrel through the vocabulary of the coopers of Jerez, 1 and 2 – "Duelas", 3 – "Calzo", 4 – "Bocacha", 5 – "Aspilla", 6 – "Venencia", 7 – "Venencia de caña", 8 – "Falsete", 9 – "tapón", 10 – "Martillo", 11 – "Mazo", 12 – "Chazo", 13 – "Bigonia", 14 – "Espolines", 15 – "Tranquilla", 16 – "Vara de mover", 17 – "Cesta de prueba", 18 – "Coqueteador", 19 – "Media caña", 20 – "Cepillo Para bocoy", 21 – "Galafate de empajar", 22 – "Media cuchilla", 23 – "Suela de carpintero", 24 – "Botas de pisar". Name of the staves; a – "Mediano", B – "Luengos", C – "Jareles", D – "Chanteles". Bottom classes; "aventado" = concave, "apandado" = convex, "normal" = flat. Hoops ("flejes") names; E – "Talugo de coronar", F – "Talugo de amasar", G – "Colete", H - "aro de en medio", I – "Aro de bojo" (mural exhibited at "Ligures" winery, mesas de Asta, Cádiz. Photo of Manuel Fernández Barcell).

2. Yeast diversity in sherry wines during biological aging

The predominant yeast species responsible for the film formation in Sherry wine aging is *Saccharomyces cerevisiae*. Once the study of flor yeasts began a few 100 years ago, it was observed that these yeasts were different from the rest of the wine yeasts belonging to the same species. Indeed, they have been able to adapt to a really hostile environment, such as the absence of easily fermentable sugars and the presence of high concentrations of ethanol and other toxic compounds such as glycerol, acetaldehyde, or acetoin [6].

These yeast strains are able to carry out an oxidative metabolism, being in contact with oxygen. This biofilm formation required genetic divergence from the *S. cerevisiae* fermentative strains, with differential expression of genes related to the regulation of sugar metabolism and tolerance to osmotic stress, crucial mechanisms in this particular winemaking context [7–9].

It is important to highlight the intrinsic genomic diversity of *S. cerevisiae* flor strains. Recent studies have revealed a remarkable genetic heterogeneity in this subspecies, evidencing a rich pool of genetic variability that contributes to their versatility and adaptability. In fact, it has been demonstrated that these yeasts have

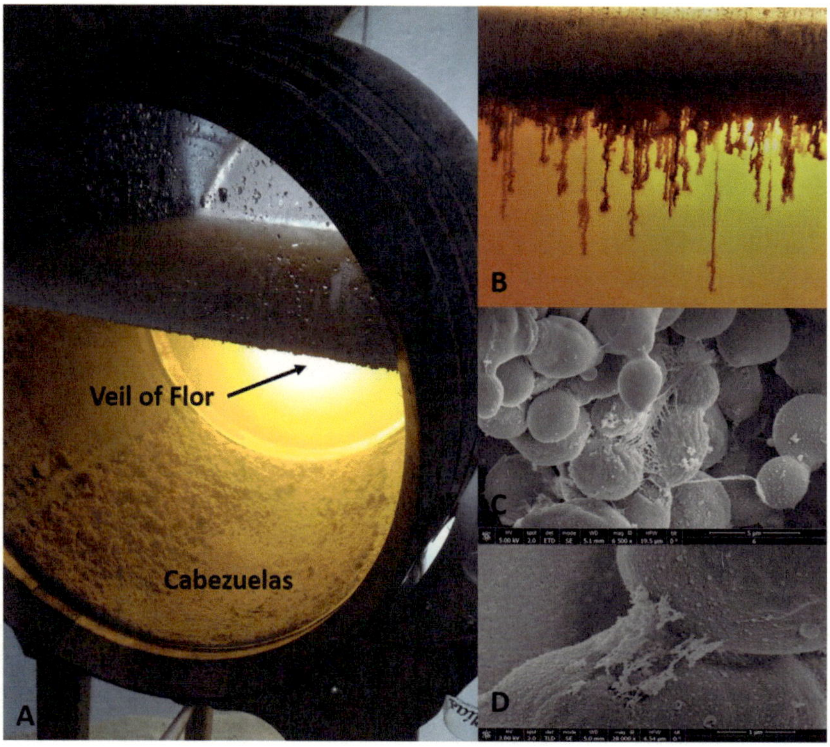

Figure 3.
(A) the "flor veil" refers to a thin layer of yeast that forms naturally on the surface of Sherry wines during aging, creating a protective film against oxidation that contributes to the wine's unique characteristics and flavors. "Cabezuelas" is yeast autolysis and its accumulation in the bottom of the barrel during the aging process. It seems to provide essential micronutrients to yeast forming the veil of flor. (B) Detail of a well-developed veil of flor. (photos of Jesús Manuel Cantoral). (C) Scanning electron microscopy (SEM) capture of the yeasts forming the veil of flor within a glycoprotein matrix. (D) Detail of the adherence of veil of flor yeasts by SEM.

a common phylogenetic origin that is distinct from the rest of the wine *S. cerevisiae* yeast strains [10–12].

Specific alleles and genetic variants have been identified in genes associated with the synthesis of key enzymes in biological aging, as well as in those related to the response to cellular stress induced by external factors, such as high ethanol concentration and nutrient limitation. For instance, a 24 bp deletion in the ITS1 region has been found in flor strains isolated in Spain and Italy, or a "C" insertion in the same region in these isolated in France [13–15].

The *FLO11* gene in *S. cerevisiae* flor yeasts plays a key role in the formation and maintenance of the flor veil, a distinctive feature of this group of yeasts. This gene codes for a surface adhesion protein that allows yeasts to bind and subsequently form a matrix that floats on the surface of the wine during biological aging. This matrix acts as a protective barrier that isolates the yeasts from the liquid medium and air, allowing them to survive in adverse conditions while maintaining their proper activity. Flor yeast strains are known for a distinct 111 bp deletion in the promoter region of the *FLO11* gene, a crucial adhesion gene responsible for cell aggregation and film formation. This deletion is common among sherry strains from Spain, France, Italy, and Hungary. Its role is to deactivate long noncoding RNAs, specifically ICR1, which represses FLO11 transcription. Disabling this negative regulator, ICR1, stimulates the production of Flo11 protein in flor yeast. The FLO11 coding sequences in *S. cerevisiae* can vary from 3 to 6 kb, primarily due to differences in the number of repeats in the central domain. Initially, it was believed that Sherry yeast strains encoded longer FLO11 variants with higher hydrophobicity. However, subsequent research revealed that the central domain of FLO11 is notably unstable under non-selective conditions, therefore it seems that one of the factors for the emergence of *S. cerevisiae* flor yeast may be due to the presence of other repeats, which in turn influence the level of expression of the *FLO11* gene [7, 16–18]. FLO11 sequence is registered in the Yeast Genome Database at Stanford University Medical School (www.yeastgenome.org). FLO11 differs from all other known cell surface flocculins in that it is located near a centromere rather than near a telomere, and its expression is regulated by mating type.

Currently, researchers employ comparative genomic and transcriptomic to unravel the molecular mechanisms behind the transition of an *S. cerevisiae* strain into a flor yeast. This approach helps identify changes in gene expression associated with unique cell adhesion, stress resistance, iron uptake, nitrogen and carbon metabolism, lipid metabolism, and the production of aromatic compounds. In this sense, other genes that have been found to be directly related to the success of S. *cerevisiae* flor yeasts include *BTN2*, *HSP12*, *YAP1*, *ACC1*, *FRE2*, *MCH2*, and *YKL222C* [11, 19, 20]. However, more research is still needed to better understand the modifications of these genes and others that still remain unexplored involved in this process to better understand the mechanism of adaptation and/or domestication of this particular yeast strains to this specific niche. All of these factors are pivotal in the distinct function of veil-forming yeast. These modern techniques are getting insights into the yeast's adaptation mechanism and enable the development of genetic markers for selecting suitable strains for sherry production [3, 21].

Traditionally, flor yeast strains have been classified based on their ability to metabolize different sugars, categorizing them into four varieties or races within the species *Saccharomyces cerevisiae*. However, this classification is considered outdated and even simplistic, as significant differences have been found in wines produced in neighboring wine casks from the same base wine [3]. Additionally, it is worth noting that traditionally, in the wine industry, as a whole and particularly in the biological aging of Sherry wines, yeast strains belonging to different species of *S. cerevisiae* capable of forming biofilm were considered undesirable and even contaminants. However, interest in these yeasts has been growing exponentially due to their potentially valuable phenotypic characteristics.

In a recent study on the diversity of the veil of flor in the Jerez region, four different species (*Wickerhamomyces anomalus*, *Pichia membranaefaciens*, *Pichia manshurica,* and *Pichia kudriavzevii*) were found in addition to *S. cerevisiae* flor strains during the biological aging of sherry wines (**Table 1**). Although they were isolated sporadically from some of the casks analyzed, their origin remains unclear [27, 28]. However, their phenotypic characteristics, such as a good capacity to aerobically assimilate ethanol, glycerol, or urea, together with their ability to form biofilm or the formation of an extracellular matrix, suggesting that they must necessarily be adapted to the specific and difficult environment of sherry wines while maintaining an active growth rate. A study carried out in the Montilla-Moriles region (Cordoba, Spain) has also confirmed the presence of some of these non-*Saccharomyces* species, as well as others not previously described in this system, such as *Candida guillermondii* and *Trichosporon ashaii*, using regrowth technique to increase their concentration prior to their identification, which was carried out using Next Generation Sequencing (NGS) techniques [12].

Species	Number of strains	Region	Substrate of isolation	Features	Reference
Saccharomyces cerevisiae: var. *beticus*	B2, B3, B16	Jerez-Xèrés-Sherry (Jerez)	Fino Sherry wine	15,5% etOH resistance, Killer resistance	[22, 23]
S. cerevisiae: var. *montuliensis*	M9, M10, M12, M17	Jerez-Xèrés-Sherry (Jerez)	Fino Sherry wine	15,5% etOH resistance, Killer resistance	[22, 23]
S. cerevisiae: var. *cheresiensis*	CH16	Jerez-Xèrés-Sherry (Jerez)	Fino Sherry wine	15,5% etOH resistance	[22, 23]
S. cerevisiae: var. *rouxii*	R13	Jerez-Xèrés-Sherry (Jerez)	Fino Sherry wine	15,5% etOH resistance	[22, 23]
S. cerevisiae: var. *beticus, montuliensis, cheresiensis, rouxii*	MY138, My91, CU2, ET7	Jerez-Xèrés-Sherry (El Puerto de Santa María)	Fino Sherry wine	15,5% etOH resistance	[7, 8]
S. cerevisiae: var. *beticus, cheresiensis*	III, A2, H2, R2, R3	Jerez-Xèrés-Sherry (Jerez)	Fino Sherry wine	15,5% etOH resistance	[24]
Pichia ssp.	n.d.	Jerez-Xèrés-Sherry (Jerez)	Fino Sherry wine	15,5% etOH resistance	[24]
S. cerevisiae (beticus, montuliensis, cheresiensis, rouxii)		Jerez-Xèrés-Sherry (Jerez)	Fino Sherry wine	15,5% etOH resistance	[13, 14]
Pichia membranaefaciens	n.d.	Jerez-Xèrés-Sherry (Jerez)	Sobretablas Fino Sherry wine	15,5% etOH resistance	[13]
P. anomala	n.d.	Jerez-Xèrés-Sherry (Jerez)	Sobretablas Fino Sherry wine	15,5% etOH resistance	[13]
C. cantarelli	n.d.	Jerez-Xèrés-Sherry (Jerez)	Fino Sherry wine	15,5% etOH resistance	[13]
D. bruxellensis	n.d.	Jerez-Xèrés-Sherry (Jerez)	Fino Sherry wine	15,5% etOH resistance	[13]
S. cerevisiae (beticus, montuliensis, cheresiensis, rouxii)	B10, BS13, BS24, CH7, CH24, CHS15, B17	Jerez-Xèrés-Sherry (Sanlúcar de Barrameda)	Manzanilla Sherry wine	15,5% etOH resistance	[25]
S. cerevisiae	FI, FII, FIII, FIV, FV, FVI, FIX, FX, MI, MII, MIII, MIV, MV, MVI and MX,	Jerez-Xèrés-Sherry (Jerez and Sanlúcar de Barrameda)	Fino and Manzanilla Sherry wines	15,5% etOH resistance	[26]
S. cerevisiae	A-E	Montilla-Moriles D.O. (Córdoba, Spain)	Fino of Montilla-Moriles	15,5% etOH resistance	[18]
S. cerevisiae (Formerly *S. cerevisiae* var. *beticus*)	ScA, ScF, ScH	Jerez-Xèrés-Sherry (Jerez and El Puerto de Santa María)	Fino Sherry wine	15,5% etOH resistance	[3]

Species	Number of strains	Region	Substrate of isolation	Features	Reference
S. cerevisiae (Formerly *S. cerevisiae* var. *cheresiensis*)	ScC, ScG	Jerez-Xèrés-Sherry (Jerez and El Puerto de Santa María)	Fino Sherry wine	15,5% etOH resistance	[3, 12, 27]
P. membranaefaciens	NsA,	Jerez-Xèrés-Sherry (Jerez and El Puerto de Santa María); Montilla-Moriles (Córdoba)	Fino Sherry wine	15,5% etOH resistance	[3, 12, 27]
Wickerhamomyces anomalus (formerly Pichia anomala)	NsB	Jerez-Xèrés-Sherry (Jerez and El Puerto de Santa María); Montilla-Moriles (Córdoba)	Fino Sherry wine	15,5% etOH resistance	[3, 12, 27]
P. kudriavzevii	NsC	Jerez-Xèrés-Sherry (El Puerto de Santa María); Montilla-Moriles (Córdoba)	Fino Sherry wine	15,5% etOH resistance	[3, 12, 27]
P. manshurica	NsD	Jerez-Xèrés-Sherry (El Puerto de Santa María)	Fino Sherry wine	15,5% etOH resistance	[3, 12, 27]
Candida guilliermondii	n.d.	Montilla-Moriles (Córdoba)	Fino wine	15,5% etOH resistance	[12]
Trichosporon ashaii	n.d	Montilla-Moriles (Córdoba)	Fino wine	15,5% etOH resistance	[12]

Table 1.
Yeasts are found in Sherry wines in the Andalusian regions and their main characteristics. Includes saccharomyces and non-saccharomyces.

The same research group has recently described the presence of 26 other previously undetected microorganisms in the system, hinting at the possibility that their presence may be due to the mites and flies present in the cellars. Regardless of their origin, and despite the growing number of studies focusing on the identification of various non-*Saccharomyces* yeast species and strains, the potential of many of these species remains largely undiscovered, and they could be of great interest not only in the context of biological aging, but also from a wider biotechnological point of view [12].

3. Other microorganisms distinct from yeasts in the veil of flor

During the biological aging process of Sherry wines, various other microorganisms can coexist with the yeast strains responsible for veil formation. These include lactic acid bacteria from the genera *Leuconostoc*, *Pediococcus*, and *Lactobacillus*, as well as opportunistic fungi like *Botrytis cinerea*, *Penicillium chrysogenum,* and fungi of the genus *Rhizopus*. The presence of lactic acid bacteria, such as *Lactobacillus hilgardii*, *Lactobacillus plantarum*, and *Lactobacillus brevis* in Sherry veils of Flor, is closely associated with the high levels of gluconic acid produced by the filamentous fungi *B. cinerea* [1, 29, 30].

The species *S. cerevisiae* and related yeasts are susceptible to various intracellular nucleic acid infectious agents, including dsRNA viruses, ssRNA viruses, LTR retrotransposons, and bicatenary DNA plasmids. These infectious agents share certain similarities with higher eukaryotic viruses. For example, yeast retrotransposons exhibit a life cycle akin to retroviruses, the structure of yeast Totiviruses mirrors the capsid of reoviruses, and yeast plasmid segregation resembles viral episome segregation strategies (**Figure 4**).

Modern experimental tools available for studying the genetics, cell biology, and evolution of *S. cerevisiae* and other yeasts are well-suited for expanding our knowledge of how these intracellular agents exploit cellular processes. Importantly, it remains unclear whether yeast-related issues in alcoholic fermentation or premature weakening of the flor veil in wines could potentially be triggered by viral infections, representing a significant area for further investigation. A recent study conducted in our laboratory revealed that flor yeasts infected by Mycoviruses develop a killer factor that facilitates their dominance in the formation of the biofilm (data not shown). This could provide an evolutionary advantage, particularly in environments with high ethanol concentrations.

The presence of this microbiota, distinct from yeast, can lead to the development of undesirable wines and an accumulation of high concentrations of biogenic amines. However, the proliferation of some yeast species like *S. cerevisiae*, can also potentially introduce sensory deviations and negatively impact the wine's quality.

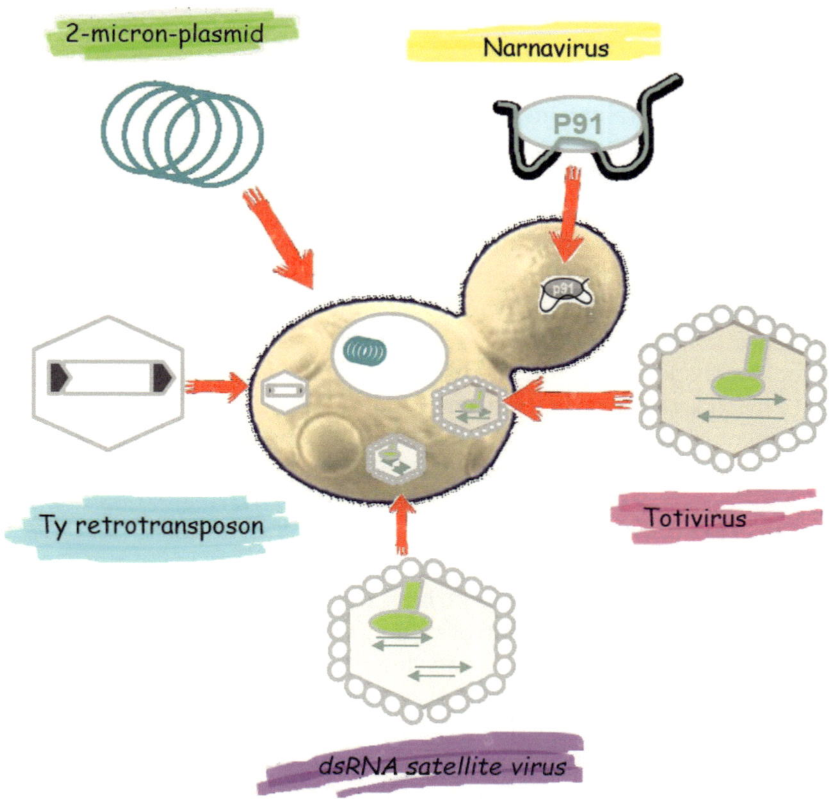

Figure 4.
Intracellular nucleic acid infectious agents described in the species S. Cereviae, including dsRNA viruses, ssRNA viruses, LTR retrotransposons, and bicatenary DNA plasmids.

4. Off-flavors and carcinogenic agents in sherry base wines; solutions to eliminate them from a microbiological and biotechnological perspective

Current studies showed that the base wine plays a more substantial role in Sherry wine production than previously believed. This suggests the potential for substantial enhancements in yeast strains that have already been chosen and are used in industrial processes. These improvements can be achieved without the need for specific genetic modifications; instead, they leverage the genetic diversity present within the yeast strain population. This way, the base wine (i.e., young white wine) used for the production of various Sherry wines can undergo not only sensory improvements but also enhance food safety, which will substantially impact the resulting young white wine and, consequently, the final product [21, 31].

Hydrogen sulfide is a metabolite associated with an aromatic defect in wines (rotten eggs) and has a very low detection threshold. Consequently, its reduction has been extensively studied in yeast but is not yet employed on an industrial scale. On the other hand, urea spontaneously reacts with ethanol to form ethyl carbamate, a genotoxic substance. The generation of this metabolite is highly favored by rising temperatures and high ethanol concentrations, so its presence in biologically aged wines is expected to increase as global warming raises temperatures [21].

In our laboratory, we conducted a study with the aim of reducing the production of hydrogen sulfide (H2S), a compound associated with an undesirable aroma defect. Two distinct improvement strategies were employed. Firstly, a mass mating technique was used, involving the hybridization of the target strain with another parental strain known for its low H2S production. Hybrid strains were selected based on complementary auxotrophic markers (inability to grow on melezitose and galactose, respectively) exhibited by both parental strains. Additionally, a distinct transposon amplification profile allowed the identification of putative hybrids with profiles intermediate between the two parental strains [21].

Furthermore, adaptive laboratory evolution was employed using ammonium molybdate as a selective pressure, an analog toxic to sulfate. Before applying the selective pressure, the yeast strain was sporulated, ascospores were dissected, and mating was allowed to increase the genetic variability within the population.

In both cases, rapid and qualitative screening of variants was made possible through the use of specific culture media (BiGGY). In total, 10 yeast variants were obtained (2 hybrids and 8 variants through adaptive evolution) that met basic oenological parameters and exhibited phenotypic characteristics similar to the parental yeast. Following pilot-scale fermentations in both synthetic and natural grape musts, three candidate variants were selected for industrial-level fermentations: two hybrids and one variant obtained through adaptive evolution.

Once produced at an industrial scale, the wines underwent analytical and sensory evaluation. Wines produced using the variants obtained through hybridization exhibited greater differences from the control strain in terms of terpenes and ester production, as expected, while the evolved strain closely resembled the parental strain. As a result, not only were wine defects reduced at an industrial level, but the wines were also noted for their enhanced floral and fruity characteristics.

On another note, efforts were made to reduce urea excretion during alcoholic fermentation, as urea is the main precursor of ethyl carbamate. For this, only the adaptive evolution technique was employed in the laboratory, similar to the first study, using L-canavanine as a selective pressure, a toxic analog of l-arginine [31].

Following screening in the laboratory, a variant was selected for industrial wine production, resulting in reduced urea production and improved fermentative capacity of the yeast strain. The wines also exhibited significant differences in both chemical composition and sensory analysis, once again being favorably rated by tasters compared to those produced with the parental strain [31].

5. Mitigating the consequences of climate change on the veil of flor formation

On the other hand, biological aging is a highly complex and delicate process. The stability of the yeast of flor biofilm depends on various factors, with the environmental conditions of the specific winery being among the most defining.

Studying yeast populations that make up the flor film using molecular techniques and analyzing their behavior during aging can provide valuable insights to better understand the process. Research is crucial for developing strategies to maintain the identity of these wines despite viticulture changes due to the effects of climate change. Additionally, it is assumed that the yeasts present in this biological aging system, whether or not they belong to the *S. cerevisiae* species, can be very interesting from a biotechnological perspective, as they have had to adapt to very adverse conditions to remain metabolically active.

The effects associated with climate change, such as rising temperatures, increased carbon dioxide levels, and a significant reduction in rainfall, are directly impacting the wine industry as a whole, with a particular focus on the Jerez region. As a consequence of rising temperatures, excessive grape ripening occurs, leading to two direct effects on the must's composition: increased sugar concentration and reduced easily assimilable nitrogen content. This imbalance can cause deviations during the base wine fermentation, even when using commercial or selected yeast strains, resulting in an increase in undesirable compounds in the wine, such as hydrogen sulfide or urea, as stated before.

Furthermore, it is known that the stability of the veil of flor largely depends on the wine's ethanol content and cellar climatic conditions, including temperature and relative humidity. Therefore, the effects of climate change directly impact the biological aging process, progressively weakening the biofilm's structure, composition, and thus its activity. This can lead to the disappearance of the flor film from the wine's surface, causing not only oxidation but also a noticeable sensory deviation in biologically aged wines.

Despite the existence of classic methods for regenerating the flor veil through direct grafts from a barrel with a healthy biofilm, it has been observed that the time it takes for this veil to re-form is often lengthy. As a consequence, there is some oxidation of wines undergoing biological aging, leading to a loss in their quality. Therefore, we aimed to apply new techniques that would contribute to the development and implementation of targeted flor veils.

We optimized a culture medium based on ethanol as the primary carbon source, allowing for the fastest possible biofilm formation. Once we obtained the culture medium that enabled faster flor veil formation, we developed a support printed with a 3D printer using polylactic acid (PLA), a biodegradable material relatively stable in acidic environments and with high ethanol concentrations. This latter prototype has resulted in the publication of a patent (ref. P202130692).

6. Searching for an alternative to wine ethanol in wine fortification

One of the key characteristics of Sherry wines is their classification within the category of fortified or "encabezados" wines. For their production, it is essential to use alcohol to increase their alcohol content above 15% vol. The relevant European legislation (Regulation [EU] 2019/934 of March 12, 2019) and the Specification of the Designation of Origin (D.O.) restrict the type of alcohol allowed for this category of products, exclusively authorizing alcohol obtained through the distillation of wines.

For the fortification, topping, or enrichment of wines protected by the D.O. Jerez-Xérès-Sherry, wine alcohol with an alcohol content ranging between 95.0 and 96.0% (v/v) is used, as stipulated by the relevant legislation. However, the impact of using alcohols derived from raw materials other than wine alcohol on the microbial populations present in wines aged under biological aging conditions (Finos) is still unknown, as well as its effect on the generation of volatile components and its influence on the sensory profile of these wines. Therefore, we proposed an alternative approach for fortifying wines using alcohols obtained from malt, grape pomace, agave, sugarcane, and cereal, as well as wine alcohols obtained through various distillation techniques. The results may reveal alternatives to the current use of grape-derived alcohol, potentially impacting both product quality and cost-effectiveness (results not shown).

Ongoing experiments at winery scale regarding the use of non-grape-based alcohols at different alcohol levels than currently authorized for fortifying base wines represent a variable that could potentially optimize the biological aging process of Fino wines. Initial results reveal significant differences in the speed of development and morphology of *S. cerevisiae* and *P. kudriavzevii* yeast biofilms for certain tests, as well as variations in the volatile composition of the aging wines [32].

It is also expected that the sensory and compositional contributions of the different alcohols will impact the organoleptic profile of Fino wines. Additionally, these alcohols provide alternative carbon sources to the flor yeast, which could be of great importance if the yeast assimilates these nutrients through metabolic pathways different from the usual ones. This could result in the formation of compounds that add distinctiveness and quality nuances to the wine produced through biological aging.

7. Conclusions

The production of biological-aged Sherry wines is a distinctive process that relies on the veil of Flor. The maintenance and storage of these cultures play a crucial role in this process, leading to the development of unique low-molecular compounds responsible for the distinct flavor and aroma of sherry wines.

Recently, numerous countries worldwide have begun adopting microbiological and food technology to produce Sherry-like wines. In our previous research, we endeavored to create a screening method for identifying promising veil-forming yeast strains to craft Sherry-like wines.

This chapter has shown that the veil of flor requires careful consideration of essential winemaking factors and contributes to our comprehension of the evolution of flor yeast. Furthermore, it aids in enhancing sherry production technology by integrating new microbial compositions into industrial strains.

On the other hand, biotechnological methods for yeast improvement focus on enhancing genetic diversity naturally through sexual reproduction. Directed evolution, sometimes combined with mutagenic agents or sexual hybridization, increases genetic and phenotypic variation for selection. Laboratory *S. cerevisiae* strains are usually preferred due to their haploid nature and better understanding. However, applying these techniques to industrial strains is challenging due to variable characteristics like sporulation efficiency and spore viability, common in the laboratory but not in industrial settings. Industrial strains are often polyploidy, lack suitable markers, etc.

In the context of climatic change and biological aging as well, the PLA-printed prototype has proven to be a powerful scaffold for veil of flor yeast at the laboratory scale. This allows for grafting to occur more quickly than free yeast cells can achieve. Nevertheless, further investigation is required to assess the stability of these scaffolds in acidic environments and under increased temperatures, among other potential factors they may encounter.

Flor yeast strains remain relatively understudied even today. Therefore, the aim of this chapter has been to provide an updated overview of these yeast strains as of the year 2023. We hope that this chapter will encourage further research and investigations into these fascinating microorganisms, particularly within the context of *Saccharomyces cerevisiae*.

Acknowledgements

We greatly appreciate the cooperation of Bodegas Fundador in providing samples for this study. We also thank Bodegas Ligures (Mesas de Asta) and Francisco Núñez for giving us access to the mural of the parts of the barrel, Jesús Pérez Chicón for his rewarding help with Mycoviruses experiments, and Manuel Fernández Barcell and Jesús Cantoral for their photos.

Conflict of interest

"The authors declare no conflict of interest."

Author details

Gustavo Cordero-Bueso[1*], Marina Ruiz-Muñoz[1], Antonio Florido-Barba[1,2] and Jesús Manuel Cantoral Fernández[1]

1 Department of Biomedicine, Biotechnology and Public Health, Campus Río San Pedro s/n, University of Cádiz, Cádiz, Spain

2 Department of Enology, Bodegas Fundador S.L.U, Jerez de la Frontera (Cádiz), Spain

*Address all correspondence to: gustavo.cordero@uca.es

References

[1] Cordero-Bueso G, Ruiz-Muñoz M, González-Moreno M, Chirino S, del Bernal-Grande MC, Cantoral JM. The microbial diversity of sherry wines. Fermentation. 2018;**4**(1):19

[2] Avdanina D, Zghun A. Sherry wines: Worldwide production, chemical composition and screening conception for flor yeasts. Fermentation. 2022;**8**(8):381

[3] Ruiz-Muñoz M, Cordero-Bueso G, Benítez-Trujillo F, Martínez S, Pérez F, Cantoral JM. Rethinking about flor yeast diversity and its dynamic in the "criaderas and soleras" biological aging system. Food Microbiology. 2020;**92**(May):103553

[4] Pliego de condiciones de la denominación de origen protegida jerez-xérès-sherry. Orden de 4 de octubre de 2022 (BOJA núm. 195 de 10/10/2022). Available from: https://www.sherry.wine/documents/274/Pliego_Jerez_normal_consolidado.pdf

[5] Sánchez CS. El Vino de Jerez en los 80 Años de la Denominación de Origen, 1935-2015: actas del Congreso científico. Jerez de la Frontera: Consejo Regulador de las Denominaciones de Origen Jerez-Xérès-Sherry; 2016

[6] Durán-Guerrero E, Castro R, de García-Moreno MV, del Rodríguez-Dodero MC, Schwarz M, Guillén-Sánchez D. Aroma of sherry products: A review. Foods. 2021;**10**(4):753

[7] Ibeas JI, Jimenez J. Mitochondrial DNA loss caused by ethanol in saccharomyces flor yeasts. Applied and Environmental Microbiology. 1997;**63**(1):7-12

[8] Ibeas JI, Lozano I, Perdigones F, Jimenez J. Dynamics of flor yeast populations during the biological aging of sherry wines. American Journal of Enology and Viticulture. 1997;**48**(1):75-79

[9] Ángeles Pozo-Bayón M, Victoria M-AM. Sherry wines. Advances in Food and Nutrition Research. 2011;**63**:17-40

[10] Legras JL, Moreno-Garcia J, Zara S, Zara G, Garcia-Martinez T, Mauricio JC, et al. Flor yeast: New perspectives beyond wine aging. Frontiers in Microbiology. 2016;7(APR):1-11

[11] Coi AL, Bigey F, Mallet S, Marsit S, Zara G, Gladieux P, et al. Genomic signatures of adaptation to wine biological ageing conditions in biofilm-forming flor yeasts. Molecular Ecology. 2017;**26**(7):2150-2166

[12] Carbonero-Pacheco J, Moreno-García J, Moreno J, García-Martínez T, Mauricio JC. Revealing the yeast diversity of the flor biofilm microbiota in sherry wines through internal transcribed spacer-metabarcoding and matrix-assisted laser desorption/ionization time of flight mass spectrometry. Frontiers in Microbiology. 2022;**12**:825756

[13] Esteve-Zarzoso B, Peris-Torán MJ, García-Maiquez E, Uruburu F, Querol A. Yeast population dynamics during the fermentation and biological aging of sherry wines. Applied and Environmental Microbiology. 2001;**67**(5):2056-2061

[14] Esteve-Zarzoso B, Fernández-Espinar MT, Querol A. Authentication and identification of *Saccharomyces cerevisiae* flor yeast races involved in sherry ageing. Antonie van

Leeuwenhoek, International Journal of General and Molecular Microbiology. 2004;**85**(2):151-158

[15] Esteve-Zarzoso B, Belloch C, Uruburu F, Querol A. Identification of yeasts by RFLP analysis of the 5.8S rRNA gene and the two ribosomal internal transcribed spacers. International Journal of Systematic and Evolutionary Microbiology. 1999;**49**(1):329-337

[16] Zara G, Goffrini P, Lodi T, Zara S, Mannazzu I, Budroni M. FLO11 expression and lipid biosynthesis are required for air-liquid biofilm formation in a *Saccharomyces cerevisiae* flor strain. FEMS Yeast Research. 2012;**12**(7):864-866

[17] Zeidan MB, Carmona L, Zara S, Marcos JF. FLO11 gene is involved in the interaction of flor strains of *Saccharomyces cerevisiae* with a biofilm-promoting synthetic hexapeptide. Applied and Environmental Microbiology. 2013;**79**(19):6023-6032

[18] Marin-Menguiano M, Romero-Sanchez S, Barrales RR, Ibeas JI. Population analysis of biofilm yeasts during fino sherry wine aging in the Montilla-Moriles D.O. Region. International Journal of Food Microbiology. 2017;**244**:67-73

[19] Zara S, Antonio Farris G, Budroni M, Bakalinsky AT. HSP12 is essential for biofilm formation by a sardinian wine strain of *S. Cerevisiae*. Yeast. 2002;**19**(3):269-276

[20] Mardanov AV, Eldarov MA, Beletsky AV, Tanashchuk TN, Kishkovskaya SA, Ravin NV. Transcriptome profile of yeast strain used for biological wine aging revealed dynamic changes of gene expression in course of flor development. Frontiers in Microbiology. 2020;**11**:538

[21] Ruiz-Muñoz M, Cordero-Bueso G, Izquierdo-Cañas PM, Mena-Morales A, Cantoral JM. Improving an industrial sherry base wine by yeast enhancement strategies. Food. 2022;**11**(8):1104

[22] Martínez P, Codón AC, Pérez L, Benítez T. Physiological and molecular characterization of flor yeasts: Polymorphism of flor yeast populations. Yeast. 1995;**11**(14):1399-1411

[23] Martínez P, Pérez Rodríguez L, Bénitez T. Factors which affect velum formation by flor yeasts isolated from sherry wine. Systematic and Applied Microbiology. 1997;**20**(1):154-157

[24] Mesa JJ, Infante JJ, Rebordinos L, Cantoral JM. Characterization of yeasts involved in the biological ageing of sherry wines. LWT - Food Science and Technology. 1999;**32**(2):114-120

[25] Rodríguez ME, Infante JJ, Mesa JJ, Rebordinos L, Cantoral JM. Enological behaviour of biofilms formed by genetically-characterized strains of sherry yeast. The Open Biotechnology Journal. 2013;**7**(1):23-29

[26] Ruíz-Muñoz M, del Bernal-Grande MC, Cordero-Bueso G, González M, Hughes-Herrera D, Manuel Cantoral J. A microtiter plate assay as a reliable method to assure the identification and classification of the veil-forming yeasts during sherry wines ageing. Fermentation. 2017;**3**(4):2-10

[27] Ruiz-Muñoz M, Hernández-Fernández M, Cordero-Bueso G, Martínez-Verdugo S, Pérez F, Cantoral JM. Non-saccharomyces are also forming the veil of flor in sherry wines. Fermentation. 2022;**8**(9):56 Available from: https://www.mdpi.com/2311-5637/8/9/456

[28] Morata A, Arroyo T, Bañuelos MA, Blanco P, Briones A, Cantoral JM, et al. Wine yeast selection in the Iberian Peninsula: *Saccharomyces* and non-*saccharomyces* as drivers of innovation in Spanish and Portuguese wine industries. Critical Reviews in Food Science and Nutrition. 2022;**2**:1-29

[29] Roldán A, Lasanta C, Caro I, Palacios V. Effect of lysozyme on "flor" velum yeasts in the biological aging of sherry wines. Food Microbiology. 2012;**30**(1):245-252

[30] Roldán AM, Lloret I, Palacios V. Use of a submerged yeast culture and lysozyme for the treatment of bacterial contamination during biological aging of sherry wines. Food Control. 2017;**71**:42-49

[31] Ruiz-Muñoz M, Cordero-Bueso G, González-García L, Izquierdo-Cañas PM, Centeno-Cuadros A, Mena-Morales A, et al. Evolved *Saccharomyces cerevisiae* strains to reduce ethyl carbamate in Sherry wines. Food Control. 2023;**153**:109958

[32] Cantoral JM, Florido-Barba A, Lopez-Molina MF, Cordero-Bueso G. Adaptación y desarrollo del velo de flor en vinos "sobretablas" de la D.O. Jerez-Xèrés-Sherry fortificados con alcoholes no vínicos. In: Roca P, editor. BIO Web Conference. Vol. 56. Les Ulis Cedex A, France: Edp Sciences; 2023. p. 02029

Chapter 10

Use of Other Species in Winemaking, and Their Interaction with *Saccharomyces cerevisiae*

Duncan Hamm and Bernardo Muñoz González

Abstract

While *Saccharomyces cerevisiae* is recognized as the yeast species that completes the process of alcoholic fermentation during winemaking, the use of starter cultures from other species has become popular in recent years. Non-*saccharomyces* yeast cultures are now widely used for their bio-protective effects and/or the contribution they make to a wine's sensory profile. Conversely, starters of wine lactic acid bacteria are also commonly utilized around the same time as commercial *Saccharomyces cerevisiae*, as an alternative to encouraging adventitious strains to proliferate. This could be either for initiating malolactic fermentation during alcoholic fermentation, or more recently for biological protection of musts prior to the fermentation process. The interactions between *S. cerevisiae* and other species are documented in the following chapter. The areas examined in more details include requirements of nutrients compared to *S. cerevisiae*, whether complimentary of symbiotic. Active bioprotective agents such as killer factors, the role of cell-to-cell contact, and the resultant effects on final wine composition when co-fermenting with *S. cerevisiae* is also discussed.

Keywords: non-*Saccharomyces* yeast, malolactic fermentation, *Oenococcus oeni*, microbial interactions, wine bio-protection, wine aroma, mixed fermentation

1. Introduction

While *Saccharomyces cerevisiae* is generally regarded as the microbial species that completes alcoholic fermentation in wine, it is far from the only species both present and thereby in some way contributing to the process. Winegrapes, the key starting material in winemaking, can contain a diverse range of microbial flora [1], so offer a natural origin for these organisms in a given fermentation. They can also arise from the winemaking environment itself [2] and will continue to be present and viable some way into the winemaking process [3, 4]. Whether colonized on the initial winegrapes or derived from the winemaking environment, typically the populations of species other than *S. cerevisiae* are naturally far more abundant in the early stages of winemaking than *S. cerevisiae* itself is [5].

More recently the use of oenological starter cultures not belonging to *S. cerevisiae* has been adopted. Often these species are inoculated with some overlap to a thriving

S. cerevisiae population. These include "non-*Saccharomyces*" yeast (NSY), effectively any yeast not part of the *Saccharomyces* spp., as well as oenological lactic-acid bacteria (LAB). Given that the vast majority of wine producing nations are observant of OIV practices, commercialized cultured are required to be natural isolates derived from a winemaking environment [6].

The benefits by applying such cultures will vary on depending on the application, but predominantly these are used to enhance the overall sensory profile in some way, or for biological protection against spoilage processes. In the case of oenological bacterial cultures, their use could be for biological protection, although in the vast majority of cases these are inoculated in order to initiate malolactic fermentation (MLF).

1.1 Non-*Saccharomyces* yeast

Naturally present in the winemaking environment, NSY make up a significant part of the microbiome of a given fermenting wine, not just in musts where indigenous alcoholic fermentation is encouraged [7], but also in those inoculated with commercial *S. cerevisiae* starters. While in these "inoculated" fermentations the inoculated strain of *S. cerevisiae* proliferates, thereby giving the winemaker a more predictable outcome, other species are present to some degree at different stages of the process. In contrast, in "un-inoculated" fermentations, NSY species will be dominant in the early stages of fermentation [3]. The question of what effects specific these species can have, has led to them being isolated, then studied in a controlled manner. Having the potential to impart positive impacts upon a given wine has led to them being evaluated for their commercial appeal, which in turn led to the launch of the first NSY cultures around 20 years ago. Initial commercialized starters were a blend of NSY and *S. cerevisiae*, followed later by single-species starters. In the years since, the NSY species have been launched as commercial wine yeasts, includes but is not limited to; *Torulaspora delbreuckii, Lachancea thermotolerans, Metschnikowia pulcherrima, Metschnikowia fructicola, Pichia kluyveri* and *Hanseniaspora vineae,* as detailed in **Table 1**.

The application of NSY species in winemaking has focused on two areas. The first of these is organoleptic impact. This could be from one of the following or a combination of; the production of volatile flavor compounds, the modulation of acidity or the production of non-volatile compounds that affect the perceived texture of a wine. In contrast, biological protection has been the other area where NSY have been studied and applied. This practice extends to applying NSY to prevent spoilage organisms

Species	No. of products	Suppliers
Hanseniaspora vineae	1	Oenobrands
Lachancea thermotolerans	9	AEB, Chr. Hansen, Lamothe-Abiet, Laffort, Lallemand, ICV
Metschnikowia fructicola	1	Lallemand
Metschnikowia pulcherrima	9	Ersblöh, Deveze, AEB, Lamothe-Abiet, Laffort, Lallemand
Pichia kluyveri	1	Chr. Hansen
Torulaspora delbreuckii	7	AEB, Agrovin, Chr. Hansen, Lamothe-Abiet, Laffort, Lallemand, ICV

Table 1.
List of commercially available NSY cultures in France, 2022 [8].

from proliferating, to preventing oxidative reactions in must, or to ensuring certain desirable organisms have a healthy environment in which to flourish.

Typically, NSY is inoculated sequentially in winemaking, as represented in **Figure 1**. This allows the NSY population to dominate the winemaking environment, before *S. cerevisiae* is later introduced complete the alcoholic fermentation.

1.2 Oenological bacteria

Oenological bacterial cultures were developed in the later part of the 20th century. Then, as with today, the two species that were initially commercialized were *Oenococcus oeni* and *Lactiplantibacillus plantarum*. The former was developed as an alternative to relying on adventitious lactic-acid bacteria to develop in a wine, whereby MLF could be managed more effectively. The first "direct inoculation" starter for MLF, Viniflora® Oenos™, was launched in 1993. Its efficacy was based on implanting a high number of viable cells into a given wine, above the threshold of 1.0E+6 cfu/ml required for MLF to start [9]. In contrast, the first oenological culture based on *L. plantarum* was for a deacidification in juice or must, by completing a partial MLF at this early stage of winemaking.

The process of MLF was seen as being distinct from alcoholic fermentation, so inoculation of *O. oeni* starters generally took place after *S. cerevisiae* had finished converting sugars to ethanol. The only exception was the *L. plantarum* mentioned above, as this was designed to be inoculated prior to alcoholic fermentation. Nevertheless, whether before or after, the two fermentations of alcoholic and malolactic fermentations in wine were seen as being distinct. In the more recent past this approach had been challenged by the adoption of the practice known as "co-inoculation". Although successfully demonstrated since 1985 [10], adopting the practice was initially very slow. Since the turn of the century, the practice has become more widespread. At the same time the risks associated with co-inoculation are better understood, and

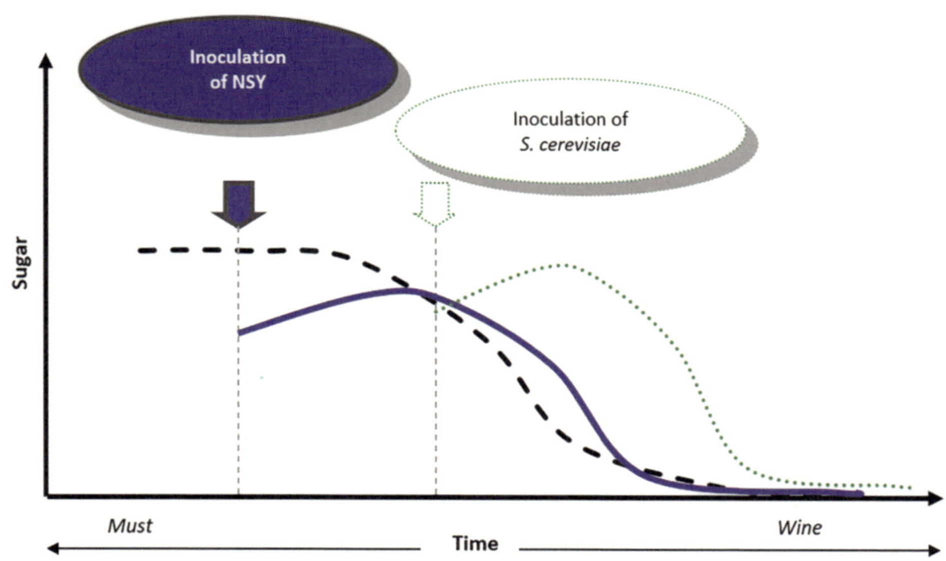

Figure 1
The depiction of sequential inoculation with NSY. NSY is introduced to must first, its population shown by the solid blue line. After a period of 2–3 days, S. cerevisiae (dotted green line) is then inoculated, so that alcoholic fermentation can complete. The sugar concentration is shown by the dashed black line.

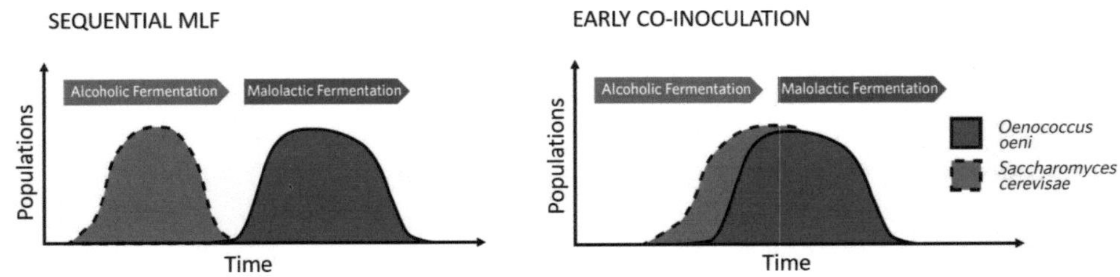

Figure 2.
The population overlap between S. cerevisiae and O. oeni when using co-inoculation, when compared to the more traditional sequential MLF. The relative population of S. cerevisiae is depicted in gray with a dashed outline, while O. oeni is depicted in blue with a solid outline.

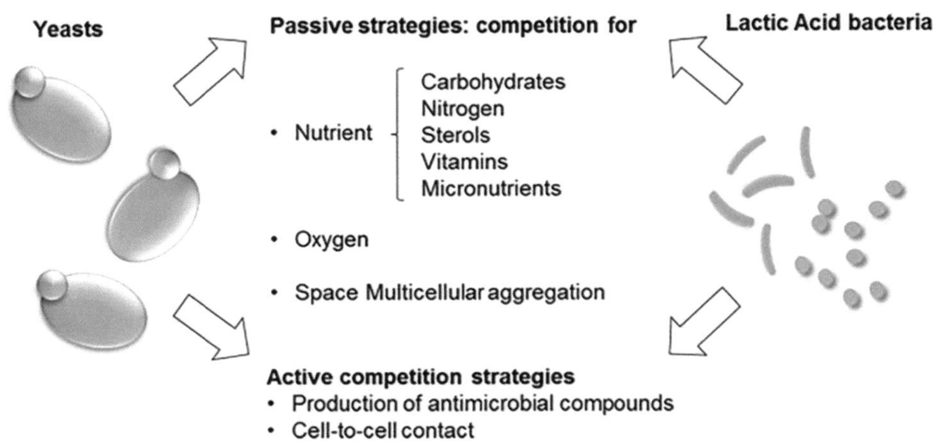

Figure 3.
The passive and active strategies used when applying culture for biological protection, taken from [15].

measures can be taken to minimize deviations. This can be done in two ways, firstly by ensuring the winemaking parameters such as pH, temperature and sulfite concentrations are conducive to both the yeast and bacteria being employed. Secondly, through selecting both yeast and bacterial strains that do not exhibit antagonism, risks of deviations are minimized [11]. The overlap of yeast and bacterial populations during co-inoculation are shown in **Figure 2**.

A further, emerging application of oenological bacteria in winemaking is by utilizing *L. plantarum* for biological protection against spoilage organisms in musts. While the topic of biological protection is relatively recent, it was initially only focused on NSY [12]. It is apparent that there is a role for bacteria to play in this as well [13, 14]. This is highlighted in **Figure 3**, summarizing the "active" and "passive" mechanisms at play when using microbial cultures for biological protection.

2. Interactions with *Saccharomyces cerevisiae*

2.1 Population dynamics

In the nutrient and oxygen poor environment of a fermenting wine, by the end of the process, *S. cerevisiae* is recognized as becoming the dominant species of yeast.

This extends to a wine fermentation where a NSY has been inoculated, although the population numbers *S. cerevisiae* can reach are generally lower than when compared with *S. cerevisiae*-only wines [16]. The converse effect has also been clearly demonstrated, where an active population of *S. cerevisiae*, can cause a significant decline in populations of NSY [17]. This may be due to competition for space, nutrients or through the production of inhibitory substances. Transcriptomics suggest that both *S. cerevisiae* and NSY strains impact upon the genetic expression of one another in a mixed fermentation [18].

Given that classical microbiological plating has been used for most of these population studies, it has been proposed more recently that NSY are not necessarily dead but could be in a viable but non-culturable state (VBNC). While these population can later be revived [19], it does not detract from the dominance seen from *S. cerevisiae*. It merely challenges the correct state of any NSY populations that could be present.

Like NSY, oenological populations of LAB can also be impacted by the presence of *S. cerevisiae*. The causes for this will be examined in more detail in the following sections. Conversely, *S. cerevisiae* populations are seldom impacted by inoculated LAB. This is unsurprising given the MLF still takes place after alcoholic fermentations for many wines, but even when co-inoculation practices care used, there has been little impact demonstrated on *S. cerevisiae* populations [20, 21].

2.1.1 Cell to cell contact

Physical contact between cells from differing yeast populations (cell-cell contact) has been demonstrated as an important aspect when assessing population dynamics during winemaking. Investigated by using double-compartment fermentation vessels (*P. kluyveri*), or by sterile-filtration between inoculating yeast populations (*L. thermotolerans*, *T. delbrueckii*), *S. cerevisiae* can become increasingly dominant over NSY when physically together [17, 22]. It has been demonstrated that this phenomenon is most associated with the presence of the FLO family genes in *S. cerevisiae*, which control flocculation [23].

Some studies have also showed that by physically removing NSY, the health of *S. cerevisiae* is subsequently affected. This most likely due to removing nutrients along with the NSY, rather than any absence of cell-cell contact [24–26].

2.2 Nutrient requirements

Wine microorganisms require nutrients to survive and thrive, so that the ultimate end-goal of producing a sound wine in a timely manner can be realized. Given that the wine microbiome changes during fermentation, whether with inoculated or native organisms, the potential for competition for nutrients exists throughout. It is clear however the requirements for different species, or even different strains within a species, will vary. In order to properly manage wine fermentations, these differences need to be considered and wherever possible be managed.

2.2.1 Assimilable nitrogen

Nitrogenous compounds, typically amino acids and ammonium ions are required by yeasts for growth. The sum of the concentrations of these are expressed as Yeast Assimilable Nitrogen (YAN). The nitrogen demand between different strains *S. cerevisiae* does vary [27], however the typical guideline for a healthy alcoholic fermentation in wine is a YAN concentration > 150 mg/L. In contrast, oenological LAB cannot utilize ammonia, with the requirement for a healthy population of *O. oeni* being 60 mg/L of amino acid nitrogen [28]. The inclusion of additional microbial species, sometime in the winemaking process, can thereby lead to competition with a higher demand for YAN when NSY and *S. cerevisiae* populations are together [29].

The nitrogen requirements of NSY will also vary between species [30]. When studying commercial strains of *L. thermotolerans*, *T. delbrueckii*, *P. kluyveri* and *M. pulcherrima*, it was shown that the first two species consume the available nitrogen, both amino acids and ammonium, at a much faster rate than the latter two. Furthermore, upon filtering samples inoculated with these two species, then inoculated with *S. cerevisiae*, primary fermentation was unable to complete. In contrast, when the same conditions were applied to both *P. kluyveri* and *M. pulcherrima*, there was much less competition for nutrients, thereby allowing primary fermentation to finish. In all cases the fermentation completed when *S. cerevisiae* was inoculated without filtration [25]. This suggests lysing of the NSY cells is required in order to supply enough nitrogen to the *S. cerevisiae*. A further facet examined in the same study is around the release of certain amino acids back into the fermenting wine by *P. kluyveri* and *M. pulcherrima* early in the first 48 hours of winemaking, as detailed in **Figure 4**. While limited scientific data exists on this, it does propose a mechanism for certain strains of NSY to be used in low-intervention winemaking, to help ensure native *S. cerevisiae* is adequately fed.

When it comes to oenological bacteria, there is also the risk of competition between *S. cerevisiae* and the bacteria being utilized. In most cases, bacteria are inoculated after yeasts are, and will have a lower nitrogen demand. Given that MLF is seen as requiring an amino nitrogen concentration of 60 mg/L, depletion of nitrogen is generally only seen against bacteria, from yeasts, not the other way around. Given that the concentration of assimilable nitrogen in a wine pre-MLF can vary, including as a result of differing YAN requirements between strains of *S. cerevisiae*, to ensure successful MLF it is often recommended to measure the amino nitrogen concentration, and address any nitrogen deficiencies, at the start of this process.

2.2.2 Thiamine and B1 vitaminers

Thiamine is necessary for successful growth of wine yeasts. While many yeasts can synthesis this compound, it has been suggested that it is not possible under the stressful conditions of a wine fermentation. Previous work has suggested that indigenous strains of *Hanseniaspora uvarum* can quickly deplete exogenous thiamine present in must, thereby depriving *S. cerevisiae* of this vitamin and leading to stuck alcoholic fermentation [31]. More recent work investigating another apiculated yeast, *H. vineae*, found no such effect on the *S. cerevisiae* present [32]. The Thiamine uptake of *T. delbreuckii* has also been investigated and found to be consumed at roughly a similar rate as *S. cerevisiae* [24]. Three other species *Starmerella bacillaris*, *M. pulcherrima* and *T. delbrueckii* demonstrated slower uptake of Thiamine and other B1 Vitamers at a lower rate than *S. cerevisiae* [33, 34].

2.2.3 Pantothenic acid and B5 vitaminers

The uptake and release of Pantothenic acid, and its B5 cofactor form Coenzyme A by NSY has been investigated. All strains studied in [32] found that both B5 vitaminers were synthesized early after inoculation, however *T. delbrueckii* was found to

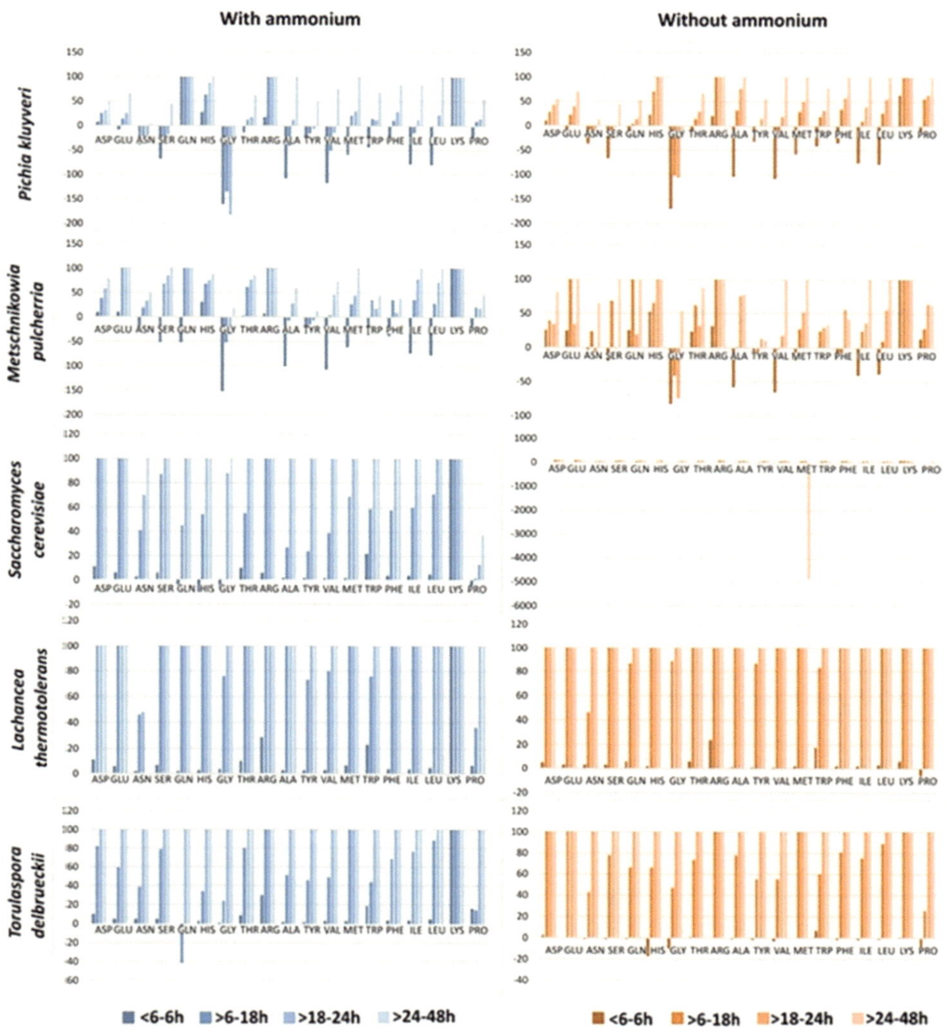

Figure 4.
The uptake and subsequent release of amino acids by S. cerevisiae and four commercial NSY in the first 48 hours of fermentation. In the cases of the P. kluyveri and M. pulcherrima there is significant release of certain amino acids back into the fermenting wine. Taken from [25].

re-consume a proportion. This has implications when later inoculating *S. cerevisiae*, as competition for these vitamins could potentially lead to decreased numbers and sluggish fermentation. In contrast, by synthesizing B5 vitaminers, *S. bacillaris* and *M. pulcherrima* could provide a beneficial environment for a later *S. cerevisiae* inoculation.

2.3 Inhibitory compounds

A range of inhibitory substances can be produced by oenological organisms, whether by *S. cerevisiae*, NSY or by LAB. For the latter, inhibitory compounds have not been reported with commercialized starters, but sluggish fermentation has been linked to adventitious spoilage strains of LAB [35]. Yeasts however are capable of producing a range of toxic compounds, the most obvious of which is ethanol. In fact the production of ethanol, and the subsequent toxicity to NSY is seen as one of the key mechanisms for *S. cerevisiae* to dominate a wine fermentation.

2.3.1 Sulfur dioxide

While sulfur dioxide (SO_2) remains the preeminent preservative used in winemaking, to control both oxidative and microbial spoilage, it is also a compound that can be synthesized by *S. cerevisiae* by the sulfate assimilation pathway. The concentrations of SO_2 produced depends on both the strain, and the winemaking conditions, and can vary widely. When expressed as Total-SO_2 (the sum of the concentrations of molecular SO_2, free bisulfate ions and bisulfate ions bound to other moieties) the concentration produced can range from 20 to 200 mg/L, [36]. This has very important implications for when applying NSY or LAB con-currently as these two families have a much lower tolerance to SO_2 than *S. cerevisiae*, or at least those strains of *S. cerevisiae* selected as oenological starters. This therefore reinforces the typically applied sequential inoculation of NSY, allowing them to proliferate before *S. cerevisiae* comes to dominate. For bacteria, to ensure successful MLF, it is therefore suggested to ensure the *S. cerevisiae* employed for the alcoholic fermentation does not produce excessive SO_2. A good rule of thumb is to keep the concentration of T- SO_2 below 40 mg/L, although specific tolerances can be cross-referenced with the supplier of MLF bacteria.

2.3.2 Medium chain fatty acids

Octanoic, decanoic and dodecanoic acids (C8, C10 and C12 respectively) are medium chain fatty acids (MCFA), always produced by wine yeasts to some degree during wine fermentation. In fact, ethyl esters of these compounds are seen as beneficial flavor compounds in wine while higher concentrations of free MCFA are seen as having a direct effect of sluggish fermentation [37]. While the toxicity of MCFA to NSY is not well studied, the production of these has been investigated. Oxygenation of mixed fermentations, with either *Lactobacillus L. thermotolerans* or *T. delbrueckii* and *S. cerevisiae,* gave lower MCFA concentrations than the same combinations fermented anaerobically, as well as the *S. cerevisiae*-only control. Low MCFA concentrations were also seen in single-strain NSY fermentations. This suggests that both of these commercial NSY strains involved in the study will naturally produce lower concentrations of MCFA [38]. These results also lead the authors to conclude with the presence of oxygen encouraged the incorporation of MCFA into longer fatty acids via Acetyl-CoA carboxylase activity.

MCFA are inhibitory to oenological bacteria, especially in their unesterified free forms. As with SO_2, it is important that that the concentration of these is kept below a certain threshold. This is below 12.5 mg/L for C10 and 2.5 mg/L for C12 [39], although individual tolerances between different strains of LAB will vary. Fermenting *S. cerevisiae* at warmer temperatures had been shown to give lower MCFA concentrations in the final wine [40], however the innovative approach of utilizing NSY combined with oxygenation could be another approach.

2.3.3 Killer toxins

The ability of yeast strains to both produce specific extracellular proteinaceous compounds that can kill other yeasts, referred to as killer toxins, has been well studied. For *S. cerevisiae* there are three killer toxins, referred to as; K1, K2 and K28, and these are only active within the species. Strains can either produce a killer toxin or be sensitive to toxins. Within oenology, the killer toxin status between different strains of commercial *S. cerevisiae* starters is therefore often published by yeast manufacturers, and expressed as 'positive', 'sensitive' or 'neutral'. The latter referring to strains where no toxin is produced and there is not sensitivity to the three known killer factors for *S. cerevisiae*.

Killer toxins have also been shown to be produced by various NSY strains, and utilizing these has been proposed as a biological protection strategy in winemaking. In [41] only a minority of these are active at wine pH, and when they are, none have been shown to be from NSY species that have been commercialized. *T. delbrueckii* is capable of producing a killer toxin referred to at TDKT, however no effect on commercial *S. cerevisiae* strains from this toxin has been observed [42]. Likewise, no inhibition of *S. cerevisiae* from killer toxin was seen in when investigating a commercial strain of *T. delbreuckii* in [24].

A similar situation exists for *L. thermotolerans*, with some studies suggesting strains of this species are positive for a killer toxin against *S. cerevisiae*. As the methodology for establishing such killer factors is carried out at pH 4.50, higher than that of wine, it can be argued that it is of limited value in oenology. Conversely multiple studies with *L. thermotolerans* in combinations with various strains of *S. cerevisiae*, repeatedly demonstrate the dominance of the latter species [43].

P. kluyveri has long been associated with producing killer toxin [44] active against *S. cerevisiae*, more recently labeled as PKKP [45]. While exhibiting less activity at the relatively low pH range of wine, the question of whether commercialized strains of this species produce this toxin should naturally be raised. It would seem unlikely, and while not directly studied in [22], it would appear that no killer toxin from the commercial strain *of P. kluyveri* used (Viniflora® FrootZen™) affected the growth of the *S. cerevisiae* employed.

M. pulcherrima has also been found to be positive for killer toxin against a sensitive strain of *S. cerevisiae* in one study [46], however the same arguments around the validity of the test being carried out at high pH apply. Interestingly, two of the authors conclude in a later paper that much of the antimicrobial activity of this species is via the production of pulcherriminic acid rather than killer factor [47]. In the same later study and using the same test for killer toxin, no inhibition was seen against the 18 commercial strains of *S. cerevisiae* tested.

2.3.4 Other biological inhibitors

Beyond killer toxins, there are other peptide-based inhibitory substances produced in mixed fermentations. Anti-microbial activity was seen against *L. thermotolerans* and *T. delbrueckii* when investigating a secreted protein fraction, 2-10 kDa in size, from a specific *S. cerevisiae* strain [48], later shown to be derive from a specific protein, abbreviated as GADPH [49].

Extracellular vesicles (EV) are another potential group of potentially inhibitory compounds in yeast-yeast interactions. EV's 100–200 nm in size were found to be produced by all six species studied in [50]; *S. cerevisiae*, *L. thermotolerans*, *H. uvarum*, *Candida sake*, *M. pulcherrima* and *T. delbrueckii*. While in this study inhibition of the *T. delbrueckii* toward *S. cerevisiae* was seen, it was not from the most abundant EV produced. No doubt more light will be shed on this topic in the coming years.

While killer toxins are seen as a yeast-yeast interaction, there is also evidence that yeast derived peptides can be inhibitory to oenological bacteria. For instance, the protein fraction derived from GADPH and studied in [49] was also found to be inhibitory toward a strain of *O. oeni*. Other studies where anti-microbial peptides derived from *S. cerevisiae* were found to inhibit bacteria suggested differing sizes of these substances. This suggests the peptides themselves, and their production, varies across the species [51].

3. Effects on final wine composition

With the goal of producing expressive, sound and fault-free wines in an economic fashion, cultures of species other than *S. cerevisiae*, whether LAB or NSY have become an indispensable tool at the winemaker's disposal. Whether utilized to ensure timely fermentation processes, to mitigate against spoilage, or to optimize the resultant organoleptic characters, it is expected that in most cases there will be an impact on the final wines composition from these species.

3.1 Mitigation of spoilage characters through biological protection

3.1.1 Ethyl phenols

The ethyl phenols 4-Ethyl Guaiacol (4-EG) and 4-Ethyl Phenol (4-EP) can arise in wine due to the presence one of the most serious of wine contaminants, *Brettanomyces bruxellensis*. While SO_2 has traditionally been used to help mitigate against the outgrowth of *B. bruxellensis*, this approach is likely to become more challenging in the future due to increasing growing season temperatures leading to increased wine pH, and the ability of *B. bruxellensis* to adapt to SO_2 [52].

Two bioprotective strategies can be employed against *B. bruxellensis*. The first is to treat must by inoculation with a pre-fermentative NSY, early in the winemaking process. Some strains of *M. pulcherrima* have been shown to have antimicrobial activity against *B. bruxellensis* [47]. The mechanism for this has been demonstrated to be due to the chelation of iron by pulcherriminic acid, thereby starving the contaminants of this essential mineral. Interestingly, this mechanism seems to have little effect on *S. cerevisiae*. Ethyl phenol development from *B. bruxellensis* contamination does however generally arise later in the winemaking process, after alcoholic fermentation is complete. Whether this early inhibition of *B. bruxellensis* persists is inconclusive.

The second and most effective microbial strategy against *B. bruxellensis* that can be utilized is to inoculate wine, or in the case of co-inoculation must, with MLF bacteria. Given that *B. bruxellensis* develops later in the winemaking process, by ensuring MLF starts and finishes in a timely manner from the inoculated bacteria, the opportunity for this species to build numbers is severely limited. In contrast, while waiting for adequate numbers of *O. oeni* to develop when relying on adventitious populations, *B. bruxellensis* has the opportunity to grow in the warm environment, largely free of sulfites. While the main protective effect seen is simply from the ability to microbially stablise a given wine sooner, this has been recent evidence to suggest *O. oeni* exhibits a fungistatic effect on *B. bruxellensis*, through cell-cell interactions. This is however, likely to be strain specific [53].

3.1.2 Volatile acidity

The excessive production of volatile acids in wine, predominantly acetic acid (AA) is a serious deviation, and limits of such acids are often mandated by law. While AA can be caused by spoilage organisms, particularly apiculated yeast and Acetic Acid Bacteria, it can also arise from the population of *S. cerevisiae* present. During alcoholic fermentation there will always be a small quantity of AA produced, so ensuring this stays at a low concentration is important. The tendency of *S. cerevisiae* to produce AA increases with more osmotic stress. Specific NSY strains of *L. thermotolerans*, *T. delbrueckii*, *S. bacillaris* and *M. pulcherrima* in comparison have often been shown to produce less AA in a single-species fermentation [46] however the lower AA effect also occurs in mixed NSY-*S. cerevisiae* fermentations with the same strains. This suggested commercial strains of NSY are an effective tool in the styles of wine most susceptible to elevated AA, for instance wines from very high-sugar musts [54].

When it comes to AA production from spoilage organisms, often oxygen is required to this to occur. Excluding oxygen during the winemaking process therefore become an important measure to limit AA production. While little scientific data exists on oxygen-uptake rates of NSY c.f. *S. cerevisiae*, this is an emerging area for further study [55].

3.1.3 Chemical oxidation

Chemical oxidation in winemaking is generally prevented by good cellar practices, that prevent oxygen exposure, combined with the use of anti-oxidants, predominantly SO_2 and to a much lesser extent ascorbic acid. Like *S. cerevisiae*, NSY will readily consume dissolved oxygen in must. Identifying candidates that do this most rapidly could have application in low-sulfite winemaking. This could be combined with the properties of *M. pulcherrima* to produce pulcherriminic acid, a substance which will chelate with iron ions [47]. Given many oxidative reactions in must and wine are catalyzed by the presence of metal ions, effectively removing these from the must/wine matrix could prevent potentially mitigate against oxidation to some degree [56].

3.1.4 Biogenic amines

Biogenic amines (BA) arise from the decarboxylation of amino acids by certain wine microorganisms. Histamine is the most predominant BA found in wine, arising from the decarboxylation of histidine, but others produced include cadaverine, putrescine and tyramine. While BA primarily arise from LAB, yeast species can also play a role. To limit their formation, it is therefore suggested to use commercial strains with little or no decarboxylase activity against histidine and the other relevant amino acids [57]. Given BA arise from the activity of adventitious flora, the use of strains for biological-protection can potentially also play a role.

3.2 Flavor compounds

In wines fermented in conjunction with species other than *S. cerevisiae*, the resultant flavor compounds produced have a significant effect on the final sensory profile.

These effects have been extensively studied in both the scientific literature and winemaking field trials. The following is by no means an exhaustive list of such work, but is a summary of the most important compounds, from a wine-aroma point of view.

3.2.1 Esters

Esters are compounds formed in wine by the combination of acids and alcohols, either by the action of microorganisms such as yeasts or bacteria (biological esterification) or during storage and aging (chemical esterification). They are produced by yeast via esterase activity, and may be volatile or non-volatile.

A wide range of yeast-derived esters are formed during a wine fermentation. This includes ethyl esters, acetates and esterified forms of fatty acids and higher alcohols. An increase in the concentrations of the sum of all esters measured was seen in [58, 59], where sequential mixed inoculations of *S. cerevisiae* and NSY were evaluated against *S. cerevisiae* only controls. In the former paper *L. thermotolerans*, *T. delbrueckii* and *H. vineae* were studied, whereas three strains each of *L. thermotolerans*, *Metschnikowia* spp. and *S. bacillaris* were the NSY species investigated in [59].

One particularly important ethyl ester to consider is ethyl lactate (EL), which is known to impart a strawberry-like aroma. While *S. cerevisiae*-only wine fermentations generally have very low concentrations, *L. thermotolerans* is able to produce L-lactic acid from grape sugars, which in turn acts as a precursor for EL production [43]. EL production is not however restricted to *L. thermotolerans* as LAB will also produce L-lactic acid, and EL. In the case of *L. plantarum* however, although somewhat strain dependent, the production of EL can far exceed that from *O. oeni*. *L. plantarum* therefore has the potential to make a positive contribution to wine aroma [59].

Acetate esters can often be more readily synthesized by NSY when compared to *S. cerevisiae*. In some cases this can be beneficial, such as for 2-phenylethyl acetate, which imparts a pleasant rose-like aroma. *H. vineae* in particular has been selected for this characteristic [60], although it has also been measured at more moderate concentrations in fermentations with *T. delbrueckii* also [61].

Isoamyl acetate is another acetate ester that has been widely studied. Giving a banana-like aroma, some strains of *S. cerevisiae* are promoted on the ability to produce this compound, using isoamyl alcohol as a precursor. While concentrations of this compound produced with respect to *S. cerevisiae* varies across studies, it is generally observed that *P. kluyveri* will increase concentrations [25].

While isoamyl acetate and 2-phenylethyl acetate are important acetate esters, without negative connotations, production of acetate esters may not always be beneficial to wine quality. One example is ethyl acetate, which while contributing positively at low concentrations, will give solvent-like characters at elevated levels. While some strains of *H. uvarum* are capable of producing very high concentrations [62], fortunately commercial NSY do not produce elevated concentrations of ethyl acetate. In fact, often lower ethyl acetate concentrations are measure in wine co-fermented with both NSY and *S. cerevisiae* compared to *S. cerevisiae* alone [61, 63].

Ethyl esters of MCFA such as ethyl hexanoate, ethyl octanoate and ethyl decanoate also play an important role in wine flavor chemistry. Research shows that co-fermenting with NSY does influence the concentration of these [46]. Whether concentrations increase or decrease for a particular ethyl ester of a MCFA depends very much on both species and strain, as well as the *S. cerevisiae* being employed.

3.2.2 Thiols

Three specific sulfur-containing flavor compounds, 3-mercaptohexan-1-ol (3MH), 3-mercaptohexyl acetate (3MHA) and 4-mercapto-4-methylpentan-2-one (4MMP) have been extensively investigated for the positive effects that they have on wine aroma. While imparting aromas of boxtree, passionfruit and grapefruit these are in contrast to much smaller mercaptans, such a dimethyl disulfide and methanethiol, which impart negative characters. As the three volatile thiols are not naturally present in wine grapes, yeasts are required to release these from their precursors, cysteinylated and glutathionylated forms of 4MMP and 3MH. Further conversing of 3MH to 3MHA then takes place from alcohol-acetyltransferases.

While many *S. cerevisiae* starters are marketed for their ability to release volatile thiols, NSY can play important role too. *P. kluyveri* has been investigated, and subsequently a commercialized NSY starter has been promoted around increasing 3MH and 3MHA concentrations. Interestingly, the interaction between this species and the strain of *S. cerevisiae* used is crucial. While the combination of *P. kluyveri* and *S. cerevisiae* can lead to an increase in both 3MH and 3MHA, for some strains of *S. cerevisiae* this led to an increase in only one of these, or in some cases neither [64].

Other NSY species shown to increase volatile thiol concentrations include *T. delbrueckii*, *L. thermotolerans*, *M. pulcherrima* and *S. bacillaris* [64, 65].

3.2.3 Terpenoids

Terpenoids such as monoterpenes, sesquiterpenes and norisoprenoids are an important group of aroma compounds in wine. Generally existing primarily in non-volatile "bound" forms in grapes, the free equivalents are released by glycosidase activity. An increase in overall total terpene concentrations from co-fermentations of NSY and *S. cerevisiae* has been demonstrated in various studies, which is unsurprising given NSY exhibit greater β-glucosidase activity when compared to *S. cerevisiae* [66]. This effect however is not seen in all cases, suggested that under winemaking conditions, some strains of *S. cerevisiae* have roughly equivalent activities [63]. Interestingly *O. oeni* can also have significant glycosidase activity, resulting in increased concentrations of free terpenoids during MLF [67].

3.2.4 Others

The production of higher alcohols has shown to vary across yeast species, which can vary further whether co-fermented with *S. cerevisiae*. This can be linked to some degree to amino acid consumption, with production via the Ehrlich pathway, but not exclusively so [25].

2,3 Butanedione, commonly referred to as diacetyl, as another biologically important flavor compound. Its production in wine derives from citrate-fermentation by LAB, primarily *O. oeni*, followed by chemical oxidation of the intermediate compound, α-acetolactate. Co-fermentation between *O. oeni* and *S. cerevisiae* will have a significant reduction on concentrations of diacetyl formed, both through viable yeast cells being able to take up this compound, and by the presence of an active population of *S. cerevisiae* ensuring oxygen concentrations are depleted.

3.3 Acidity modulation

In winemaking, the biological process which has the largest impact upon a wine's acidity in MLF. When carrying out this process with *O. oeni* in conjunction with alcoholic fermentation, some studies have shown a slight increase in pH or titratable acidity, compared to running both fermentations sequentially [21]. The bacterial genus that offers the most oenological promise for carrying out such acidification however is *Lactobacillus*, utilizing the production of lactate from grape sugars. Due to the generally low tolerance to ethanol from Lactobacilli, this process needs to take place early in the winemaking process, so most likely with an overlap with the presence of *S. cerevisiae*.

One particular species of NSY that offers an alternative to LAB for MLF is *Schizosaccharomyces pombe*, through maloalcoholic fermentation [68]. While two examples have been commercialized, their use to date has been very limited.

L. thermotolerans is the most high-profile NSY species investigated for the effect that it can have on wine acidity. Like *Lactobacillus* spp., it can synthesize lactate from sugars. The immediate application for such yeasts is for musts lacking in acidity [43], so is not surprising that it is one of the most utilized NSY species in winemaking [8], producing up to 8 g/L of Lactic acid. While helping to balance low acidity, further benefits include a potential shift to higher concentrations of molecular SO_2 from unbound sulfites, as well as potentially lower final ethanol concentrations. Both are depicted in **Figure 5**.

3.4 Textural elements

A long-held belief among some parts of the winemaking community is that wines fermented using spontaneous fermentation, so by inference using adventitious NSY to a large degree, give a fuller mouthfeel than ones simply inoculated with *S. cerevisiae*. Research into mixed fermentations of NSY and *S. cerevisiae* has been able to clearly demonstrate a connection between a more diverse yeast ecology and increased

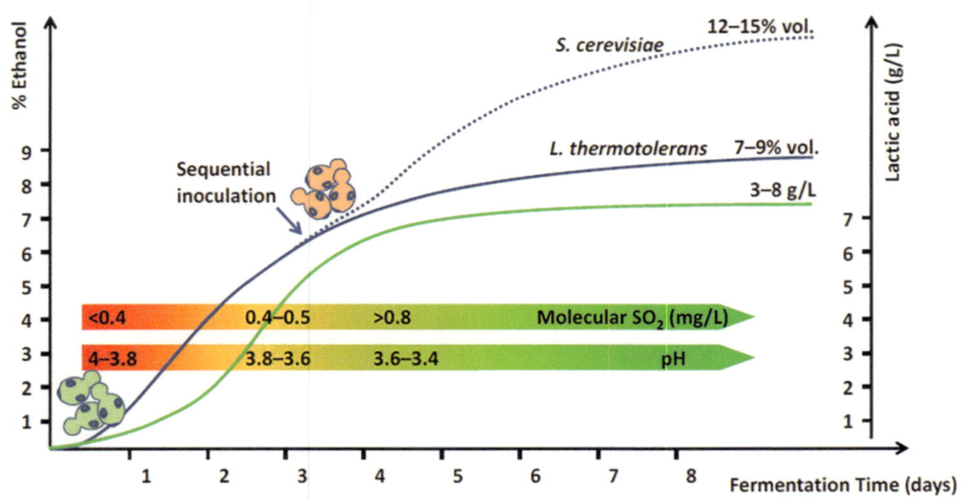

Figure 5.
The effect on lactic acid production (green line), alcohol content (blue line), pH, and molecular SO_2 when running sequential fermentation, first with L. thermotolerans (green yeast) and then with S. cerevisiae (pink yeast). Taken from [12].

palate-weight, with NSY often producing levels of polysaccharide well in excess of *S. cerevisiae* alone [46], as detailed in **Table 2**. It is unsurprising that yeast cell walls are largely composed of polysaccharides, which can be released into the medium upon cell lysis. Polysaccharides are compounds soluble in alcohol, and the combination of polysaccharides with tannins and anthocyanins increases viscosity and fullness in the mouth, provides complexity and aromatic persistence, reduces astringency, thus improving the structure, density and texture of the wine and stabilizes the coloring matter [69]. Although increases have been observed with many species of NSY,

Inoculum, (cells ml^{-1})	Ethanol (% v/v)	pH	Total acidity ($g\,l^{-1}$)	Volatile acidity ($g\,l^{-1}$)	Glycerol ($g\,l^{-1}$)	Δ Polysaccharides ($mg\,l^{-1}$)
S. cerevisiae 10^7	13.93 ± 0.06^{a}	3.20 ± 0.04^{a}	7.05 ± 0.04^{a}	0.46 ± 0.01^{a}	6.23 ± 0.54^{a}	97 ± 10^{a}
S. cerevisiae 10^5	13.87 ± 0.00^{a}	3.16 ± 0.05^{a}	7.12 ± 0.11^{a}	0.47 ± 0.01^{a}	6.65 ± 0.05^{a}	98 ± 11^{a}
S. cerevisiae 10^3	13.88 ± 0.03^{a}	3.17 ± 0.21^{a}	7.02 ± 0.05^{a}	0.50 ± 0.06^{a}	6.46 ± 0.52^{a}	68 ± 3.0^{a}
C. zemplinina + *S. cerevisiae* 10^7	13.83 ± 0.04^{a}	3.21 ± 0.01^{a}	6.88 ± 0.27^{a}	0.43 ± 0.04^{a}	6.25 ± 0.30^{a}	123 ± 42^{a}
C. zemplinina + *S. cerevisiae* 10^5	13.78 ± 0.05^{a}	3.15 ± 0.05^{a}	6.84 ± 0.02^{a}	0.44 ± 0.06^{a}	7.18 ± 1.30^{b}	140 ± 42^{a}
C. zemplinina + *S. cerevisiae* 10^3	13.64 ± 0.04^{b}	3.08 ± 0.18^{b}	6.88 ± 0.04^{a}	0.52 ± 0.01^{a}	7.95 ± 1.28^{b}	181 ± 48^{a}
L. thermotolerans + *S. cerevisiae* 10^7	13.80 ± 0.02^{a}	3.16 ± 0.01^{a}	7.30 ± 0.07^{a}	0.38 ± 0.01^{b}	6.95 ± 0.20^{b}	133 ± 1.0^{a}
L. thermotolerans + *S. cerevisiae* 10^5	13.80 ± 0.01^{a}	2.97 ± 0.03^{b}	9.00 ± 1.96^{b}	0.40 ± 0.00^{a,b}	7.29 ± 0.96^{b}	139 ± 10^{a}
L. thermotolerans + *S. cerevisiae* 10^3	13.70 ± 0.18^{a}	2.90 ± 0.01^{b}	9.20 ± 1.93^{b}	0.40 ± 0.00^{a,b}	7.58 ± 0.46^{b}	158 ± 3.0^{a}
T. delbrueckii + *S. cerevisiae* 10^7	13.90 ± 0.04^{a}	3.19 ± 0.01^{a}	7.12 ± 0.02^{a}	0.38 ± 0.01^{b}	5.88 ± 0.04^{a}	157 ± 16^{a}
T. delbrueckii + *S. cerevisiae* 10^5	13.85 ± 0.08^{a}	3.10 ± 0.08^{a,b}	7.36 ± 0.51^{a}	0.40 ± 0.04^{a,b}	6.14 ± 0.22^{a}	269 ± 44^{b}
T. delbrueckii + *S. cerevisiae* 10^3	13.76 ± 0.04	3.08 ± 0.11^{a,b}	7.34 ± 0.49^{a}	0.41 ± 0.01^{a,b}	6.29 ± 0.61^{a}	308 ± 42^{b}
M. pulcherrima + *S. cerevisiae* 10^7	13.87 ± 0.01^{a}	3.40 ± 0.08^{c}	6.33 ± 0.27^{a}	0.30 ± 0.04^{b}	6.53 ± 0.27^{a}	120 ± 10^{a}
M. pulcherrima + *S. cerevisiae* 10^5	13.79 ± 0.13^{a}	3.39 ± 0.14^{c}	6.50 ± 0.16^{a}	0.34 ± 0.07^{b}	6.98 ± 0.00^{b}	126 ± 10^{a}
M. pulcherrima + *S. cerevisiae* 10^3	13.65 ± 0.19^{b}	3.40 ± 0.00^{c}	6.64 ± 0.37^{a}	0.33 ± 0.01^{b}	7.25 ± 0.25^{b}	154 ± 17^{a}

Data are means ± standard deviations of two independent experiments. Values displaying different superscript letters ($^{a,\,b,\,c}$) within each column are significantly different according to the Duncan test (0.05%).

Table 2.
Analytical profiles of wines fermented as mixed culture of S. cerevisiae and four different NSY. While an increase in polysaccharide concentration is measured for all four NSY species, the increase is significantly greater for T. delbrueckii. Taken from [46].

T. delbrueckii is seen as the species which can give the largest increase to the final polysaccharide concentration of a given wine [70, 71].

4. Conclusion

The microbiome of grapes, must and wine during winemaking is complex, with multiple interactions taking place between populations, whether naturally present or inoculated. While *S. cerevisiae* is dominant at the later stages of alcoholic fermentation, there are key interactions between this and other microbial species, whether NSY or bacteria.

The use of oenological bacterial cultures and NSY has grown significantly in recent years, as winemakers strive to make expressive and fault-free wines, which are true to their origins. This could be to bring biological protection, enhance the organoleptic profile, or in the case of oenological bacteria, to induce MLF. In order to ensure success however, it is important to understand the potential relationships between inoculated populations.

The benefits of utilizing species other than *S. cerevisiae* are multiple, whether increasing volatile flavor compounds, helping to prevent the effects of spoilage or impacting the acidic profile of the final wine.

Acknowledgements

Beatriz Buenafe, Chr. Hansen, S.L.
Jesper Kjeldgaard Andersen, Chr. Hansen A/S.
Anne-Claire Bauquis, Chr. Hansen SA.

Conflict of interest

The authors declare no conflict of interest.

Author details

Duncan Hamm[1*] and Bernardo Muñoz González[2]

1 Chr. Hansen A/S, Hoersholm, Denmark

2 Chr. Hansen, S.L., Madrid, Spain

*Address all correspondence to: gbdha@chr-hansen.com

References

[1] Renouf V, Claisse O, Lonvaud-Funel A. Inventory and monitoring of wine microbial consortia. Applied Microbiology and Biotechnology. 2007;**75**:149-164. DOI: 10.1007/s00253-006-0798-3

[2] Ocón E, Gutiérrez AR, Garijo P, López R, Santamaría P. Presence of non-*Saccharomyces* yeasts in cellar equipment and grape juice during harvest time. Food Microbiology. 2010;**27**(8):1023-1027. DOI: 10.1016/j.fm.2010.06.012

[3] Goddard MR. Quantifying the complexities of *Saccharomyces cerevisiae*'s ecosystem engineering via fermentation. Ecology. 2008;**89**(8):2077-2082. DOI: 10.1890/07-2060.1

[4] Guzzon R, Widmann G, Settanni L, Malacarne M, Francesca N, Larcher R. Evolution of yeast populations during different biodynamic winemaking processes. South African Journal for Enology and Viticulture. 2011;**32**:242. DOI: 10.21548/32-2-1384

[5] Pinto C, Pinho D, Cardoso R, Custódio V, Fernandes J, Sousa S, et al. Wine fermentation microbiome: A landscape from different Portuguese wine appellations. Frontiers in Microbiology. 2015;**6**:905. DOI: 10.3389/fmicb.2015.00905

[6] International Oenological Codex. 2022. Available from: https://www.oiv.int/sites/default/files/publication/2022-10/international-oenological-codex.pdf

[7] Liu PT, Lu L, Duan CQ, Yan G. The contribution of indigenous non-*saccharomyces* wine yeast to improved aromatic quality of cabernet sauvignon wines by spontaneous fermentation. LWT—Food Science and Technology. 2016;**71**:356-363. DOI: 10.1016/j.lwt.2016.04.031

[8] Vitisphere. 2023. Available from: https://www.vitisphere.com/index-oenologie-vinification.html

[9] Rice AC. The malo-lactic fermentation in New York state wines. American Journal of Enology and Viticulture. 1965;**16**(2):62-68. DOI: 10.5344/ajev.1965.16.2.62

[10] Beelman RB, Kunkee RE. Inducing simultaneous malolactic-alcoholic fermentation in red table wines. In: Proceedings of the Australian Society of Viticulture and Oenology Seminar on Malolactic Fermentation. South Australia: Australian Wine Research Institute Urrbrae; 1985

[11] Alexandre H, Costello PJ, Remize F, Guzzo J, Guilloux-Benatier M. *Saccharomyces cerevisiae–Oenococcus oeni* interactions in wine: Current knowledge and perspectives. International Journal of Food Microbiology. 2004;**93**:2. DOI: 10.1016/j.ijfoodmicro.2003.10.013

[12] Morata A, Loira I, González C, Escott C. Non-*saccharomyces* as biotools to control the production of off-flavors in wines. Molecules. 2021;**26**(15):4571. DOI: 10.3390/molecules26154571

[13] Rubio-Bretón P, Gonzalo-Diago A, Iribarren M, Garde-Cerdán T, Pérez-Álvarez EP. Bioprotection as a tool to free additives winemaking: Effect on sensorial, anthocyanic and aromatic profile of young red wines. LWT. 2018;**98**:458-464. DOI: 10.1016/j.lwt.2018.08.050

[14] Englezos V, Di Gianvito P, Peyer L, Giacosa S, Segade SR, Edwards N, et al.

Bioprotective effect of *Pichia kluyveri* and *Lactiplantibacillus plantarum* in winemaking conditions. American Journal of Enology and Viticulture. 2022;**73**(4):294-307

[15] Di Gianvito P, Englezos V, Rantsiou K, Cocolin L. Bioprotection strategies in winemaking. International Journal of Food Microbiology. 2022;**2**(364):109532. DOI: 10.1016/j.ijfoodmicro.2022.109532

[16] Zhu X, Torija M-J, Mas A, Beltran G, Navarro Y. Effect of a multistarter yeast inoculum on ethanol reduction and population dynamics in wine fermentation. Food. 2021;**10**(3):623. DOI: 10.3390/foods10030623

[17] Nissen P, Arneborg N. Characterization of early deaths of non-saccharomyces yeasts in mixed cultures with *Saccharomyces cerevisiae*. Archives of Microbiology. 2003;**180**:257-263. DOI: 10.1007/s00203-003-0585-9

[18] Shekhawat K, Patterton H, Bauer F, Setati M. RNA-seq based transcriptional analysis of *Saccharomyces cerevisiae* and *Lachancea thermotolerans* in mixed-culture fermentations under anaerobic conditions. BMC Genomics. 2019;**20**:145. DOI: 10.1186/s12864-019-5511-x

[19] Wang C, Mas A, Esteve-Zarzoso B. The interaction between *Saccharomyces cerevisiae* and non-saccharomyces yeast during alcoholic fermentation is species and strain specific. Frontiers in Microbiology. 2016;**7**:502. DOI: 10.3389/fmicb.2016.00502

[20] Huang YC, Edwards CG, Peterson JC, Haag KM. Relationship between sluggish fermentations and the antagonism of yeast by lactic acid bacteria. American Journal of Enology and Viticulture. 1996;**47**(1):1-0. DOI: 10.5344/ajev.1996.47.1.1

[21] Abrahamse CE, Bartowsky EJ. Timing of malolactic fermentation inoculation in shiraz grape must and wine: Influence on chemical composition. World Journal of Microbiology and Biotechnology. 2012;**28**:255-265. DOI: 10.1007/s11274-011-0814-3

[22] Hu K, Zhao H, Edwards N, Peyer L, Tao Y, Arneborg N. The effects of cell-cell contact between *Pichia kluyveri* and *Saccharomyces cerevisiae* on amino acids and volatiles in mixed culture alcoholic fermentations. Food Microbiology. 2022;**103**:103960. DOI: 10.1016/j.fm.2021.103960

[23] Rossouw D, Meiring S, Bauer F. Modifying *Saccharomyces cerevisiae* adhesion properties regulates yeast ecosystem dynamics. mSphere. 2018;**3**(5):e00383-e00318. DOI: 10.1128/mSphere.00383-18

[24] Roca-Mesa H, Delgado-Yuste E, Mas A, Torija MJ, Beltran G. Importance of micronutrients and organic nitrogen in fermentations with *Torulaspora delbrueckii* and *Saccharomyces cerevisiae*. International Journal of Food Microbiology. 2022;**381**:109915. DOI: 10.1016/j.ijfoodmicro.2022.109915

[25] Prior K, Bauer F, Divol B. The utilisation of nitrogenous compounds by commercial non-*saccharomyces* yeasts associated with wine. Food Microbiology. 2019;**79**:75-84. DOI: 10.1016/j.fm.2018.12.002

[26] Luyt NA, Beaufort S, Divol B, Setati ME, Taillandier P, Bauer FF. Phenotypic characterization of cell-to-cell interactions between two yeast species during alcoholic fermentation. World Journal of Microbiology and Biotechnology. Nov 2021;**37**(11):186. DOI: 10.1007/s11274-021-03154-8

[27] Brice C, Cubillos FA, Dequin S, Camarasa C, Martínez C. Adaptability of the *Saccharomyces cerevisiae* yeasts to wine fermentation conditions relies on their strong ability to consume nitrogen. PLoS One. 12 Feb 2018;**13**(2):e0192383. DOI: 10.1371/journal.pone.0192383

[28] Remize F, Gaudin A, Kong Y, Guzzo J, Alexandre H, Krieger S, et al. *Oenococcus oeni* preference for peptides: Qualitative and quantitative analysis of nitrogen assimilation. Archives of Microbiology. 2006;**185**:459-469. DOI: 10.1007/s00203-006-0116-6

[29] Andorrà I, Berradre M, Mas A, Esteve-Zarzoso B, Guillamón JM. Effect of mixed culture fermentations on yeast populations and aroma profile. LWT. 1 Nov 2012;**49**(1):8-13. DOI: 10.1016/j.lwt.2012.04.008

[30] Gobert A, Tourdot-Maréchal R, Morge C, Sparrow C, Liu Y, Quintanilla-Casas B, et al. Non-*Saccharomyces* yeasts nitrogen source preferences: Impact on sequential fermentation and wine volatile compounds profile. Frontiers in Microbiology. 6 Nov 2017;**8**:2175. DOI: 10.3389/fmicb.2017.02175

[31] Bataillon M, Rico A, Sablayrolles JM, Salmon JM, Barre P. Early thiamin assimilation by yeasts under enological conditions: Impact on alcoholic fermentation kinetics. Journal of Fermentation and Bioengineering. 1 Jan 1996;**82**(2):145-150. DOI: 10.1016/0922-338X(96)85037-9

[32] Medina K, Boido E, Dellacassa E, Carrau F. Growth of non-Saccharomyces yeasts affects nutrient availability for *Saccharomyces cerevisiae* during wine fermentation. International Journal of Food Microbiology. 2012;**157**(2):245-250. DOI: 10.1016/j.ijfoodmicro.2012.05.012

[33] Evers MS, Ramousse L, Morge C, Sparrow C, Gobert A, Roullier-Gall C, et al. To be or not to be required: Yeast vitaminic requirements in winemaking. Food Microbiology. 1 Oct 2023;**115**:104330. DOI: 10.1016/j.fm.2023.104330

[34] Evers MS, Roullier-Gall C, Morge C, Sparrow C, Gobert A, Vichi S, et al. Thiamine and biotin: Relevance in the production of volatile and non-volatile compounds during *Saccharomyces cerevisiae* alcoholic fermentation in synthetic grape must. Foods. 24 Feb 2023;**12**(5):972. DOI: 10.3390/foods12050972

[35] Bisson L, Walker G, Ramakrishnan V, Luo Y, Fan O, Wiemer E, et al. Two faces of *Lactobacillus kunkeei*: Wine spoilage agent and bee probiotic catalyst: Discovery into practice. Catalyst: Discovery into Practice. 2017;**1**:1-11. DOI: 10.5344/catalyst.2016.16002

[36] Costello PJ, Kolouchova R, McCarthy J, Espinase Nandorfy D, Likos D, Schmidt SA. Transient acetaldehyde production by SO2-producing *Saccharomyces cerevisiae* promotes the survival of *Oenococcus oeni* during co-fermentation. OENO One. 2023;**57**(2):399-415. DOI: 10.20870/oeno-one.2023.57.2.7306

[37] Bardi L, Cocito C, Marzona. *Saccharomyces cerevisiae* cell fatty acid composition and release during fermentation without aeration and in absence of exogenous lipids. International Journal of Food Microbiology. 1999;**47**(1-2). DOI: 10.1016/S0168-1605(98)00203-7

[38] Shekhawat K, Bauer FF, Setati ME. Impact of oxygenation on the performance of three non-*saccharomyces* yeasts in co-fermentation

with *Saccharomyces cerevisiae*. Applied Microbiology and Biotechnology. 2017;**101**(6):2479-2791. DOI: 10.1007/s00253-016-8001-y

[39] Capucho I, San Romao MV. Effect of ethanol and fatty acids on malolactic activity of *Leuconostoc oenos*. Applied Microbiology and Biotechnology. 1994;**42**:391-395. DOI: 10.1007/BF00902747

[40] Torija MJ, Beltran G, Novo M, Poblet M, Guillamón JM, Mas A, et al. Effects of fermentation temperature and *Saccharomyces* species on the cell fatty acid composition and presence of volatile compounds in wine. International Journal of Food Microbiology. 2003;**85**(1-2):127-136. DOI: 10.1016/s0168-1605(02)00506-8

[41] Yap NA, de Barros LM, Langridge P, Henschke PA. The incidence of killer activity of non-*Saccharomyces* yeasts towards indigenous yeast species of grape must: Potential application in wine fermentation. Journal of Applied Microbiology. 2000;**89**(3):381-389. DOI: 10.1046/j.1365-2672.2000.01124.x

[42] Villalba ML, Sáez JS, Del Monaco S, Lopes CA, Sangorrín MP. TdKT, a new killer toxin produced by *Torulaspora delbrueckii* effective against wine spoilage yeasts. International Journal of Food Microbiology. 2016;**217**:94-100. DOI: 10.1016/j.ijfoodmicro.2015.10.006

[43] Hranilovic A, Albertin W, Capone DL, Gallo A, Grbin PR, Danner L, et al. Impact of *Lachancea thermotolerans* on chemical composition and sensory profiles of merlot wines. Food Chemistry. 2021;**349**:129015. DOI: 10.1016/j.foodchem.2021.129015

[44] Middelbeek EJ, Stumm C, Vogels GD. Effects of *Pichia kluyveri* killer toxin on sensitive cells. Antonie Van Leeuwenhoek. 1980;**46**:205-220. DOI: 10.1007/bf00444075

[45] Labbani FZ, Turchetti B, Bennamoun L, Dakhmouche S, Roberti R, Corazzi L, et al. A novel killer protein from *Pichia kluyveri* isolated from an Algerian soil: Purification and characterization of its in vitro activity against food and beverage spoilage yeasts. Antonie Van Leeuwenhoek. 2015;**107**:961-970. DOI: 10.1007/s10482-015-0388-4

[46] Comitini F, Gobbi M, Domizio P, Romani C, Lencioni L, Mannazzu I, et al. Selected non-*saccharomyces* wine yeasts in controlled multistarter fermentations with *Saccharomyces cerevisiae*. Food Microbiology. 2011;**28**(5):873-882. DOI: 10.1016/j.fm.2010.12.001

[47] Oro L, Ciani M, Comitini F. Antimicrobial activity of *Metschnikowia pulcherrima* on wine yeasts. Journal of Applied Microbiology. 2014;**116**(5):1209-1217. DOI: 10.1111/jam.12446

[48] Albergaria H, Francisco D, Gori K, Arneborg N, Gírio F. *Saccharomyces cerevisiae* CCMI 885 secretes peptides that inhibit the growth of some non-*saccharomyces* wine-related strains. Applied Microbiology and Biotechnology. 2010;**86**:965-972. DOI: 10.1007/s00253-009-2409-6

[49] Branco P, Francisco D, Chambon C, Hébraud M, Arneborg N, Almeida MG, et al. Identification of novel GAPDH-derived antimicrobial peptides secreted by *Saccharomyces cerevisiae* and involved in wine microbial interactions. Applied Microbiology and Biotechnology. 2014;**98**:843-853. DOI: 10.1007/s00253-013-5411-y

[50] Mencher A, Morales P, Valero E, Tronchoni J, Patil KR, Gonzalez R.

Proteomic characterization of extracellular vesicles produced by several wine yeast species. Microbial Biotechnology. 2020;**13**(5):1581-1596. DOI: 10.1111/1751-7915.13614

[51] Nehme N, Mathieu F, Taillandier P. Impact of the co-culture of *Saccharomyces cerevisiae–Oenococcus oeni* on malolactic fermentation and partial characterization of a yeast-derived inhibitory peptidic fraction. Food Microbiology. 2010;**27**(1):150-107. DOI: 10.1016/j.fm.2009.09.008

[52] Bartel CE, Borneman AR, Curtin CD, Varela CA. Is *Brettanomyces bruxellensis* becoming more SO2 tolerant in industry? In: 17th Australian Wine Industry Technical Conference; Adelaide, South Australia; 21-24 July 2019. 2020. p. 50

[53] Lauderdale AD. The Impact of Malolactic Fermentation Conducted by *Oenococcus oeni* on *Brettanomyces bruxellensis* Growth and Volatile Phenol Production [thesis]. Corvallis: Oregon State University; 2021

[54] Bely M, Stoeckle P, Masneuf-Pomarède I, Dubourdieu D. Impact of mixed *Torulaspora delbrueckii–Saccharomyces cerevisiae* culture on high-sugar fermentation. International Journal of Food Microbiology. 2008;**122**(3):312-320. DOI: 10.1016/j.ijfoodmicro.2007.12.023

[55] Windholtz S, Nioi C, Coulon J, Masneuf-Pomarède I. Bioprotection by non-*Saccharomyces* yeasts in oenology: Evaluation of O_2 consumption and impact on acetic acid bacteria. International Journal of Food Microbiology. 2023;**405**:110338. DOI: 10.1016/j.ijfoodmicro.2023.110338

[56] Giménez P, Just-Borras A, Pons P, Gombau J, Heras JM, Sieczkowski N, et al. Biotechnological tools for reducing the use of sulfur dioxide in white grape must and preventing enzymatic browning: Glutathione; inactivated dry yeasts rich in glutathione; and bioprotection with *Metschnikowia pulcherrima*. European Food Research and Technology. 2023;**249**(6):1491-1501. DOI: 10.1007/s00217-023-04229-6

[57] RESOLUTION OIV-CST 369-2011. OIV Code of Good Vitivinicultural Practices in Order to Minimise the Presence of Biogenic Amines in Vine-Based Products. 2011. Available from: https://www.oiv.int/public/medias/2766/oiv-cst-369-2011-en.pdf [Accessed: July 21, 2023]

[58] Zhang B, Tang C, Yang D, Liu H, Xue J, Duan C, et al. Effects of three indigenous non-*Saccharomyces* yeasts and their pairwise combinations in co-fermentation with *Saccharomyces cerevisiae* on volatile compounds of Petit Manseng wines. Food Chemistry. 2022;**368**:130807. DOI: 10.1016/j.foodchem.2021.130807

[59] Pozo-Bayón MA, G-Alegríaolo E, Polo MC, Tenorio C, Martín-Álvarez PJ, Calvo De La Banda MT, et al. Wine volatile and amino acid composition after malolactic fermentation: Effect of *Oenococcus oeni* and *Lactobacillus plantarum* starter cultures. Journal of Agricultural and Food Chemistry. 2005;**53**(22):8729-8735. DOI: 10.1021/jf050739y

[60] Viana F, Belloch C, Vallés S, Manzanares P. Monitoring a mixed starter of *Hanseniaspora vineae–Saccharomyces cerevisiae* in natural must: Impact on 2-phenylethyl acetate production. International Journal of Food Microbiology. 2011;**151**(2):235-240. DOI: 10.1016/j.ijfoodmicro.2011.09.005

[61] Zhang B, Ivanova-Petropulos V, Duan C, Yan G. Distinctive chemical

and aromatic composition of red wines produced by *Saccharomyces cerevisiae* co-fermentation with indigenous and commercial non-*saccharomyces* strains. Food Bioscience. 2021;**41**:100925. DOI: 10.1016/j.fbio.2021.100925

[62] Plata C, Millan C, Mauricio JC, Ortega JM. Formation of ethyl acetate and isoamyl acetate by various species of wine yeasts. Food Microbiology. 2003;**20**(2):217-224. DOI: 10.1016/S0740-0020(02)00101-6

[63] Dutraive O, Benito S, Fritsch S, Beisert B, Patz CD, Rauhut D. Effect of sequential inoculation with non-*saccharomyces* and *saccharomyces* yeasts on Riesling wine chemical composition. Fermentation. 2019;**5**(3):79. DOI: 10.3390/fermentation5030079

[64] Anfang N, Brajkovich M, Goddard MR. Co-fermentation with *Pichia kluyveri* increases varietal thiol concentrations in Sauvignon Blanc. Australian Journal of Grape and Wine Research. 2009;**15**(1):1-8. DOI: 10.1111/j.1755-0238.2008.00031.x

[65] Zott K, Thibon C, Bely M, Lonvaud-Funel A, Dubourdieu D, Masneuf-Pomarede I. The grape must non-*Saccharomyces* microbial community: Impact on volatile thiol release. International Journal of Food Microbiology. 2011;**151**(2):210-215. DOI: 10.1016/j.ijfoodmicro.2011.08.026

[66] Rosi I, Vinella M, Domizio P. Characterization of β-glucosidase activity in yeasts of oenological origin. Journal of Applied Bacteriology. 1994;**77**(5):519-527. DOI: 10.1111/j.1365-2672.1994.tb04396.x

[67] Ugliano M, Moio L. The influence of malolactic fermentation and *Oenococcus oeni* strain on glycosidic aroma precursors and related volatile compounds of red wine. Journal of the Science of Food and Agriculture. 2006;**86**(14):2468-2476. DOI: 10.1002/jsfa.2650

[68] Benito Á, Calderón F, Palomero F, Benito S. Combine use of selected *Schizosaccharomyces pombe* and *Lachancea thermotolerans* yeast strains as an alternative to the traditional malolactic fermentation in red wine production. Molecules. 2015;**20**(6):9510-9523. DOI: 10.3390/molecules20069510

[69] Fleet GH. Yeast interactions and wine flavour. International Journal of Food Microbiology. 2003;**86**(1-2):11-22. DOI: 10.1016/S0168-1605(03)00245-9

[70] Domizio P, Liu Y, Bisson LF, Barile D. Use of non-*saccharomyces* wine yeasts as novel sources of mannoproteins in wine. Food Microbiology. 2014;**43**:5-15. DOI: 10.1016/j.fm.2014.04.005

[71] Belda I, Navascués E, Marquina D, Santos A, Calderon F, Benito S. Dynamic analysis of physiological properties of *Torulaspora delbrueckii* in wine fermentations and its incidence on wine quality. Applied Microbiology and Biotechnology. 2015;**99**:1911-1922. DOI: 10.1007/s00253-014-6197-2

Section 3

Beer Biotechnology

Chapter 11

Diacetyl Production during Brewing and Its Management through Process Optimization and Molecular Evolution of Yeast

Dagnew Bitew and Berhanu Andualem

Abstract

Diacetyl is butter-tasting off-flavor compound produced as by-product of yeast valine metabolism during brewery fermentation. Yeasts produce diacetyl during primary fermentation and then reabsorb it in secondary fermentation. This causes a non-productive lengthy maturation period, which is costly. Several strategies have been proposed to manage diacetyl and improve the productivity of brewery industries. This review aimed to assess diacetyl production and proposed strategies to manage diacetyl production during brewing. Diacetyl production and its amount in the green beer are influenced by brewing condition and type of strain used. Green beer conditioning and brewing process optimization are regarded as simple and feasible approaches. However, these have their own inherent drawbacks. On the other hand, a plethora of researches declared that genetic manipulation of yeasts is an effective strategy in reducing diacetyl amount and ultimately to shorten the maturation period and thereby maximize profitability of brewery industries. But the applicability genetic engineering limited, due to firm regulation of utilization of genetically modified organisms in food processing industries. Therefore, though extensive research was done on identifying and understanding factors which influence yeast diacetyl formation and reduction, diacetyl management is persisting as a challenge in brewing systems.

Keywords: beer, brewer's yeast, diacetyl, process optimization, strain improvement

1. Introduction

The art of brewing accounts for a longer period of time and seems to be a well-established process. However, the brewery industry and researchers are still striving for optimization of fermentation conditions. The development of high-performing yeasts and, in general, affirming consumers' claims are crucial to the sustained profitability of the brewery industry. Indeed, nowadays, brewery industries are facing increasing product demand along with enhanced product quality and optimal profitability [1, 2]. One of the bottlenecks to ensuring the increased availability of beer and brewery economics is the longer beer maturation period, which is in turn associated with diacetyl formation during brewery fermentation [1].

Diacetyl is one of the vicinal diketones produced by yeast and imparts a distinct buttery/butterscotch flavor to fermented foods and beverages [3]. In fermented beverages, diacetyl content may be perceived positively or negatively depending on the product type, style, and concentration. In lager beers, diacetyl is considered a serious off-flavor compared to ale beers [3, 4]. The reported acceptable diacetyl threshold values in a final beer are around 0.1–0.2 mg/l in lager-style beers and 0.1–0.4 mg/l in ale-style beers [5], although threshold values as low as 0.017 mg/l [6] and 0.014–0.016 mg/l [7] have also been reported. The variation in flavor threshold value is due to variations in a taster's geographical background, ethnicity, diet, and smoking practices [8].

As result, brewers always attempt various strategies to manage the diacetyl level below the threshold [2, 4]. Beer stabilization, or keeping the green beer for an extended maturation period, is the most commonly used method to reduce the diacetyl amount in the final beer. In lager beer, the green beer has to be stored for 2 to 3 weeks at a temperature close to the freezing point until the diacetyl concentration declines below its taste threshold value. However, this prolonged maturation demands cost for additional storage facilities and ultimately makes the process economically unfeasible [9].

Consequently, considering the off-flavor property of diacetyl and the prolonged maturation period as a result of it, a number of alternative strategies are being used and have been proposed. This includes selection of yeast strain, optimization of fermentation conditions such as increased fermentation temperature [10], normal wort specific gravity and pitching rate [11], wort valine content [12], decreased wort pH value [13] and optimal wort aeration and oxygen availability [14], addition of exogenous enzyme [15], passing the beer through a column of immobilized yeast [16], and genetic modification of yeast [17, 18]. However, most strategies have inherent drawbacks and lack of social acceptability in the case of genetic modification [3, 4, 19].

Though the brewing process is a highly complex series of chemical reactions, identifying, understanding, and optimizing factors that influence the formation and reabsorption of diacetyl is critical to managing the diacetyl content in the final beer. This in turn shortens the maturation time and consequently increases the productivity, profitability, and competitiveness of the brewery industry in the global market. Therefore, the purpose of this review was to perform a critical analysis of diacetyl production and its management strategies during the brewing process.

2. Production of diacetyl in brewery fermentation

It is a fact that understanding the diacetyl biosynthesis pathway, identifying the enzymes and coding genes involved, and other influencing brewing conditions play a tremendous role in reducing the diacetyl amount in final beer. Diacetyl is produced by the brewer's yeast during fermentation. It is formed extracellularly through spontaneous nonenzymatic oxidative decarboxylation of α-acetolactate, which is an intermediate metabolite in the valine biosynthesis pathway, as indicated in **Figure 1** [4].

The conversion of α-acetolactate to 2,3-dihydroxy-isovalerate is rate-limiting; hence, during fermentation, excessively produced α-acetolactate is secreted out through the cell's plasma membrane into the wort, where it is spontaneously

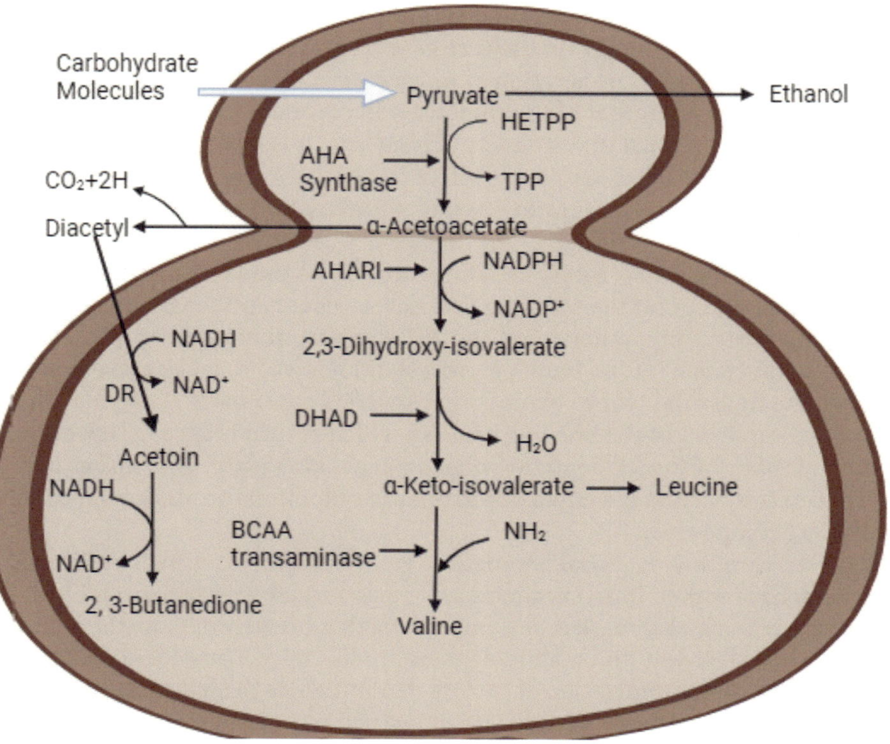

Figure 1.
Graphic outline of diacetyl formation and reduction process in valine biosynthesis pathways of Saccharomyces spp.

decarboxylated to diacetyl. Likely, the spontaneous decarboxylation of α-acetolactate to diacetyl is also rate-limiting. Therefore, during fermentation, the concentrations of free diacetyl in the wort are usually low, whereas α-acetolactate exists at higher concentrations. As time goes on, it decarboxylates into diacetyl [20]. Hence, the total diacetyl concentration in beer is expressed as the sum of free diacetyl and α-acetolactate (potential diacetyl) [4, 20].

Where: AHA synthase (α-acetolactate synthase), AHARI (acetohydroxyacid reductoisomerase), DHAD (dihydroxyacid dehydratase), BCAA transaminase (branched-chain amino acid transaminase), and DR (diacetyl reductase).

2.1 Green beer conditioning and brewing process optimization

2.1.1 Optimizing fermentation process

Fermentation is the central and most important step in alcoholic beverage production. The progression rate of fermentation is affected by the viability and vitality of brewery yeast strains, which in turn are affected by fermentation conditions, such as temperature [4], wort pH [21], dissolved oxygen content [22], and osmotic pressure (associated with wort specific gravity and pitching rate) [11].

Indeed, the flavor profile of the final beer is greatly affected by the fermentation conditions. Diacetyl is a normal metabolic byproduct of yeast during brewery fermentation. But its amount, rate of formation, and subsequent removal rate from the green beer (immature beer) are influenced by the above-mentioned factors [8].

For instance, brewing conditions, such as lower fermentation temperature, deviation from the normal amount (in the case of wort gravity and pitching rate), higher wort pH, and excessive wort aeration or oxygenation, have a considerable positive effect on the diacetyl content of the final beer. In contrast, increased fermentation temperature, normal wort gravity and pitching rate, lower wort pH, and optimal wort oxygenation have a recognizable effect on reducing diacetyl content in beer and then minimizing the maturation period without altering other beer parameters [4, 11, 21–23].

The steps in modern brewing procedures are almost the same as in ancient craftsmanship, except for the technological development and the implementation of modern industrial equipment [2, 4, 24, 25]. But several reports complain that the technical advancement of the brewing process did not lead to an improvement in the quality of the final product in a manner that satisfied consumers' demands. This is due to the complexity of the brewing process, which in turn induces a lack of understanding of much of the fermentation phenomena taking place. This makes it challenging to exactly define the effect of process alterations on the processing condition and product composition [24, 25].

However, the growing global population, along with the continual increase in consumers, has resulted in an ever-increasing demand for beer products. Thus, to be competitive in the global market, it is imperative that brewers operate their production processes effectively and maximize their productivity. Thus, identification, understanding, and optimization of factors that influence the formation and reabsorption of diacetyl without affecting other beer quality parameters are required to be studied. This in turn shortens the maturation time and consequently increases the productivity, profitability, and competitiveness of the brewery industry in the global market.

2.1.2 Lowering preferred amino acids and improving wort valine content

Studies reported that wort amino acids are divided into four groups (Groups A, B, C, and D) depending on their preference and uptake rate by yeast, as indicated in **Table 1**. Groups A, B, C, and D [12].

Since diacetyl is an intermediate metabolite in the biosynthesis of valine, its amount produced during fermentation is dependent on the wort valine content. The level of diacetyl content is reduced when the amount of valine in the wort is increased and the amount of other preferred amino acids is lowered. The study of Krogerus and Gibson [12] showed that the increment of valine supplementation to the wort from 100 to 300 mg/L resulted in 37 and 33% diacetyl concentration reductions during primary fermentation and at the end of fermentation, respectively, without affecting the quality of the beer. Moreover, lowering the amount of highly preferred amino acids (those taken up faster than valine) will create an opportunity for increased uptake of valine because of less competition for amino acid transporter proteins (permease) interactions. This implies a reduced total diacetyl amount in green beer [12, 26–28].

On the other hand, malt type and its processing conditions (such as steeping, germination, and kilning) greatly affect the amino acid profile of wort. The study by Nie et al. [29] found that different varieties of barley have different amino acid profiles. The same study also showed that different malting conditions resulted in different wort amino acid profiles. Likewise, the study by Samaras et al. [30] revealed that high malt kilning temperatures can decrease the concentration of amino acids in the malt, whereas malts kilned at lower temperatures contained a higher concentration of

Category of amino acids	Degree of absorption/ preference	Amino acids
Group A	Highly preferred and absorbed	Glutamic acid, aspartic acid, asparagine, glutamine, serine, threonine, lysine, and arginine
Group B	Moderately preferred and absorbed	Valine, methionine, leucine, isoleucine, and histidine
Group C	Poorly preferred and absorbed	Glycine, phenylalanine, tyrosine, tryptophan, and alanine
Group D	No/Least preferred and absorbed	Proline

Table 1.
The classification of wort amino acids based on their uptake rate in S. cerevisiae.

both valine and the total amount of amino acids. Thus, it is possible to conclude that diacetyl can be managed by adjusting the amino acid content of wort, selecting the best barley variety, and optimizing malting conditions.

2.1.3 Green beer conditioning

2.1.3.1 Addition of exogenous enzyme to the green beer

The brewing process is based on a myriad of endogenous and exogenous enzymatic activities. The use of commercial exogenous enzymes is a common practice in modern breweries. These enzymes help breweries be competitive by increasing process efficiency, enhancing product quality, extending product ranges, and resolving intermittent process problems, such as incomplete fermentation, turbidity/haze, and increased diacetyl content. Endogenous enzymes are needed to convert starch into sugars, complex proteins into simple amino acids, and breakdown and solubilize plant cell wall materials for yeast nutrition. In addition, brewery industries add exogenous enzymes such as α-acetolactate decarboxylase and β-1,4-glucanase to hasten the fermentation process and avoid incomplete fermentation, respectively [15, 31].

Alpha-Acetolactate decarboxylase (ALDC) is an enzyme that directly converts α-acetolactate to acetoin by bypassing diacetyl formation. This avoids the slow oxidative decarboxylation of α-acetolactate into diacetyl and the subsequent yeast diacetyl reabsorption stage and/or maturation period [15]. However, ALDC could not be produced by the pitching yeast. But this enzyme is produced by different species of bacteria, such as *Bacillus subtilis*, *Klebsiella ternigena*, *Enterobacter aerogenes*, *Lactobacillus casei*, *Bacillus brevis*, and *Streptococcus diacetylactis*. Hence, safe production of these enzymes from the aforementioned producer organisms and the addition of ALDC to the fermentation accelerate the maturation of beer, resulting in quick and improved vessel utilization and thus increasing the brewing capacity of brewery industries [32–34].

The study of Choi et al. [15] aimed to assess the efficacy of ALDC in reducing diacetyl content in two different types of beer (Jinyang and Dahyang) and showed a remarkable diacetyl reduction. The diacetyl content of Jinyang decreased from 5.17 to 1.31 ppm, while the content of Dahyang decreased from 9.75 to 1.54 ppm. These values showed that the diacetyl content was reduced by 25% in Jinyang and 15% in Dahyang compared to the controls.

2.1.3.2 Treatment of green beer with immobilized yeast cell

Nowadays, a variety of immobilized cell technologies are available and have been used in bioprocess industries such as the brewery industry. As compared with traditional fermentation with freely suspended cells and operated in batch mode, immobilized cell systems offer many advantages, such as a faster fermentation rate, increased volumetric productivity, and the possibility of continuous operation. Hence, immobilized cell technology has been attracting the attention of fermentation industries and is used for different stages in the beer fermentation process [16].

Immobilized cell technology is being used in secondary fermentation to remove diacetyl and then shorten the maturation period [16, 35]. In this process, immobilized yeast cells absorb diacetyl from the green beer and are then converted into acetoin, which in turn is converted into 2,3-butanediol. Diacetyl removal from green beer involves rapid heating of the beer after the primary fermentation to facilitate the conversion of α-acetolactate into diacetyl. Then the heat-treated green beer will be passed through a column of immobilized yeast, in which the diacetyl reduction will take place (**Figure 2**) [36].

2.2 Molecular evolution of yeasts

It is worth mentioning that the brewer's yeasts played a major role in diacetyl production and removal from green beer. The molecular evolution of yeast could be the right and effective diacetyl management strategy. However, blind accusations and frustration with genetically modified yeast are common phenomena in breweries. The applicability of genetically modified brewery yeasts is still not allowed and is continuing to be under strict control due to illogical perceptions and ethical and legislative barriers. Moreover, brewers are hesitant to use engineered yeast because they suspect that it might alter the branding characteristics of their beer. Hence, though genetically

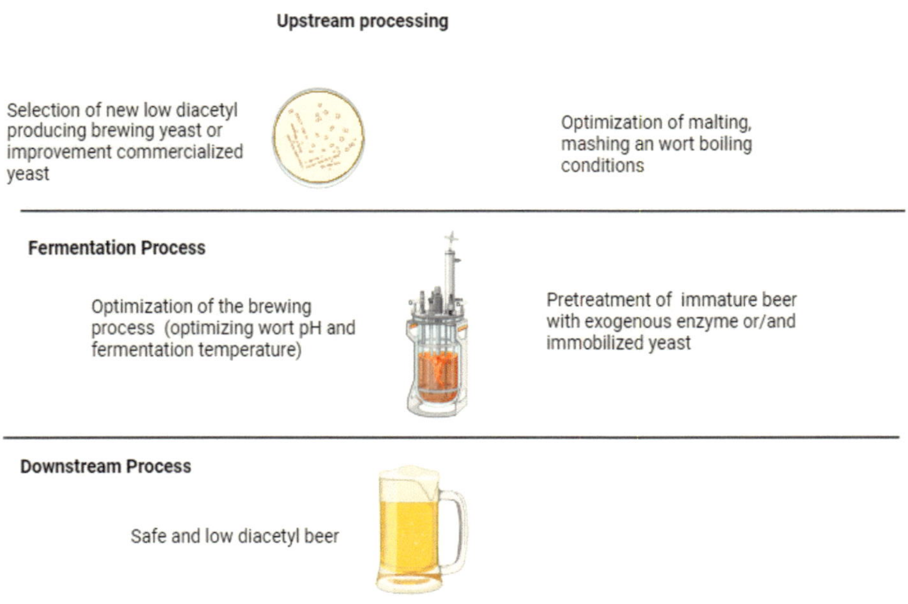

Figure 2.
Schematic presentation suggested optimization in the brewing process for lowering diacetyl production.

modified microbes are used in medical aspects and foods originated from genetically modified crops are consumed, the use of genetically modified brewer's yeast in the brewery industry is lagging behind [17].

Genetic engineering is a recent, faster, more accurate, and very effective technique, and considerable outcomes are obtained in the improvement of microbes, crops, and livestock [17, 18, 37]. Brewer's yeast strains that show improved ethanol tolerance, fermentation speed, and attenuation produce very specific flavors, less diacetyl, ferment starch, and/or dextrin are developed through genetic engineering. Therefore, breweries are starting to appreciate the applications of genetically modified yeasts to increase the productivity of breweries and afford the increasing consumer demand [17, 18, 37].

Moreover, different reports revealed that genetically modified brewer's yeast strains produce beer with less undesirable flavor compounds, such as diacetyl and 2,3-pentanedione, which increase the productivity of breweries and satisfy consumer demand. For instance, a yeast strain that can degrade some off-flavor molecules was patented in the USA [38]. Genetic modification approaches involving manipulation of the genes associated with diacetyl production and reduction are stated as effective, though they have limited applicability [1, 18, 37].

2.2.1 Altering genes involved in diacetyl synthesis pathway

Reports showed that manipulating genes involved in the diacetyl production pathway can effectively manage diacetyl content to the threshold limit. *S. cerevisiae* has four genes (*Sc-ILV2*, *ILV3*, *ILV5*, *ILV6*, and *BDH2*) that are associated with diacetyl production. *Sc-ILV2* and *ILV6* encode AHA synthase and its regulatory subunit. Whereas *ILV3*, *ILV5*, and *BDH2* code for dihydroxyacid dehydratase, acetohydroxyacid reductoisomerase, and diacetyl reductase, respectively, as indicated in **Figure 1**. Extensive research conducted so far has declared that *Sc-ILV2* and *ILV6* are directly associated with diacetyl production. But *ILV3*, *ILV5*, and *BDH2* are inversely related to diacetyl production and diacetyl content in the final beer. Previous studies suggested that reduction in the gene copy numbers of *Sc-ILV2* and *ILV6* and overexpression of *ILV3*, *ILV5*, and *BDH2* play a tremendous role in minimizing diacetyl production and then managing its amount below the threshold level in the final beer [18, 37, 39, 40].

2.2.2 Deletion of Sc-ILV2 and ILV6 genes

The results of different studies claimed that *Sc-ILV2* and *ILV6* gene copy numbers and expression levels are positively associated with diacetyl content in the final beer. Multiple copy numbers and overexpression of these genes lead to higher diacetyl production during beer production. The *Sc-ILV2* gene encodes the enzyme AHA synthase, which facilitates the conversion of pyruvate to α-acetolactate and leads to a higher diacetyl content in the final beer. The increased expression of this gene results in increased AHA synthase production and activity, which in turn increase the production of α-acetolactate. The increased production of α-acetolactate results in an increased diacetyl level in the final beer [37, 39].

Reports showed that deletion of one more of this gene reduced diacetyl production and then shortened the nonproductive, lengthy maturation period [39, 41]. The study of Zhang et al. [39] showed that disruption of one allele of the *Sc-ILV2* gene decreased AHAS activity by 59 and 41% in T1 and Q9 recombinant yeasts, respectively, as compared with their normal parent. Moreover, the acetoin amount at the end of fermentation is decreased by 35 and 40% in T1 and Q9 than that of their parent, as a result of limited diacetyl production (precursor metabolite of acetoin) reduction [39]. Furthermore, the study by Gibson et al. [42] indicated that cells treated with a sublethal amount of chlorsulfuron (specifically, that inhibits AHA synthase production) produce beer with a 60% lower diacetyl content as compared with nontreated cells of the same species.

The other gene that is responsible for diacetyl production is *Sc-ILV6*, which encodes a subunit of the AHAS multisubunit enzyme. The study of Duong et al. [41] found that lowered expression of *Sc-ILV6* during fermentation in a lager yeast strain produced beer with a low diacetyl level as compared with other yeast strains, in which *Sc-ILV6* was overexpressed. The same study also observed lower diacetyl production by a lager yeast strain with a disrupted *Sc-ILV6* gene compared with the unmodified strain, suggesting that the *Sc-ILV6* encodes an AHA synthase enzyme subunit. In this study, a *Sc-ILV6* double deletion mutant brewery yeast resulted in a 65% reduction in the final diacetyl concentration [41].

2.2.3 Overexpression of Sc-ILV5, ILV-3, and BDH2 genes

Contrary to the *Sc-ILV2* and *ILV6* genes, the higher gene copy number and expression level of *Sc-ILV5*, *ILV3*, and *BDH2* are inversely related to diacetyl production and total diacetyl concentration at the end of fermentation. *ILV3* and *ILV5* genes encode for dihydroxyacid dehydratase and acetohydroxyacid reductoisomerase, which catalyze the second and third reactions in the valine biosynthesis pathway, respectively. Whereas BDH2 encodes diacetyl reductase that catalyzes the reduction of diacetyl into acetoin within the yeast cell after it is absorbed [18, 37, 40].

Overexpression of these genes increases the activity of the respective encoded enzymes. The increased activity of dihydroxyacid dehydratase and acetohydroxyacid reductoisomerase leads to an increased anabolic flux of α-acetoacetate to the intermediate metabolites of valine. This situation consequently decreases the amount of α-acetolactate leaking out of the yeast cells during fermentation. On the other hand, the greater the activity of diacetyl reductase, the faster the conversion of leaked diacetyl to acetoin, a compound with a high threshold value, ultimately resulting in a low total diacetyl concentration at the end of fermentation [18, 37, 40].

The study of Qin and Park [37] reported that recombinant brewery yeast strains that have taken up the *ILV5* gene showed 3.7 times higher diacetyl reduction as compared with the not-transformed parent strain. The utilization of those transformed yeast strains for wine fermentation using Campbell Early and Muscat Baily A grapes brought approximately 35 to 39% diacetyl content reduction as compared with nontransformed control yeasts. Moreover, the same study also assessed the activity of cell-free acetohydroxyacid reductoisomerase. Acetohydroxyacid reductoisomerase extracted from it showed approximately 4 to 5 times higher activity than the control yeasts.

On the other hand, studies that involved deletion of *Sc-ILV2* and overexpression of *Sc-ILV5* and *BDH2* played a remarkable role in diacetyl reduction. Three strains, WY1, WY1-2, and WY1-12, transformed with *BDH1*, *BDH2*, and both *BDH1* and *BDH2*, produced beer with diacetyl concentrations that decreased by 39.81, 33.42, and 46.71%, respectively, compared with the control strain [43]. The study of Kusunoki and Ogata [40], involving construction of self-cloning lager yeast via insertion of *Sc-ILV5* DNA upstream of *ILV2* in bottom-fermenting yeast to increase the copy number of *Sc-ILV5*, achieved an approximately 60% reduction in the diacetyl amount at the end of fermentation and shortened the maturation period during which diacetyl was reduced to within the threshold value, with no impact on beer quality. Similarly, the report of [18] claimed a 55.65% diacetyl reduction in the recombinant strain, where *BDH2* was an overexpressed gene and one *Sc-ILV2* allelic gene was deleted. Moreover, a strain in which the *Sc-ILV5* gene was overexpressed and one *Sc-ILV2* allelic gene knockout showed a 69.13% diacetyl reduction.

There are also studies that indicate the role of co-expression of those genes in reducing the diacetyl amount to the threshold level at the end of fermentation. The study of [44] reported that the co-expression of *Sc-ILV3* and *Sc-ILV5* in brewer's yeast resulted in a 60% diacetyl reduction in beer. The same study revealed that overexpression of *ILV3* and *ILV5* in the S1 yeast strain resulted in a 40 and 70% diacetyl reduction, respectively, when compared to that of the normal parent strain.

2.2.4 *Introduction of α-acetolactate decarboxylase enzyme gene to brewer's yeast*

Alpha-acetolactate decarboxylase (ALDC) is an enzyme that catalyzes the nonoxidative decarboxylation of α-acetolactate into acetoin directly without the formation of diacetyl [15]. However, the *ALDC* gene is not found in brewery yeast strains, and they did not produce ALDC. In contrast, ALDC is produced by different species of bacteria, such as *Bacillus subtilis*, *Klebsiella ternigena*, *Enterobacter aerogenes*, *Lactobacillus casei*, *Bacillus brevis*, and *Streptococcus diacetylactis*. The results of different studies confirmed that the integration of *ALDC* genes originated from different bacterial species into brewery yeasts plays a crucial role in the reduction of total diacetyl content [33, 34].

The study by Blomqvist et al. [32] on the integration of *Klebsiella ternigena* and *Enterobacter aerogenes* genes encoding α-acetolactate decarboxylase into brewery yeasts revealed that the diacetyl amount was reduced even below the threshold value at the end of primary fermentation. Similarly, *Saccharomyces cerevisiae* with a truncated active ALDC from *Acetobacter aceti* ssp. xylinum attached to the cell wall using the C-terminal anchoring domain of α-agglutinin decreased the diacetyl concentration by 30% as compared with control yeasts displaying only the anchoring domain [45].

2.2.5 *Overexpression of BAP2 and BAP3 genes*

It is mentioned that the amount of diacetyl produced in the green beer is influenced by the wort amino acid composition and yeast uptake rate [12]. Apart from this, the uptake rate of amino acids via the yeast cell is determined by the expression level of amino acid transporter protein (permease) encoding genes [4, 12, 46]. In yeast, various amino acid permeases are involved in the transport of amino acids across the plasma membrane with different affinities, specificities, capacities, and regulations. Most of the branched-chain amino acids are transported by Bap2p, Bap3p, and Tat1p. These amino acid permease membrane proteins indirectly play a critical role in the flavor profile of the final beer. For instance, Bap2p and Bap3p are amino acid permeases involved in the uptake of leucine, isoleucine, and valine (branched-chain amino acids) [12, 46]. The overexpression of these amino acid permease genes increases the uptake level of valine and isoleucine. When cells have sufficient valine uptake, diacetyl levels are reduced due to feedback inhibition in the valine biosynthesis pathway. Similarly, the increased uptake rate of isoleucine resulted in a remarkable 2,3-pentanedione reduction [12, 46]. This implies that inducing the transcription of

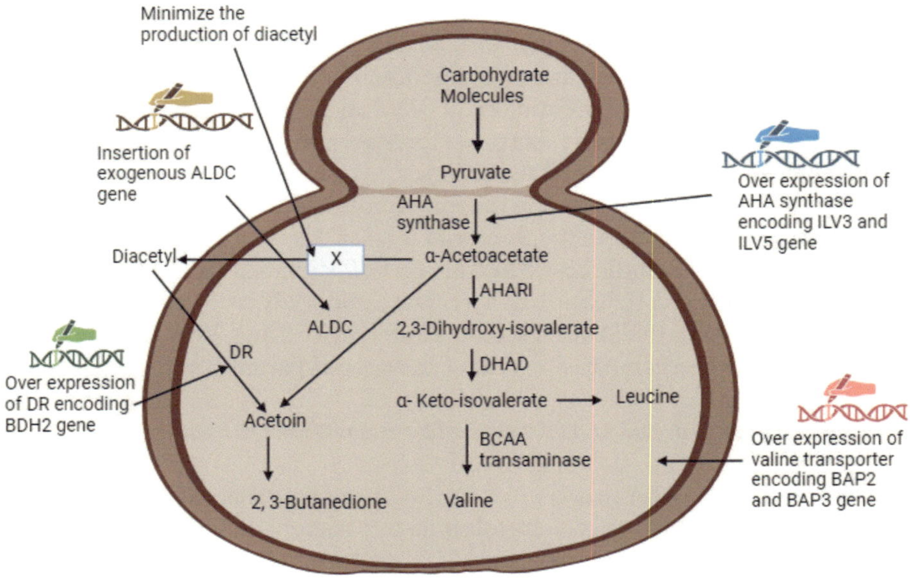

Figure 3.
Schematic presentation of expression of genes involved in valine biosynthesis in brewer's yeast targeting lowering diacetyl amount in the final beer.

those genes increases the uptake of branched-chain amino acids such as valine, which then reduces the diacetyl amount through feedback inhibition (**Figure 3**).

The study of Kodama [27] found that the constitutive expression of *BAP2* in a brewing yeast strain accelerated the assimilation rate of branched-chain amino acids, while the disruption of *BAP2* did not affect assimilation rates for these amino acids during the brewing process. These could be Bap3p, Tat1p, and/or other branched-chain amino acid permease homologs, which exist in lager brewing yeasts. The transcription of the branched-chain amino acid permease genes, particularly *BAP2* and *BAP3*, is induced by some amino acids, such as leucine and phenylalanine [47], in the medium, and this induction requires Ssy1p as a sensor for external amino acids [48].

3. Conclusion

Brewing is the oldest and most well-established process, and beer is the most economical and widely produced beverage all over the world. Nowadays, as a result of the increasing global population and the advancement of human life, the brewery industry is facing increasing product demand and consumer's quality claims, along with sustaining their profitability. One of the bottlenecks to ensuring increased availability and avoiding quality claims is associated with a longer maturation period. This in turn allied with diacetyl production, a buttery-tasting off-flavor compound produced by yeasts during brewery fermentation as byproduct of valine metabolism.

Though yeast diacetyl reduction is not well understood, the diacetyl amount in green beer is affected by the brewing conditions and genetic constituents of the yeasts used. As a result, green beer conditioning, brewing process optimization, and genetic modification of yeasts have been used and proposed to manage the diacetyl amount below the detectable or threshold amount. However, these approaches have an inherent problem that affects other beer quality parameters and increases the production cost.

Whereas genetic modification of yeast strains, particularly manipulation of genes responsible for diacetyl production and removal, is believed to be an efficient and economically feasible approach. But it has been inapplicable due to the stringent regulation of utilizing genetically modified organisms in the food processing industry. Considerable research has been conducted and huge publications are available, but diacetyl management is still a logjam to brewery productivity. Therefore, there must be an amalgamated pressure against the illogical frustration of breweries and societies with new and/or engineered yeasts.

Acknowledgements

The current study has not received external funding.

Compliance with ethical standards

Ethical statement

This manuscript is in compliance with Ethical Standards. Since it does not involve any human or animals participant.

Conflict of interest

The authors declare that they have no any competing interests.

Author details

Dagnew Bitew[1,2]* and Berhanu Andualem[2]

1 Department of Biology, Mizan-Tepi University, Ethiopia

2 Department of Biotechnology, University of Gondar, Ethiopia

*Address all correspondence to: btdagne@gmail.com

References

[1] Wang Z, He X, Liu N, Zhang B. Construction of self-cloning industrial brewing yeast with high-glutathione and low diacetyl production. International Journal of Food Science and Technology. 2008;**43**:989-994

[2] Karabin M, Jelinek L, Kotrba P, Cejnar R, Dostalek P. Enhancing the performance of brewing yeasts. Biotechnology Advancement. 2018;**36**:691-706

[3] Gibson B, Krogerus K, Ekberg J, Monroux A, Mattinen L, Rautio J, et al. Variation in α-acetolactate production within the hybrid lager yeast group *Saccharomyces pastorianus* and affirmation of the central role of the *ILV6* gene. Yeast. 2015;**32**:301-316

[4] Krogerus K, Gibson BR. 125th Anniversary Review: Diacetyl and its control during brewery fermentation. Journal of the Institute of Brewing. 2013a;**119**:86-97

[5] Meilgaard MC. Beer flavor terminology. Journal of the Institute of Brewing. 1979;**85**:38-42

[6] Saison D, de Schutter D, Uyttenhove B, Delvaux F, Delvaux FR. Contribution of staling compounds to the aged flavor of lager beer by studying their flavour thresholds. Food Chemistry. 2009;**114**:1206-1215

[7] Kluba R, de Banchs N, Fraga A, Jansen G, Langstaff S, Meilgaard M, et al. Sensory threshold determination of added substances in beer. Journal of the American Society of Brewing Chemist. 1993;**51**:181-183

[8] Stewart GG. The production of secondary metabolites with flavor potential during brewing and distilling wort fermentations. Fermentation. 2017;**3**(63):1-27

[9] Puligundla P, Smogrovicova D, Obulam VSR, SanghoonKo S. Very high gravity (VHG) ethanolic brewing and fermentation: A research update. Journal of Industrial Microbiology & Biotechnology. 2011;**38**:1133-1144. DOI: 10.1007/s10295-011-0999-3

[10] Kucharczyk K, Tuszynski T. The effect of temperature on fermentation and beer volatiles at an industrial scale. Journal of the Institute of Brewing. 2018;**124**:230-235

[11] Verbelen P, Saerens S, Mulders S, Delvaux F, Delvaux R. Impact of pitching rate on yeast fermentation performance and beer flavor. Applied Microbiology and Biotechnology. 2009a;**82**:155-167

[12] Krogerus K, Gibson BR. Influence of valine and other amino acids on total diacetyl and 2,3-pentanedione levels during fermentation of brewer's wort. Applied Microbiology and Biotechnology. 2013b;**97**:6919-6930

[13] Stewart GG. The influence of high gravity wort on the stress characteristics of brewer's yeast and related strains. Cerevisia. 2007;**32**:37-48

[14] Kucharczyk K, Tuszyński T. The effect of wort aeration on fermentation, maturation and volatile components of beer produced on an industrial scale. Journal of the Institute of Brewing. 2017;**123**(1):31-38

[15] Choi EJ, Ahn HW, Kim WJ. Effect of α-acetolactate decarboxylase on diacetyl content of beer. Food Science and Biotechnology. 2015;**24**(4):1373-1380

[16] Nedovic V, Gibson B, Mantzouridou TF, Bugarski B, Djordjevic V, Kaluševic A, et al. Aroma formation by immobilized yeast cells in fermentation processes. Yeast. 2015;**32**:173-216

[17] Steensels J, Snoek T, Meersman E, Nicolino MP, Aslankoohi E, et al. Selecting and generating superior yeasts for the brewing industry. Cerevisia. 2012;**37**:63-67

[18] Shi T, Li P, Chen S, Chen Y, Guo X, Xiao D. Reduced production of diacetyl by overexpressing *BDH2* gene and *ILV5* gene in yeast of the lager brewers with one *ILV2* allelicgene deleted. Journal of Industrial Microbiology & Biotechnology. 2017;**44**:397-405

[19] Canonico L, Comitini F, Ciani M. Dominance and influence of selected Saccharomyces cerevisiae strains on the analytical profile of craft beer refermentation. Journal of the Institute of Brewing. 2014;**120**:262-267

[20] Olaniran AO, Hiralal L, Mokoena MP, Pillay B. Flavour-active volatile compounds in beer: Production, regulation and control. Journal of the Institute of Brewing. 2017;**123**:13-23

[21] Kobayashi K, Kusaka K, Takahashi T, Sato K. Method for the simultaneous assay of diacetyl and acetoin in the presence of α-acetolactate: Application in determining the kinetic parameters for the decomposition of α-acetolactate. Journal of Bioscience and Bioengineering. 2005;**99**:502-507

[22] Verbelen P, Saerens S, Mulders S, Delvaux F, Delvaux R. The role of oxygen in yeast metabolism during high cell density brewery fermentations. Applied Microbiology Cell Physiology. 2009b;**82**:1143-1156

[23] Sigler K, Matoulkova D, Dienstbier M, Gabriel P. Net effect of wort osmotic pressure on fermentation course, yeast vitality, beer flavor, and haze. Applied Microbiology and Biotechnology. 2009;**82**:1027-1035

[24] Rodman AD, Gerogiorgis DI. Multi-objective process optimisation of beer fermentation via dynamic simulation. Food and Bioproducts Processing. 2016;**100**:255-274. DOI: 10.1016/j.fbp.2016.04.002

[25] Rodman AD, Gerogiorgis DI. Dynamic optimization of beer fermentation: Sensitivity analysis of attainable performance vs. product flavor constraints. Computers and Chemical Engineering. 2017;**106**:582-595. DOI: 1016/j.compchemeng.2017.06.024

[26] James N. The influence of an amino acid permease on diacetyl production during beer fermentation [MSc thesis] Berlin, Germany. 2014

[27] Kodama Y, Omura F, Ashikari T. Isolation and characterization of a gene specific to lager brewing yeast that encodes a branched-chain amino acid permease. Applied and Environmental Microbiology. 2001;**67**:3455-3462

[28] Lei H, Zheng L, Wang C, Zhao H, Zhao M. Effects of worts treated with proteases on the assimilation of free amino acids and fermentation performance of lager yeast. International Journal of Food Microbiology. 2013;**161**:76-83

[29] Nie C, Wang C, Zhou G, Dou F, Huang M. Effects of malting conditions on the amino acid compositions of final malt. African. Journal of Biotechnology. 2010;**9**(53):9018-9025

[30] Samaras T, Camburn P, Chandra S, Gordon M, Ames MJ.

Antioxidant properties of kilned and roasted malts. Journal of Agricultural and Food Chemistry. 2005;**53**:8068-8074

[31] Maradyn D. Enzymes in brewing: Theory and practical considerations. In: 74th Annual Meeting of the American Society of Brewing Chemists. Fort Myers, FL, USA. 11-15 June 2011

[32] Blomqvist K, Suihko TML, Knowles T, Penttila M. Chromosomal integration and expression of two bacterial α-acetolactate decarboxylase genes in brewer's yeast. Applied and Environmental Microbiology. 1991;**57**(10):2796-2803

[33] Yamano S, Tanaka J, Inoue T. Construction of a brewer's yeast having alpha-acetolactate decarboxylase gene from *Acetobacter aceti* ssp. xylinum integrated in the genome. Journal of Biotechnology. 1994a;**32**:173-178

[34] Yamano S, Tanaka J, Inoue T. Cloning and expression of the gene encoding alpha-acetolactate decarboxylase from *Acetobacter aceti* ssp. xylinum in brewer's yeast. Journal of Biotechnology. 1994b;**32**:165-171

[35] Verbelen PJ, Van Mulders S, Saison D, Van Laere S, Delvaux F, Delvaux FR. Characteristics of high cell density fermentations with different lager yeast strains. Journal of the Institute of Brewing. 2008;**114**(2):127-133

[36] Krogerus K, Gibson B, Hytonen E. An improved model for prediction of fermentation and total diacetyl profile during brewery fermentation. Journal of American Society. 2015;**73**:90-99

[37] Qin Y, Park H. Overexpressed acetohydroxyacid reductoisomerase (ILV5) gene in *Saccharomyces cerevisiae* reduces diacetyl contents in Korean Campbell early and Muscat Bailey a grape wines. Journal of Korean Society for Applied Biological Chemistry. 2012;**55**:799-801

[38] Chambers PJ, Pretorius IS. Fermenting knowledge: The history of winemaking, science and yeast research. EMBO Reports. 2010;**11**(12):914-920

[39] Zhang Y, Wang Z, He X, Liu N, Zhang B. New industrial brewing yeast strains with ILV2 disruption and LSD1 expression. International Journal of Food Microbiology. 2008;**123**:18-24

[40] Kusunoki K, Ogata T. Construction of self-cloning bottom-fermenting yeast with low vicinal diketone production by the homo-integration of *ILV5*. Yeast. 2012;**29**:435-442

[41] Duong CT, Strack L, Futschik M, Katou Y, Nakao Y, Fujimura T, et al. Identification of *Sc* type- *ILV6* as a target to reduce diacetyl formation in lager brewers' yeast. Metabolism Engineering. 2011;**13**:638-647

[42] Gibson B, Vidgren V, Peddinti G, Krogerus K. Diacetyl control during brewery fermentation via adaptive laboratory engineering of the lager yeast *Saccharomyces pastorianus*. Journal of Indu Microbiology and Biotechnology. 2018;**45**:1103-1112

[43] Li P, Guo X, Shi T, Hu Z, Chen Y, Du L, et al. Reducing diacetyl production of wine by overexpressing *BDH1* and *BDH2* in *Saccharomyces uvarum*. Journal of Industrial Microbiology & Biotechnology. 2017;**44**:1541-1550

[44] Villa KD, Lee S, Goossens E, et al. Control of vicinal diketone production by brewer's yeast. I. Effects of ILV5 and ILV3 gene amplification on vicinal diketone production and ILV enzyme activity. Journal of American Society Brew Chemistry. 1995;**53**:49-53

[45] Cejnar R, Hlozkova K, Kotrba P, Dostalek P. Surface-engineered Saccharomyces cerevisiae displaying a-acetolactate decarboxylase from *Acetobacter aceti ssp xylinum*. Biotechnology Letters. 2016;**38**:2145-2151. DOI: 10.1007/s10529-016-2205-1

[46] Gibson B, Boulton C, Box W, Graham N, Lawrence S, Linforth R, et al. Amino acid uptake and yeast gene transcription during industrial brewery fermentation. Journal of American Society Brew Chemistry. 2009;**67**(3):157-165. DOI: 10.1094/ASBCJ-2009-0720-01

[47] de Boer M, Jan-Paul Bebelman J, Paula M, Gonaives PM, Maat J, van Heerikhuizen H, et al. Regulation of expression of the amino acid transporter gene *BAP3* in *Saccharomyces cerevisiae*. Molecular Microbiology. 1998;**30**(3):603-613

[48] Didion T, Regenberg B, Jorgensen MU, Killand-Brandt MC, Andersen HA. The permease homolog Ssy1p controls the expression of amino-acid and peptide transporter genes in *Saccharomyces cerevisiae*. Molecular Microbiology. 1998;**27**:643-650

Chapter 12

Non-Conventional *Saccharomyces* Yeasts for Beer Production

Vanesa Postigo, Margarita García and Teresa Arroyo

Abstract

Beer is a world-famous beverage, second only to tea and coffee, where the yeasts traditionally used are *Saccharomyces cerevisiae* and *Saccharomyces pastorianus* for the production of ale and lager beer, respectively. Their production, especially craft beer production, has grown in recent years, as has the development of new products. For this reason, research has focused on the selection of yeasts with good fermentation kinetics, as well as beers with outstanding aromatic profiles. The final flavor and aroma of beer is a combination of hundreds of active aroma compounds produced mostly during fermentation as a result of yeast metabolism (higher alcohols, esters, aldehydes, and vicinal diketones). Likewise, several studies have demonstrated the potential of wild yeasts of the genus *Saccharomyces*, both in aromatic production and in the production of healthy compounds of interest such as melatonin. This chapter therefore focuses on non-conventional *Saccharomyces* yeasts as they have the capacity to produce outstanding aroma compounds, as well as compounds that can provide health benefits, under moderate consumption.

Keywords: non-conventional yeast, *Saccharomyces*, beer, aromatic compounds, functional beer

1. Introduction

Nowadays, there is a growing global interest in craft beer, as well as in the production of new beers that meet market needs. For this reason, brewers have focused especially on the use of yeasts, especially non-conventional yeast, to innovate in the brewing sector [1–5]. Wild yeast strains can provide different aroma and flavor characteristics with which to obtain varieties and styles of beer that are alternative and new to existing ones [6]. However, the use of these domesticated yeasts may show variable fermentative characteristics and affect the consistency and quality of the beers produced [4, 7]. Unlike the current commercial strains of *Saccharomyces* yeast, their domestication has occurred over millennia, resulting in brewers obtaining consistent beers with established aromas, flavours and fermentation conditions [4, 8]. Nowadays, however, all the steps leading to yeast domestication can be done in a faster way [2, 9], since the metabolism pathways of yeast are better known, as well as the production pathways of the different aromatic compounds (higher alcohols, esters, phenols, acids, and monoterpenes) [2, 10–13]. It is for this reason that brewers and researchers have focused on the search for new yeasts from different environments using current techniques [14]. Determining the suitability of new yeast strains can be done by evaluation by expert panels; however, this must be coupled with their analysis using physicochemical analytical techniques to show correlation with the aroma compounds produced. Such compounds are produced in smaller proportions than ethanol and carbon dioxide, which are the main products of fermentation, but are nevertheless of great importance [10, 15]. Secondary metabolites can be divided into different categories and include sulfur-containing aroma compounds, undesirable carbonyl compounds, volatile phenols, organic acids, fusel alcohols, esters, and monoterpene alcohols [10, 15–17]. These groups of aromatic compounds play a number of roles in the impact of flavor on beer, as many of them can act synergistically, thus increasing flavors, despite being below their threshold [11, 15, 16].

The present review focuses on the use of non-conventional yeasts of the genus *Saccharomyces* for fermenting beer wort, obtaining an outstanding organoleptic profile, as well as the production of secondary metabolites with healthy characteristics for potential use by brewers.

2. Beer ingredients

Beer is a worldwide-known beverage and the most popular after tea and water. However, people do not focus their attention on how it is brewed, as well as on its ingredients, which are barley as the main cereal, water, hops, and yeast. Other cereals or adjuncts may also be added [18, 19].

Water is the main component of beer, accounting for 85–92% of beer, as well as serving as the basis for brewing the beer wort. Water quality is an important factor to take into account since the presence of any odor, taste, and impurity is unacceptable because of its influence on the sensory properties of the beer. The enzymatic and colloidal reactions that occur during brewing are influenced by the content of dissolved ions found in water such as Ca^{2+}, Mg^{2+}, Na^{+1}, K^{+1}, SO_4^{-2}, Cl^-, HCO_3^{-1}, and CO_3^{-2} [20]. Calcium (Ca^{2+}) is involved in protein coagulation and polyphenol and carbohydrate formation, forms and precipitates calcium oxalate and calcium carbonate, reduces pH during mashing and boiling, and promotes yeast metabolism and flocculation, and its concentration is key to residual hardness and alkalinity. Magnesium (Mg^{2+}) is a vital nutrient for yeast and necessary for the isomerization of humulone to isohumulone in hops. Sodium (Na^+) can contribute to the beer body. Chloride (Cl^-) in low amounts enhances malt sweetness. Bicarbonate (HCO_3^-) functions as a buffering agent, buffering pH variations [21].

The basis of beer production is **malted barley**. The malting process is carried out for the yeast fermentation, as well as proteins that promote and enhance foam stability [22]. Malt can provide the beer with aromas of biscuit, honey, cinnamon, bread, chocolate, cocoa, coffee, and roasted and/or smoked, among others [23].

Hops (*Humulus lupulus* L.) is the ingredient that will mainly provide the bitterness of the beer due to the iso-α-acids (iso-normal-humulone, iso-co-humulone, and iso-ad-humulone) obtained by the isomerization of the α-acids (normal-humulone, co-humulone, and ad-humulone) during the wort boiling and whirlpool process in brewing. These α-acids are contained in the resins of hop flowers [24]. In addition to bitterness, hop also contributes to beer aromas, such as citrus, floral (geraniol, linalool), herbaceous, and fruity (limonene, myrcene, β-pyrene), due to the resins, proteins, and essential oils they possess [25]. Hops are sensitive to oxidation in contact with air, so it is important that they are stored under vacuum in a cool, dry place; otherwise they can transmit unpleasant aromas to the beer (such as isovaleric acid with rancid cheese aroma).

Yeast is an important ingredient in beer brewing, as it is responsible for fermentation and thus for the conversion of the sugars in the beer wort into ethanol and carbon dioxide (CO_2). However, it will also produce different by-products, the main ones being aromatic compounds that will characterize the aromatic profile of the beers.

In the past, these four ingredients were not well-known, but it was known that yeast was necessary for the production of fermented beverages. It was not until the development of lager beers that the exact composition of yeast was discovered. Although yeasts have been used for centuries to brew beer, they were first identified as responsible for the fermentation of malted barley water in the 19th century. The first principles of yeast function were discovered during the 17th and 18th centuries [26]; however, it was not until the mid-19th century that the French scientist Louis Pasteur was able to demonstrate that yeast is made up of living cells responsible for the fermentation process [27]. In the same vein, sugars were also not known to be another essential ingredient and can be classified according to the type of fermented beverage, whether from sugar cane (sucrose), milk (lactose), fruit or honey (fructose and glucose), or cereals (maltose) [19].

3. Beginnings in the use of yeast for beer production

The beginnings of beer production date back to the Neolithic civilization. Beer is a traditional product and valued for its physicochemical properties that give it quality. Therefore, the history of brewing is not only the history of scientific and technological developments but also the history of people: their government, their culture, and their daily life [28]. Early brewers, winemakers, and bakers realized that by using small portions of finished products that had already been fermented, it was possible to obtain products with faster and more predictable fermentation. Thus, the ability of *Saccharomyces cerevisiae* yeast to convert sugars into ethanol, as well as into other compounds that provide organoleptic characteristics to foods, in addition to alcohol to beverages, began to be exploited [2, 29]. However, the use of pure yeast cultures did not begin until sometime after the pioneering work of Pasteur and Hansen in the 19th century [27]. It should be noted that it has not yet been determined whether the yeast lineages used industrially originated from the loss of contact with their natural niches and limited dispersal or from adaptation to the industrial niche (domestication) [30, 31]. Domestication therefore is about artificially selecting and breeding wild species with improved characteristics adapted to artificial environments. However, this domestication can lead to genome breakdown, polyploidy, chromosomal rearrangements, gene amplifications and deletions, horizontal gene transfer, and loss of genetic diversity due to bottlenecking [32, 33].

The most studied yeast at the industrial level is *Saccharomyces cerevisiae*, where its diversity lies both in genetic drift by bottleneck and small isolated populations as well as in selection and niche adaptation. In yeasts used for winemaking, horizontal gene transfer events [34–36] and copy number variations [37–40] have been described that increase sugar and nitrogen metabolic activity, which favors fermentation of grape must, as well as better tolerance to chemicals used in vineyards and in wine [41]. In contrast, the most significant changes are found in the yeasts used for brewing, as yeasts are generally reused for several fermentation batches. This continuous growth in a very specific industrial niche has resulted in a continuous selection imposed by the brewing environment. Therefore, the yeasts currently used at the industrial level could be considered as the result of a centuries-long evolutionary experiment in a highly selective niche.

4. Brewing potential of wild *Saccharomyces*

The increase in the consumption of craft beer [42], as well as consumers' interest in trying new beer styles [43], has encouraged the application of new yeasts in brewing [44, 45]. These yeasts include the wild yeast *Saccharomyces eubayanus* (ancestor of today's lager yeast hybrid), as well as various species such as *Torulaspora delbrueckii, Lachancea thermotolerans*, and *Mrakia gelida* [46–48]. These non-conventional yeasts have functional advantages over commercial yeasts as they provide novel flavor profiles, production of flavorful beers with low alcohol content, reduced caloric content, and primary acidification, in addition to their use to improve existing beer styles or create new styles. Currently, the use of new yeasts focuses not only on the isolation of wild yeasts but also on mutagenesis and breeding. However, the most attractive option for brewers remains the isolation of yeast strains from different niches [2]. The advantages offered by wild yeast isolation are extensive:

- The natural biodiversity of microorganisms representing the habitat of a geographic region can be tapped.
- These new species can bring added value to beer in terms of organoleptic qualities.
- Some yeasts isolated from nature already have the status of Generally Recognized as Safe/Qualified Presumption of Safety.
- Current regulations favor the use of unmodified genetic stocks, as they add identity and uniqueness to differentiate the production line.

On an industrial level, not only ethanol and glycerol production but also the utilization of available sugars in the wort, hop tolerance and resistance to low temperatures, and the relative production of aromatic compounds such as esters and higher alcohols, as well as low levels of acetic acid and hydrogen sulfide, are important [2, 45, 49]. Not all species are able to ferment the sugars present in the wort (glucose, fructose, maltose, and maltotriose). In the case of *S. eubayanus*, it seems highly efficient at utilizing maltose in addition to presenting an efficient production of biomass at low temperature (ideal for lager brewing), thus balancing with the non-capacity to ferment maltotriose. Wild yeasts are noted for their potential to produce remarkable aroma compounds in brewing [50]. The type strain of *Saccharomyces arboricola* has been found to produce high levels of ethyl esters [50]. However, it should also be noted that numerous *Saccharomyces* wild yeast strains tend to produce significant perceivable levels of unpleasant phenolic aroma (POF), mainly through the production of 4-vinyl guaiacol from ferulic acid [51], imparting a clove-like aroma to beer. This is considered undesirable in most beer styles. In some cases, POF aromas do not represent a problem, but are desired in the final fermented product, such as wheat beer, where clove aroma is part of the normal flavor profile [52]. Therefore, the ability of Saccharomyces wild yeasts to ferment at low or high temperatures, to obtain new aromatic profiles, high or low ethanol levels and a reduction in H_2S production, as well as to supporting stressful wort conditions, represents a potential for new processing strategies. Cubillos et al. [53] collect in their study the application of various strains of wild yeast for application in beer.

Another advantage of wild yeasts is that they can be subjected to different fermentation conditions to observe their behavior. In this case, from a technological point of view, the application of aeration during fermentation is an interesting tool for controlling yeast metabolism during fermentation [54]. First, due to Crabtree-negative yeasts, when the oxygen concentration in the medium is saturated, the yeast metabolism begins to be predominantly oxidative, thus reducing the ethanol content and increasing the yeast biomass. On the other hand, aeration may also affect aromatic compounds by reducing or increasing their concentration in beer (the acetaldehyde content may decrease, and the concentration of higher alcohols increases) [55–57]. Within the non-conventional strains of *Saccharomyces*, there are strains that may present the Kluyver effect with aeration during fermentation, as in the study conducted by Postigo et al. [58]. In this study, during fermentation with continuous aeration, it was observed that one of the tested strains (CLI 1057) that under anaerobic conditions only fermented glucose, fructose, and sucrose, when changing to aeration conditions, began to ferment part of the maltose present in the must [59, 60]. This behavior resulted in a variation in the concentration of aromatic compounds where isoamyl acetate, DMS, diacetyl, and 2,3-pentanedione decreased considerably, as observed in other studies such as Mauricio et al. [61]. However, the rest of the compounds increased, resulting in a more phenolic beer than without aeration.

5. Aroma production by non-conventional yeast

The beverage industry has focused on the search for fruity and floral aromas, which is why consumers demand beers with fruitier aromatic profiles. The ingredients in beer that can provide such aromas and flavors are hops, but mostly yeast during the fermentation process, as it will provide a fruitier organoleptic profile to beers. Various aroma compounds can be found in beer, although studies mainly focus on the alcohols and esters produced by yeast as they will provide the main aromas found in beer [10, 62, 63].

Meilgaard elaborated the "beer aroma wheel," where all important beer aromas were included, including all yeast-derived aromas, as well as all other raw materials used during brewing [64].

Main aromatic compounds relevant in beer:

Sulfur compounds: yeasts can produce sulfur dioxide (SO_2) and hydrogen sulfide (H_2S) during fermentation [65, 66]. SO_2 has several functionalities, including acting as an antimicrobial agent, bleaching agent, oxygen scavenger, reducing agent, and enzyme inhibitor [67]. However, the maximum levels of SO_2 allowed in beer depend on the regulations of each country, always taking into account its perception threshold (20 mg/L), as it can impart unpleasant aromas to beer [68]. Although it should be noted that it is a positive characteristic desired in some bottom-fermented beers [69], this compound is usually found in concentrations of less than 10 mg/L, being a secondary or feedback inhibition product of amino acid anabolism [65, 69]. On the other hand, H_2S (rotten egg aroma) has a lower perception threshold (0.005 mg/L) and can mask desired aromas in beer [70, 71]. Its production occurs mainly during the cell maturation cycle, but it is assimilated later in the budding cycle [66].

Carbonyl compounds: within this group, the most prominent compounds that produce undesirable aromas in beer are vicinal diketones and acetaldehyde, produced by yeasts during the fermentation process. Vicinal diketones include diacetyl and 2,3-pentanedione. Both have low thresholds (0.15 mg/L and 0.9 mg/L, respectively [72]) and a buttery or toffee-like aroma. Diacetyl is a by-product of valine anabolism during glucose metabolism. Later, during the beer maturation phase, yeasts take up diacetyl and reduce it to 2,3-butanediol, which has no unpleasant aroma [73]. 2,3-pentanedione, on the other hand, is produced during the synthesis of the amino acid isoleucine in the mitochondria of yeast cells, but like diacetyl, its content is reduced during maturation [73, 74]. Acetaldehyde is produced during the yeast growth phase as a result of sugar metabolism and is subsequently converted to ethanol [75]. It has an undesirable green apple or grassy taste, with a threshold of 30 mg/L [72, 76, 77].

Phenolic compounds: these compounds are generally referred to as phenolic off-flavors (POFs) for most beers but are accepted in Belgian-style beers or wheat beers. These compounds contribute clove, smoky, spicy, medicinal, and burnt aromas [78, 79]. The compounds responsible for these aromas are generally 4-vinylguaiacol, 4-vinylphenol, 4-ethylguaiacol, 4-ethylphenol, 4-vinylsyringol, styrene, eugenol, guaiacol, and vanillin [80]. Although they depend on the presence of certain precursors in the wort for their production, generally the type of yeast used is largely responsible for their formation. The precursors that can be found in the wort are phenolic acids with a high flavor threshold, such as ferulic, coumaric, and cinnamic acids, which are derived from malt [51, 80]. These compounds usually have low thresholds: 4-ethylphenol, 0.9 mg/L (phenolic aroma, astringent); 4-ethylguaiacol, 0.13 mg/L (phenolic aroma, sweet); 4-vinylguaiacol, 0.3 mg/L (phenolic aroma, bitter, clove); guaiacol, 3.88 μg/L (smoky aroma); and 4-vinylphenol, 0.2 mg/L (phenolic aroma, smoky) [72, 81].

Organic acids: within the organic acids are non-volatile and volatile acids, the latter being the most prominent in beer as they generally contribute along with the rest of the off-flavors [82]. In beer, the main organic acids that can be found are propionic, isobutyric, butyric, isovaleric, valeric, caproic, acetic, caprylic, caprylic, capric, and lauric, the last four being the ones that most influence the flavor of beer [83, 84]. Their concentration plays an important role since in high concentrations, they can contribute acidic, salty, or even cheesy and sweaty flavors [16, 83–85]. Acetic acid, with a vinegar-like aroma, has a high threshold of 175 mg/L, whereas caprylic acid, with a goat-like aroma, has a much lower threshold of 15 mg/L. Capric acid contributes waxy-like aroma with a threshold of 10 mg/L, and lauric acid is described as soapy above the threshold of 6.1 mg/L [16]. On the other hand, the most prominent non-volatile organic acids are oxalic acid, citric acid, malic acid, fumaric acid, succinic acid, lactic acid, and pyruvic acid, with thresholds higher than those of the volatile acids, ranging from 220 to 700 mg/L, most of which provide acidic flavors [16, 83, 84, 86–88]. This group of organic acids is dependent on the yeast strain used and is a by-product of glycolysis, the citric acid cycle, and the metabolism of amino acids and fatty acids [88].

Higher alcohols: higher alcohols are produced by yeasts as a by-product of amino acid metabolism and catabolism. During catabolism, yeasts take up the amino acids and they are transaminated by four transaminase enzymes [89]. The resulting product is an α-keto acid, which through an irreversible reaction forms the higher alcohols, being known as the Ehrlich route [90, 91]. The main aromas provided by the higher alcohols are floral, fruity, or herbaceous, the most important in beer being n-propanol, isobutanol, isoamyl alcohol, and 2-phenylethanol [10, 16, 92]. N-butanol provides a sweet taste, but its threshold is high (600 mg/L) [92]. Isobutanol and amyl alcohol have solvent-like aromas, with thresholds of 100 and 50–70 mg/L, respectively. Isoamyl alcohol and 2-phenylethanol give fruitier aromas, whereas isoamyl alcohol shows more banana and alcohol flavors with a threshold of 50–65 mg/L. The 2-phenylethanol compound has a gummy bear and rose flavor at a threshold of 40 mg/L [16, 92].

Esters: in addition to higher alcohols, esters are also an important group of aromatic compounds because they have low thresholds and provide fruity aromas to beer [10, 11]. The esters can be divided into two groups, the acetate esters whose concentration is majority in beers and the medium-chain fatty acid ethyl esters [10, 11]. The most important esters that can be found in beer are ethyl acetate (solvent flavor with a threshold of 33 mg/L), isoamyl acetate (banana flavor with a threshold of 1.6 mg/L), isobutyl acetate (fruity and sweet flavor with a threshold of 1.6 mg/L), and phenylethyl acetate (aroma of rose, apple, and honey with a threshold of 3.8 mg/L) [16, 93, 94]. Moreover, medium-chain fatty acid ethyl esters are formed from a medium-chain fatty acid and an ethanol radical. The most prominent are ethyl hexanoate (apple, anise flavor, with a threshold of 0.23 mg/L) and ethyl octanoate (acid apple flavor, with a threshold of 0.9 mg/L) [10, 11, 16].

The use of non-conventional yeasts in winemaking has been extensively studied in both *Saccharomyces* and non-*Saccharomyces*, as they have shown great potential in the production of aromatic esters during vinification, in both mixed and sequential inoculation [95–97]. However, despite the great aromatic potential of non-conventional yeasts, there are fewer studies on beer than on wine [6, 98]. In *Saccharomyces* species, mainly the Ehrlich pathway for fusel alcohols or the enzymes responsible for ester formation have been studied as main pathways for the formation of aroma compounds [99–101]. Gamero et al. [102] also carried out their studies with non-*Saccharomyces* yeasts with a prominent organoleptic profile, so these non-conventional yeasts should also present such metabolic pathways involved in the formation of this type of flavor compounds, such as the Ehrlich pathway, or the specific enzymes responsible for the synthesis of esters (Atf1p, Atf2p in *S. cerevisiae*). However, the regulation of their expression or the functionality of these enzymes may vary from non-*Saccharomyces* yeast to *Saccharomyces* yeast.

The study carried out by Postigo et al. [103] with 114 *Saccharomyces cerevisiae* yeast strains isolated from wine environment resulted in the production of aromatic compounds similar to those produced by commercial strains such as SafAle S-04 (Fermentis, Lesaffre, France). However, the relative proportions between the specific volatiles, or the flavor profiles, were different. This diversity of aromatic profiles was immense, and, interestingly, some of these strains could produce greater amounts of aromatic compounds than the commercial *Saccharomyces* strain. The strains studied were characterized by the production of higher concentrations of alcohols, such as isoamyl alcohol, isobutanol, and β-phenylethanol, as well as esters such as isoamyl acetate and ethyl butyrate compared to the commercial strain. On the other hand, it was also observed that some of these strains showed guaiacol production above the threshold (3.88 ppb) (smoked flavor) [104], giving a phenolic character to beers. A particularity of non-conventional yeasts is the production of POFs, generally unwanted in some beer styles but appreciated in wheat beers [52]. In other studies, such as the one carried out by Rossi et al. [105] with 12 yeast strains isolated from environments such as grape must, bakery, wine, and apple stillage showed behaviors similar to those observed in the previous study. **Table 1** shows the different yeast strains isolated from non-conventional environments and the production

Yeast strain	Species	Source of isolation	Apparent attenuation (%)	Flavor compound	Concentration (mg/L)	Reference
DBVPG 1058	*Saccharomyces cerevisiae*	Baker's yeast	71.25	Ethyl acetate	32.00	[105]
				Isoamyl acetate	1.90	
				Ethyl hexanoate	0.60	
M4	*S. cerevisiae*	Craft beer Ale	69.85	Acetaldehyde	164.92	[106]
				Isoamyl alcohol	161.04	
MT-15	*S. cerevisiae*	Sourdoughs	60.29	Acetaldehyde	43.14	[106]
				Isoamyl alcohol	101.88	
Granvin2	*S. cerevisiae*	Norwegian kveik	n.a.	Ethyl caproate	0.37	[107]
				Ethyl caprilate	4.56	
				Ethyl decanoate	0.46	
Grancin6	*S. cerevisiae*	Norwegian kveik	n.a.	Ethyl caproate	0.37	[107]
				Ethyl caprilate	5.01	
				Ethyl decanoate	0.88	
G 520	*S. cerevisiae*	Organic cellar	72.00	Isoamyl acetate	2.34	[103]
				Guaiacol	0.05	
CLI 1109	*S. cerevisiae*	Vineyard	71.00	Isoamyl acetate	1.90	[103]
				Ethyl hexanoate	0.60	
				Guaiacol	0.03	

n.a.: not appear.

Table 1.
Aromatic compounds found in different beers fermented with non-conventional Saccharomyces yeasts whose concentrations are above their threshold.

of different aromatic by-products, the most important of which are shown below. Some of the strains studied exhibited fermentative behaviors close to commercial strains (Safbrew-S33 and Nottingham), with concentrations of aromatic compounds higher in some cases than those in commercial ones. Another peculiarity of the non-conventional yeasts of *Saccharomyces* is that there are some strains that will only ferment in the wort glucose, fructose, and sucrose but not maltose [58, 103]. This makes them ideal candidates for brewing low ethanol beers, as they can produce levels of aromatic compounds similar to those of other *Saccharomyces* strains that do ferment maltose, which can cause the characteristic wort flavor of some beers with low ethanol content [108, 109]. This also prevents the use of mechanical methods to remove ethanol content. Furthermore, residual sugars that may remain in beer can influence aspects such as the viscosity that contributes to the sensation in the mouth and body of beer [110, 111]. Likewise, this type of strains can be used in cofermentation with other strains of *Saccharomyces* to improve the aroma of beer [106].

6. Functional beer

Functional beer is defined as a beer that can provide health benefits under moderate consumption. Functional beers include those that have a low ethanol content, as well as those that provide high concentrations of compounds such as fiber, vitamins, minerals, polyphenols, and probiotics [1].

Recent years have seen an increase in consumption and interest in low-alcohol beers. This is mainly due to health and safety reasons, in addition to an increase in strict social regulations [112]. Low-ethanol beers can have health benefits due to the healthy components they contain, besides a lower energy intake and the total absence of negative effects of alcohol consumption.

Several studies have shown that with the use of non-conventional yeast, functional beers can be obtained, as they not only possess the ability to produce remarkable aromatic compounds, as well as other by-products such as melatonin [6, 113].

Functional beers also include beers that are gluten-free, thus covering consumers suffering from coeliac disease, which is a gluten-sensitive, immune-mediated enteropathy.

Probiotic beers are also included in functional beers. Probiotics include those live microorganisms that are added to food and that under certain doses can be potentially beneficial for human health, especially for the intestinal microbial balance [114]. Therefore, a probiotic beer is one obtained by using probiotic microorganisms during the fermentation process. The best known microorganisms used for their probiotic characteristics are lactic acid bacteria (*Lactobacillus, Bifidobacterium, Enterococcus*, or *Streptococcus*) [115]. However, studies have also been extended to other microorganisms such as yeasts, where experiments have been carried out with *Saccharomyces cerevisiae* var. *boulardii*, since it has been found to have the ability to prevent diarrhea associated with the use of antibiotics, as well as helping with infections caused by *Escherichia coli* or *Clostridium difficile* [116, 117]. In addition, beer trials have also been carried out with *S. cerevisiae* var. *boulardii*, where the results have shown an increase in the antioxidant capacity due to metabolite secretion by the yeasts in the craft beers obtained [118]. However, studies have focused not only on the study of non-conventional yeasts of the *Saccharomyces* genus but also on the use of other non-*Saccharomyces* genera (*Lachancea, Kluyveromyces, Torulaspora, Metschnikowia, Kazachstania, Brettanomyces, Pichia, Candida, Hanseniaspora, Rhodotorula, Rodosporidobolus*) since they have been found to have great potential for use in the production of craft beers with probiotic characteristics [119]. Craft beer is generally not unfiltered or unpasteurized; thus, the yeast cells remaining after fermentation can provide a probiotic character to the beer and can be considered beneficial to health. This could be seen in the studies carried out by Mulero et al. [118] comparing unfiltered beers fermented with *S. cerevisiae* var. *boulardii* and a commercial *S. cerevisiae* strain. According to the studies of Hill et al. [120], the minimum recommended dose of live probiotic cell count per product sample is 9 Log colony forming units (CFU). On the other hand, industrial beers, after being subjected to high temperatures during pasteurization, may eliminate such probiotics. For this reason, craft rather than industrial brewing would be more appropriate since viability is crucial for the efficacy of probiotics [106]. However, it should be noted that the presence of yeasts in beer can give beer a unique taste but reduce its shelf life as during storage, cell lysis can negatively affect its quality. This is why yeast strain selection is important in the production of craft beers [121].

6.1 Melatonin production

Melatonin is a mammalian hormone that regulates sleep and has antioxidant properties. It is produced by yeasts during fermentation and can therefore be a source of exogenous melatonin for the body, since as people age, less melatonin is produced in the body. In addition, it provides beer with antioxidant, antiaging, anti-inflammatory, antitumor, and immunomodulatory capacities [122–124].

Much of the food and beverages we consume on a daily basis contain melatonin. Therefore, their intake helps to increase the melatonin level in the body and its antioxidant status in human serum, which is the reason that this molecule is absorbed in the gastrointestinal tract [123, 125, 126] and readily crosses all morphophysiological barriers and tissue and cell membranes [124–129]. Likewise, melatonin interacts with toxic reagents, generating other metabolites that are in turn direct free radical scavengers. The combined actions of melatonin and its derivatives greatly enhance the efficacy of melatonin in protecting against free radical damage and reducing the likelihood of human disease [130].

Melatonin is a by-product of yeast that is produced in the final stages of fermentation and is excreted into the medium during the stationary phase of the yeast fermentation [131]. In the studies carried out by Postigo et al. [103, 132] with different strains of wild *Saccharomyces* and non-*Saccharomyces* yeast, it was generally observed that *Saccharomyces* yeasts produce melatonin in a higher percentage than non-*Saccharomyces* yeasts. However, the concentration of non-*Saccharomyces* yeasts (ranged between 6.69 and 102.98 ng/mL) is in some cases double that of *Saccharomyces* yeasts (between 5.04 and 56.51 ng/mL). Furthermore, in studies carried out with sequential fermentations, where both yeast genera participate, melatonin concentrations were detected in almost all beers (values from 33.63 to 66.57 ng/mL) [133]. On the other hand, in the studies carried out by García-Moreno et al. [122], the melatonin values found in different commercial beers ranged from 58 to 169 pg/mL. As Valera et al. [134] determined in their studies, the presence of *Saccharomyces* seems to promote a higher presence of melatonin in the medium. In the studies carried out by Fernandez-Cruz et al. [135] at the intracellular level with wine yeasts, melatonin was only detected in non-*Saccharomyces* yeasts. Also, Juhnevica-Radenkova et al. [131] mention in their work that melatonin is produced in the final stages of fermentation and is excreted into the medium during the stationary phase of the yeast fermentation. In comparison with other types of food such as bread, tomato, or yogurt, we can find different melatonin values of 28.9, 138.1, and 126.7 pg/mL, respectively, being therefore lower than what we could find in some beers fermented with non-conventional yeasts [136]. The comparison of the different concentrations can be seen in **Figure 1**.

These levels of melatonin that can be found in various foods, as well as in beer, are concentrations below those studied that have supposed positive effects on health (1–10 mg) [137, 138]; however, if ingested together with other foods, they can contribute to increase its concentration in the human serum. Studies carried out by Maldonado et al. [123] determined that the intake of beer with high melatonin content (169.7 pg/ml) contributed to increase the antioxidant properties of human serum. However, although melatonin can act as a strong antioxidant, it can be degraded in the presence of oxygen, light, or free radicals that are present during the aging process [139]. This fact could be observed in the studies of lambic beer carried out by Postigo et al. [140], where the analysis at different maturation times of the lambic beers brewed determined their degradation over time. It should also be taken into account whether the final product is subjected to final heat treatments to prolong

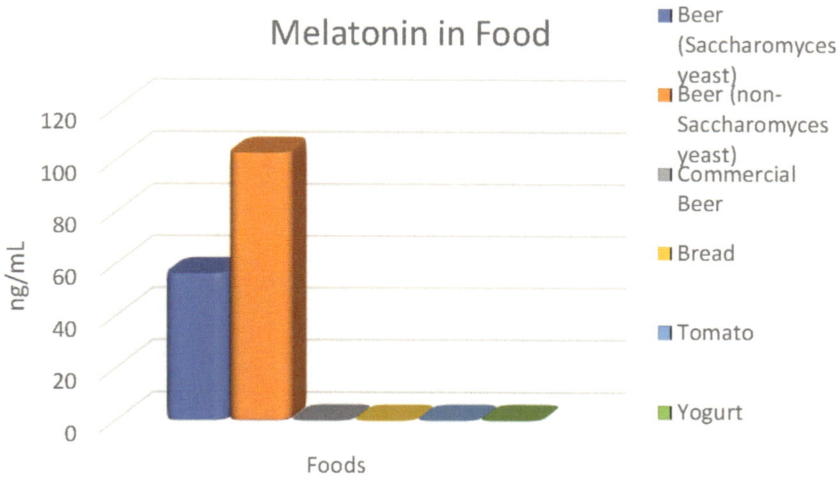

Figure 1.
Melatonin concentrations found in different foods. Beer (Saccharomyces yeast) [103], beer (non-Saccharomyces yeast) [132], commercial beer [122], bread, tomato, and yogurt [136].

the stability of the product (such as pasteurization), since it has been shown that high temperatures can also degrade it and reduce its concentration in food [136].

6.2 Antioxidant capacity

Different compounds such as phenolics can be found in beer [141]. These substances can also be found naturally in fruits, vegetables, nuts, seeds, and beverages [142]. When studying the antioxidant fraction of beer, phenolic compounds are the most studied, which are found in hops (20–30%) as well as in malt (70–80%) [141]. Hops provide the beer with phenolic acids, prenylated chalcones, flavonoids, catechins, and proanthocyanidins [141]. Malt contains an overall phenolic mass of 1.0–1.9 mg g^{-1} dry matter [143]. Likewise, the yeast used in brewing can also influence the phenolic composition and antioxidant capacity of the final product [106]. Several studies, such as the one carried out by Viana et al. [144], showed that the use of certain yeast strains for the production of Pale Ale beer significantly influences the antioxidant capacity of the beers.

Antioxidant capacity is related to parameters such as total phenolic and flavonoid content; 2,2-Diphenyl-1-picrylhydrazyl (DPPH) radical scavenging activity; and 2,2′-azinobis-(3-ethylbenzothiazoline-6-sulfonic acid) (ABTS) radical cation scavenging activity [145]. In the studies carried out by Postigo et al. [103] with different strains of wild yeast, it was observed that there were no major differences between the yeasts of *Saccharomyces* (9.50 to 13.67 mmol TE/L) and the commercial strain S-04 (11.18 mmol TE/L), as well as in sequential fermentations (9.63 to 13.70 mmol TE/L) (**Table 2**). However, in those strains where maltose was not fermented, or where there was competition between species in sequential culture, the antioxidant capacity was lower [133]. In contrast, in the studies carried out by Granato et al. [113], the values found in beers (ale and lager) were lower, between 424.77 and 10,508.47 μmol TE/L. These antioxidant capacity values may also vary depending on the method used for their analysis, since by the Oxygen Radical Absorption Capacity (ORAC) method in another study, values between 3.70 and 29.11 mmol TE/L were obtained. The ORAC method is based on the absorption capacity of oxygen radicals, whereby the decrease

Beer	Concentration	Method	Reference
Ale (*Saccharomyces* yeast)	9.50 to 13.67 mmol TE/L	TEAC	[103]
Ale (sequential fermentation)	9.63 to 13.70 mmol TE/L	TEAC	[133]
Commercial Ale (S-04)	11.18 mmol TE/L	TEAC	[103]
Commercial Ale	0.76 to 10.51 mmol TE/L	ORAC	[113]
Commercial Lager	0.44 to 7.72 mmol TE/L	ORAC	[113]
Commercial Ale	3.70 to 29.11 mmol TE/L	ORAC	[113]
Commercial Lager	0.58 to 1.02 mmol TE/L	TEAC	[146]
Commercial Ale	0.73 to 1.19 mmol TE/L	TEAC	[146]
Commercial Lager	3.70 to 14.79 mmol TE/L	ORAC	[146]
Commercial Ale	10.65 to 29.11 mmol TE/L	ORAC	[146]

ORAC (Oxygen Radical Absorption Capacity), TEAC (Trolox equivalent antioxidant capacity).

Table 2.
Antioxidant capacity of different ale and lager beers analyzed by different methods.

in fluorescence emission is measured [146, 147]. Another method used to determine the antioxidant capacity is that using the ferric-reducing antioxidant power, whose principle is the determination of the reduction of a ferric-tripyridyltriazine complex to its ferrous form, colored, in the presence of antioxidant components, where values of 3125 μmol Fe^{2+}/L can be found in ale beer.

7. Conclusions

The use of non-conventional *Saccharomyces* yeasts provides beer with additional natural flavor variants that differentiate it from other commercial beers. These characteristics can be obtained either in pure culture, in sequential culture, or with aeration during the fermentation, which makes them very versatile. Furthermore, such wild yeasts can provide bio-healthy properties to beer under moderate consumption, as they may present production of compounds such as melatonin, with concentrations higher than those of commercial strains and even those of other foods.

Acknowledgements

Work funded by IMIDRA under project FP22-CERVEZAM.

Conflict of interest

The authors declare no conflict of interest.

Author details

Vanesa Postigo*, Margarita García and Teresa Arroyo
Department of Agrifood Research, Madrid Institute for Rural, Food and Agriculture Research and Development (IMIDRA), Spain

*Address all correspondence to: vanesa.postigo@madrid.org

References

[1] Yeo HQ, Liu SQ. An overview of selected specialty beers: Developments, challenges and prospects. International Journal of Food Science and Technology. 2014;**49**(7):1607-1618. DOI: 10.1111/ijfs.12488

[2] Steensels J, Verstrepen KJ. Taming wild yeast: Potential of conventional and nonconventional yeasts in industrial fermentations. Annual Review of Microbiology. 2014;**68**(1):61-80. DOI: 10.1146/annurev-micro-091213-113025

[3] James SA, Stratford M. *Zygosaccharomyces*. In: The Yeasts. London: Elsevier; 2011. pp. 937-947. DOI: 10.1016/B978-0-444-52149-1.00084-7

[4] Lentz M, Putzke T, Hessler R, Luman E. Genetic and physiological characterization of yeast isolated from ripe fruit and analysis of fermentation and brewing potential. Journal of the Institute of Brewing. 2014;**120**(4):559-564. DOI: 10.1002/jib.154

[5] Petruzzi L, Rosaria Corbo M, Sinigaglia M, Bevilacqua A. Brewer's yeast in controlled and uncontrolled fermentations, with a focus on novel, nonconventional, and superior strains. Food Review International. 2016;**32**(4):341-363. DOI: 10.1080/87559129.2015.1075211

[6] Basso RF, Alcarde AR, Portugal CB. Could non-*Saccharomyces* yeasts contribute on innovative brewing fermentations? Food Research International. 2016;**86**:112-120. DOI: 10.1016/j.foodres.2016.06.002

[7] Vanderhaegen B, Neven H, Coghe S, Verstrepen KJ, Derdelinckx G, Verachtert H. Bioflavoring and beer refermentation. Applied Microbiology and Biotechnology. 2003;**62**(2-3):140-150. DOI: 10.1007/s00253-003-1340-5

[8] Piskur J, Rozpedowska E, Polakova S, Merico A, Compagno C. How did Saccharomyces evolve to become a good brewer? Trends in Genetics. 2006;**22**(4):183-186. DOI: 10.1016/j.tig.2006.02.002

[9] Lodolo EJ, Kock JLF, Axcell BC, Brooks M. The yeast *Saccharomyces cerevisiae*-the main character in beer brewing. FEMS Yeast Research. 2008;**8**(7):1018-1036. DOI: 10.1111/j.1567-1364.2008.00433.x

[10] Pires EJ, Teixeira JA, Brányik T, Vicente AA. Yeast: The soul of beer's aroma—A review of flavour-active esters and higher alcohols produced by the brewing yeast. Applied Microbiology and Biotechnology. 2014;**98**(5):1937-1949. DOI: 10.1007/s00253-013-5470-0

[11] Verstrepen KJ, Derdelinckx G, Pierre DJ, Winderickx J, Thevelein JM, Pretorius IS, et al. Flavor-active esters: Adding fruitiness to beer. Journal of Bioscience and Bioengineering. 2003;**96**(2):110-118. DOI: 10.1016/S1389-1723(03)90112-5

[12] Procopio S, Qian F, Becker T. Function and regulation of yeast genes involved in higher alcohol and ester metabolism during beverage fermentation. European Food Research and Technology. 2011;**233**(5):721-729. DOI: 10.1007/s00217-011-1567-9

[13] Procopio S, Brunner M, Becker T. Differential transcribed yeast genes involved in flavour formation and its associated amino acid metabolism during brewery fermentation. European

Food Research and Technology. 2014;**239**(3):421-439. DOI: 10.1007/s00217-014-2236-6

[14] Steensels J, Snoek T, Meersman E, Picca Nicolino M, Aslankoohi E, Christiaens JF, et al. Selecting and generating superior yeasts for the brewing industry. Cerevisia. 2012;**37**(2):63-67. DOI: 10.1016/j.cervis.2012.08.001

[15] Verstrepen KJ, Derdelinckx G, Delvaux FR. Esters in beer-part 1: The fermentation process: More than ethanol formation. Cerevisia. 2003;**28**(3):41-49

[16] Meilgaard MC. Flavor chemistry in beer: Part I: Flavor interaction between principal volatiles. Technical Quarterly—Master Brewers Association. 1975;**12**(2):107-117

[17] Kollmannsberger H, Biendl M, Nitz S. Occurrence of glycosidically bound flavour compounds in hops, hop products and beer. Monatsschrift fur Brauwissenschaft. 2006;**5**(6):83-89

[18] Bogdan P, Kordialik-Bogacka E. Alternatives to malt in brewing. Trends in Food Science and Technology. 2017;**65**:1-9. DOI: 10.1016/j.tifs.2017.05.001

[19] Nelson M. The Barbarian's Beverage [Internet]. New York: Routledge; 2005. DOI: 10.4324/9780203309124

[20] Belitz H, Grosch W. Alcoholic beverages. In: Food Chemistry. Berlin, Heidelberg: Springer Berlin Heidelberg; 2004. pp. 892-937. DOI: 10.1007/978-3-540-69934-7_21

[21] Eumann M. Water in brewing. In: Brewing. London: Elsevier; 2006. pp. 183-207. DOI: 10.1533/9781845691738.183

[22] Niu C, Han Y, Wang J, Zheng F, Liu C, Li Y, et al. Malt derived proteins: Effect of protein Z on beer foam stability. Food Bioscience. 2018;**25**:21-27. DOI: 10.1016/j.fbio.2018.07.003

[23] Parr H, Bolat I, Cook D. Modelling flavour formation in roasted malt substrates under controlled conditions of time and temperature. Food Chemistry. 2021;**337**:127641. DOI: 10.1016/j.foodchem. 2020.127641

[24] Kishimoto T, Teramoto S, Fujita A, Yamada O. Evaluation of components contributing to the international bitterness unit of wort and beer. Journal of the American Society of Brewing Chemists. 2022;**80**(1):53-61. DOI: 10.1080/03610470.2021.1878684

[25] Astray G, Gullón P, Gullón B, Munekata PES, Lorenzo JM. *Humulus lupulus* L. as a natural source of functional biomolecules. Applied Sciences. 2020;**10**(15):5074. DOI: 10.3390/app10155074

[26] Barnett JA. Beginnings of microbiology and biochemistry: The contribution of yeast research. Microbiology. 2003;**149**(3):557-567. DOI: 10.1099/mic.0.26089-0

[27] Barnett JA. A history of research on yeasts 2: Louis Pasteur and his contemporaries, 1850-1880. Yeast. 2000;**16**(8):755-771. DOI: 10.1002/1097-0061 (20000615)16:8<755::AID-YEA587>3.0.CO;2-4

[28] Meussdoerffer FG. A comprehensive history of beer brewing. In: Handbook of Brewing. Weinheim, Germany: Wiley-VCH Verlag GmbH & Co. KGaA; 2009. pp. 1-42. DOI: 10.1002/ 9783527623488.ch1

[29] Michel RH, McGovern PE, Badler VR. Chemical evidence for ancient beer. Nature. 1992;**360**(6399):24-24

[30] Goddard MR, Greig D. *Saccharomyces cerevisiae*: A nomadic yeast with no niche? FEMS Yeast Research. 2015;**15**(3):1-6

[31] Warringer J, Zörgö E, Cubillos FA, Zia A, Gjuvsland A, Simpson JT, et al. Trait variation in yeast is defined by population history. PLoS Genetics. 2011;7(6):e1002111

[32] Driscoll CA, Macdonald DW, O'Brien SJ. An evolutionary view of domestication. PNAS. 2009;**106**:9971-9978

[33] Purugganan MD, Fuller DQ. The nature of selection during plant domestication. Nature. 2009;**457**(7231):843-848

[34] Marsit S, Mena A, Couloux A, Guy J, Luc LJ, Barrio E, et al. Evolutionary advantage conferred by an eukaryote-to-eukaryote gene transfer event in wine yeasts. Molecular Biology and Evolution. 2015;**32**(7):1695-1707

[35] Marsit S, Sanchez I, Galeote V. Horizontally acquired oligopeptide transporters favour adaptation of *Saccharomyces cerevisiae* wine yeast to oenological environment. Environmental Microbiology. 2016;**18**:1148-1161

[36] Almeida P, Barbosa R, Zalar P, Imanishi Y, Shimizu K, Turchetti B, et al. A population genomics insight into the Mediterranean origins of wine yeast domestication. Molecular Ecology. 2015;**24**(21):5412-5427

[37] Marsit S, Dequin S, Montpellier F, Supagro M, Montpellier F, Montpellier F. Diversity and adaptive evolution of Saccharomyces wine yeast: A review. FEMS Yeast Research. 2015;**15**:fov067

[38] Dunn B, Richter C, Kvitek DJ, Pugh T, Sherlock G. Analysis of the *Saccharomyces cerevisiae* pan-genome reveals a pool of copy number variants distributed in diverse yeast strains from differing industrial environments. Genome Research. 2012;**22**:908-924

[39] Steenwyk J, Rokas A. Extensive copy number variation in fermentation-related genes among *Saccharomyces cerevisiae* wine strains. G3 (Bethesda). 2017;7(May):1475-1485

[40] Strope PK, Skelly DA, Kozmin SG, Mahadevan G, Stone EA, Magwene PM, et al. The 100-genomes strains, an *S. cerevisiae* resource that illuminates its natural phenotypic and genotypic variation and emergence as an opportunistic pathogen. Genome Research. 2015;**25**:762-774

[41] Almeida P, Barbosa R, Bensasson D, Lisboa UN De. Adaptive divergence in wine yeasts and their wild relatives, Molecular Ecology. 2017;26(7):2167-2182

[42] CJS G. Economics of the craft beer revolution: A comparative international perspective. In: Garavaglia C, Swinnen J, editors. Economic Perspectives on Craft Beer. Cham: Springer International Publishing; 2018. pp. 3-54. DOI: 10.1007/978-3-319-58235-1

[43] Aquilani B, Laureti T, Poponi S, Secondi L. Beer choice and consumption determinants when craft beers are tasted: An exploratory study of consumer preferences. Food Quality and Preference. 2015;**41**:214-224. DOI: 10.1016/j.foodqual.2014.12.005

[44] Hittinger CT, Steele JL, Ryder DS. Diverse yeasts for diverse fermented beverages and foods. Current Opinion in Biotechnology. 2018;**49**:199-206. DOI: 10.1016/j.copbio.2017.10.004

[45] Gibson B, Geertman JMA, Hittinger CT, Krogerus K, Libkind D,

Louis EJ, et al. New yeasts-new brews: Modern approaches to brewing yeast design and development. FEMS Yeast Research. 2017;**17**(4):1-13

[46] De FG, Sannino C, Sileoni V, Marconi O, Filippucci S, Tasselli G, et al. Mrakia Gelida in brewing process: An innovative production of low alcohol beer using a psychrophilic yeast strain. Journal of Food Microbiology. 2018;**76**(April):354-362. DOI: 10.1016/j.fm.2018.06.018

[47] Domizio P, House JF, Joseph CML, Bisson LF, Bamforth CW. *Lachancea thermotolerans* as an alternative yeast for the production of beer. Journal of the Institute of Brewing. 2016;**122**(4):599-604. DOI: 10.1002/jib.362

[48] Michel M, Kopecká J, Meier-Dörnberg T, Zarnkow M, Jacob F, Hutzler M. Screening for new brewing yeasts in the non-*Saccharomyces* sector with *Torulaspora delbrueckii* as model. Yeast. 2016;**33**(4):129-144. DOI: 10.1002/yea.3146

[49] Gallone B, Mertens S, Gordon JL, Maere S, Verstrepen KJ, Steensels J. Origins, evolution, domestication and diversity of *Saccharomyces* beer yeasts. Current Opinion in Biotechnology. 2018;**49**:148-155. DOI: 10.1016/j.copbio.2017.08.005

[50] Nikulin J, Krogerus K, Gibson B. Alternative *Saccharomyces* interspecies hybrid combinations and their potential for low-temperature wort fermentation. Yeast. 2018;**35**(1):113-127. DOI: 10.1002/yea.3246

[51] Coghe S, Benoot K, Delvaux F, Vanderhaegen B, Delvaux FR. Ferulic acid release and 4-vinylguaiacol formation during brewing and fermentation: Indications for feruloyl esterase activity in *Saccharomyces cerevisiae*. Journal of Agricultural and Food Chemistry. 2004;**52**(3):602-608. DOI: 10.1021/jf0346556

[52] Gallone B, Steensels J, Prahl T, Soriaga L, Saels V, Herrera-Malaver B, et al. Domestication and divergence of *Saccharomyces cerevisiae* beer yeasts. Cell. 2016;**166**(6):1397-1410. DOI: 10.1016/j.cell.2016.08.020

[53] Cubillos FA, Gibson B, Grijalva-Vallejos N, Krogerus K, Nikulin J. Bioprospecting for brewers: Exploiting natural diversity for naturally diverse beers. Yeast. 2019;**36**(6):383-398. DOI: 10.1002/yea.3380

[54] Rodrigues AJ, Raimbourg T, Gonzalez R, Morales P. Environmental factors influencing the efficacy of different yeast strains for alcohol level reduction in wine by respiration. LWT. 2016;**65**:1038-1043. DOI: 10.1016/j.lwt.2015.09.046

[55] Kirsop BH. Oxygen in brewery fermentation. Journal of the Institute of Brewing. 1974;**80**(3):252-259. DOI: 10.1002/j.2050-0416.1974.tb03614.x

[56] Virkajärvi I, Lindborg K, Jukka Kronlöf EP. Effects of aeration on flavor compounds in immobilized primary fermentation. Monatsschrift für Brauwiss. 1999;**52**(1):9-28

[57] Kucharczyk K, Tuszyński T. The effect of wort aeration on fermentation, maturation and volatile components of beer produced on an industrial scale. Journal of the Institute of Brewing. 2017;**123**(1):31-38. DOI: 10.1002/jib.392

[58] Postigo V, O'Sullivan T, Elink Schuurman T, Arroyo T. Non-conventional yeast: Behavior under pure culture, sequential and aeration conditions in beer fermentation. Food. 2022;**11**(22):3717. DOI: 10.3390/foods11223717

[59] Fukuhara H. The Kluyver effect revisited. FEMS Yeast Research. 2003;**3**(4):327-331. DOI: 10.1016/S1567-1356(03)00112-0

[60] Sims AP, Barnett JA. The requirement of oxygen for the utilization of maltose, cellobiose and D-galactose by certain anaerobically fermenting yeasts (Kluyver effect). Journal of General Microbiology. 1978;**106**(2):277-288. DOI: 10.1099/00221287-106-2-277

[61] Mauricio JC, Moreno J, Zea L, Ortega JM, Medina M. The effects of grape must fermentation conditions on volatile alcohols and esters formed by *Saccharomyces cerevisiae*. Journal of the Science of Food and Agriculture. 1997;**75**(2):155-160. DOI: 10.1002/(SICI)1097-0010(199710)75:2<155::AID-JSFA853>3.0.CO;2-S

[62] Saerens SMG, Delvaux FR, Verstrepen KJ, Thevelein JM. Production and biological function of volatile esters in *Saccharomyces cerevisiae*. Microbial Biotechnology. 2010;**3**(2):165-177. DOI: 10.1111/j.1751-7915.2009.00106.x

[63] Vanderhaegen B, Neven H, Coghe S, Verstrepen KJ, Verachtert H, Derdelinckx G. Evolution of chemical and sensory properties during aging of top-fermented beer. Journal of Agricultural and Food Chemistry. 2003;**51**(23):6782-6790. DOI: 10.1021/jf034631z

[64] Meilgaard MC, Dalgliesh CE, Clapperton JF. Beer flavour terminology. Journal of the Institute of Brewing. 1979;**85**(1):38-42. DOI: 10.1002/j.2050-0416.1979.tb06826.x

[65] Dvořák J, Dostálek P, Štěrba K, Čejka P, Kellner V, Čulík J, et al. Determination of total sulphur dioxide in beer samples by flow-through chronopotentiometry. Journal of the Institute of Brewing. 2006;**112**(4):308-313. DOI: 10.1002/j.2050-0416.2006.tb00736.x

[66] Nagami K. Hydrogen sulfide in brewing. MBAA Technical Quarterly. 1980;**17**:64-68

[67] Guido LF. Sulfites in beer: Reviewing regulation, analysis and role. Science in Agriculture. 2016;**73**(2):189-197. DOI: 10.1590/0103-9016-2015-0290

[68] Meilgaard MC. Individual differences in sensory threshold for aroma chemicals added to beer. Food Quality and Preference. 1993;**4**(3):153-167. DOI: 10.1016/0950-3293(93)90158-3

[69] Ilet DR. Aspects of the analysis, role, and fate of sulphur dioxide in beer—A review. Tech Quaterly. Master Brewers Association of the America. 1995;**32**:213-221

[70] Oka K, Hayashi T, Matsumoto N, Yanase H. Decrease in hydrogen sulfide content during the final stage of beer fermentation due to involvement of yeast and not carbon dioxide gas purging. Journal of Bioscience and Bioengineering. 2008;**106**(3):253-257. DOI: 10.1263/jbb.106.253

[71] Anderson RJ, Howard GA. The origin and occurrence of volatile sulphur compounds in British ales and lagers. Journal of the Institute of Brewing. 1974;**80**(4):357-370. DOI: 10.1002/j.2050-0416.1974.tb03630.x

[72] Meilgaard M, Elizondo A, Moya E. A study of carbonyl compounds in beer, part II. Flavor and flavor thresholds of aldehydes and ketones added to beer. Technical Quarterly - Master Brewers Association. 1970;**7**(3):143-149

[73] Krogerus K, Gibson BR. 125th anniversary review: Diacetyl and its control during brewery fermentation.

Journal of the Institute of Brewing. 2013;**119**(3):86-97. DOI: 10.1002/jib.84

[74] Ryan ED, Kohlhaw GB. Subcellular localization of isoleucine-valine biosynthetic enzymes in yeast. Journal of Bacteriology. 1974;**120**(2):631-637. DOI: 10.1128/jb.120.2.631-637.1974

[75] Liu SQ. Impact of yeast and bacteria on beer appearance and flavor [Internet]. In: Hill AE, editor. Brewing Microbiology. Oxford, UK: Woodhead Publishing; 2015. pp. 357-374. DOI: 10.1016/B978-1-78242-331-7.00017-4

[76] Canonico L, Agarbati A, Comitini F, Ciani M. *Torulaspora delbrueckii* in the brewing process: A new approach to enhance bioflavour and to reduce ethanol content. Food Microbiology. 2016;**56**:45-51. DOI: 10.1016/j.fm.2015.12.005

[77] Eßlinger HM. Handbook of Brewing Processes Technology Markets-Wiley. Weinheim: Wiley; 2009. p. 779

[78] Thurston P. The phenolic off-flavour test: A method for confirming the presence of wild yeasts. Journal of the Institute of Brewing. 1986;**92**(1):9-10

[79] Scholtes C, Nizet S, Collin S. Guaiacol and 4-Methylphenol as specific markers of torrefied malts. Fate of volatile phenols in special beers through aging. Journal of Agricultural and Food Chemistry. 2014;**62**(39):9522-9528. DOI: 10.1021/jf5015654

[80] Vanbeneden N, Van Roey T, Willems F, Delvaux F, Delvaux FR. Release of phenolic flavour precursors during wort production: Influence of process parameters and grist composition on ferulic acid release during brewing. Food Chemistry. 2008;**111**(1):83-91. DOI: 10.1016/j.foodchem.2008.03.029

[81] Thurston PA, Tubb RS. Screening yeast strains for their ability to produce phenolic off-flavours: A simple method for determining phenols in wort and beer. Journal of the Institute of Brewing. 1981;**87**(3):177-179. DOI: 10.1002/j.2050-0416.1981.tb04012.x

[82] Rodrigues JEA, Erny GL, Barros AS, Esteves VI, Brandão T, Ferreira AA, et al. Quantification of organic acids in beer by nuclear magnetic resonance (NMR)-based methods. Analytica Chimica Acta. 2010;**674**(2):166-175. DOI: 10.1016/j.aca.2010.06.029

[83] Clapperton JF, Brown DGW. Caprylic flavour as a feature of beer flavour. Journal of the Institute of Brewing. 1978;**84**(2):90-92. DOI: 10.1002/j.2050-0416.1978.tb03844.x

[84] Siebert KJ. Modeling the flavor thresholds of organic acids in beer as a function of their molecular properties. Food Quality and Preference. 1999;**10**(2):129-137. DOI: 10.1016/S0950-3293(98)00059-7

[85] Montanari L, Perretti G, Natella F, Guidi A, Fantozzi P. Organic and phenolic acids in beer. LWT - Food Science and Technology. 1999;**32**(8):535-539. DOI: 10.1006/fstl.1999.0593

[86] Amata AI, Germain P. The effect of pitching yeast aeration on the production of acetic acid during fermentations with brewers' yeast: An enzymatic approach. Journal of the Institute of Brewing. 1990;**96**(3):131-134. DOI: 10.1002/j.2050-0416.1990.tb01023.x

[87] Li H, Liu F. The chemistry of sour taste and the strategy to reduce the sour taste of beer. Food Chemistry. 2015;**185**:200-204. DOI: 10.1016/j.foodchem.2015.03.135

[88] Coote N, Kirsop BH. The content of some organic acids in beer and

other fermented media. Journal of the Institute of Brewing. 1974;**80**(5):474-483. DOI: 10.1002/j.2050-0416.1974.tb06797.x

[89] Eden A, Simchen G, Benvenisty N. Two yeast homologs of ECA39, a target for c-myc regulation, code for cytosolic and mitochondrial branched-chain amino acid aminotransferases. The Journal of Biological Chemistry. 1996;**271**(34):20242-20245. DOI: 10.1074/jbc.271.34.20242

[90] Hazelwood LA, Daran JM, van Maris AJA, Pronk JT, Dickinson JR. The ehrlich pathway for fusel alcohol production: A century of research on *Saccharomyces cerevisiae* metabolism. Applied and Environmental Microbiology. 2008;**74**(8):2259-2266. DOI: 10.1128/AEM.02625-07

[91] Ehrlich F. Über die Bedingungen der Fuselölbildung und über ihren Zusammenhang mit dem Eiweissaufbau der Hefe. Berichte der Dtsch Chem Gesellschaft. 1907;**40**(1):1027-1047

[92] Engan S. Organoleptic threshold values of some organic acids in beer. Journal of the Institute of Brewing. 1974;**80**(2):162-163. DOI: 10.1002/j.2050-0416.1974.tb03598.x

[93] Engan S. Organoleptic threshold values of some alcohols and esters in beer. Journal of the Institute of Brewing. 1972;**78**(1):33-36. DOI: 10.1002/j.2050-0416.1972.tb03425.x

[94] Peddie HA. Ester formation in brewery fermentations. Journal of the Institute of Brewing. 1990;**96**(5):327-331. DOI: 10.1002/j.2050-0416.1990.tb01039.x

[95] Ciani M, Comitini F, Mannazzu I, Domizio P. Controlled mixed culture fermentation: A new perspective on the use of non-*Saccharomyces* yeasts in winemaking. FEMS Yeast Research. 2010;**10**(2):123-133. DOI: 10.1111/j.1567-1364.2009.00579.x

[96] Ciani M, Comitini F. Non-*Saccharomyces* wine yeasts have a promising role in biotechnological approaches to winemaking. Annales de Microbiologie. 2011;**61**(1):25-32. DOI: 10.1007/s13213-010-0069-5

[97] Jolly NP, Varela C, Pretorius IS. Not your ordinary yeast: Non-*Saccharomyces* yeasts in wine production uncovered. FEMS Yeast Research. 2014;**14**(2):215-237. DOI: 10.1111/1567-1364.12111

[98] Michel M, Meier-Dörnberg T, Jacob F, Methner F, Wagner RS, Hutzler M. Review: Pure non-*Saccharomyces* starter cultures for beer fermentation with a focus on secondary metabolites and practical applications. Journal of the Institute of Brewing. 2016;**122**(4):569-587. DOI: 10.1002/jib.381

[99] Carrau FM, Medina K, Boido E, Farina L, Gaggero C, Dellacassa E, et al. De novo synthesis of monoterpenes by *Saccharomyces cerevisiae* wine yeasts. FEMS Microbiology Letters. 2005;**243**(1):107-115. DOI: 10.1016/j.femsle.2004.11.050

[100] Stribny J, Gamero A, Pérez-Torrado R, Querol A. *Saccharomyces kudriavzevii* and *Saccharomyces uvarum* differ from *Saccharomyces cerevisiae* during the production of aroma-active higher alcohols and acetate esters using their amino acidic precursors. International Journal of Food Microbiology. 2015;**205**:41-46. DOI: 10.1016/j.ijfoodmicro.2015.04.003

[101] Janssens L, De Pooter HL, Schamp NM, Vandamme EJ. Production of flavours by microorganisms. Process Biochemistry. 1992;**27**(4):195-215. DOI: 10.1016/0032-9592(92)80020-4

[102] Gamero A, Quintilla R, Groenewald M, Alkema W, Boekhout T, Hazelwood L. High-throughput screening of a large collection of non-conventional yeasts reveals their potential for aroma formation in food fermentation. Food Microbiology. 2016;**60**:147-159. DOI: 10.1016/j.fm.2016.07.006

[103] Postigo V, García M, Cabellos JM, Arroyo T. Wine *Saccharomyces* yeasts for beer fermentation. Fermentation. 2021;**7**(4):290. DOI: 10.3390/fermentation7040290

[104] Sterckx FL, Missiaen J, Saison D, Delvaux FR. Contribution of monophenols to beer flavour based on flavour thresholds, interactions and recombination experiments. Food Chemistry. 2011;**126**(4):1679-1685. DOI: 10.1016/j.foodchem.2010.12.055

[105] Rossi S, Turchetti B, Sileoni V, Marconi O, Perretti G. Evaluation of *Saccharomyces cerevisiae* strains isolated from non-brewing environments in beer production. Journal of the Institute of Brewing. 2018;**124**(4):381-388. DOI: 10.1002/jib.503

[106] Capece A, Romaniello R, Pietrafesa A, Siesto G, Pietrafesa R, Zambuto M, et al. Use of *Saccharomyces cerevisiae var. boulardii* in co-fermentations with *S. Cerevisiae* for the production of craft beers with potential healthy value-added. International Journal of Food Microbiology. 2018;**284**(March):22-30. DOI: 10.1016/j.ijfoodmicro.2018.06.028

[107] Preiss R, Tyrawa C, Krogerus K, Garshol LM, Van Der Merwe G. Traditional Norwegian Kveik are a genetically distinct group of domesticated *Saccharomyces cerevisiae* brewing yeasts. Frontiers in Microbiology. 2018;**9**:2137

[108] De Francesco G, Turchetti B, Sileoni V, Marconi O, Perretti G. Screening of new strains of *Saccharomycodes ludwigii* and *Zygosaccharomyces rouxii*. Journal of the Institute of Brewing. 2015;**121**(1):113-121. DOI: 10.1002/jib.185

[109] Saison D, De Schutter DP, Vanbeneden N, Daenen L, Delvaux F, Delvaux FR. Decrease of aged beer aroma by the reducing activity of brewing yeast. Journal of Agricultural and Food Chemistry. 2010;**58**(5):3107-3115. DOI: 10.1021/jf9037387

[110] Hughes PS, Baxter ED. Beer: Quality, Safety and Nutritional Aspects. Cambridge: Royal Society of Chemistry; 2001. pp. 40-73

[111] Langstaff SA, Lewis MJ. The mouthfeel of beer—A review. Journal of the Institute of Brewing. 1993;**99**(1):31-37. DOI: 10.1002/j.2050-0416.1993.tb01143.x

[112] Brányik T, Silva DP, Baszczyňski M, Lehnert R, Almeidae Silva JB. A review of methods of low alcohol and alcohol-free beer production. Journal of Food Engineering. 2012;**108**(4):493-506. DOI: 10.1016/j.jfoodeng.2011.09.020

[113] Granato D, Branco GF, Faria JDAF, Cruz AG. Characterization of Brazilian lager and brown ale beers based on color, phenolic compounds, and antioxidant activity using chemometrics. Journal of the Science of Food and Agriculture. 2011;**91**(3):563-571. DOI: 10.1002/jsfa.4222

[114] Moslehi-Jenabian S, Lindegaard L, Jespersen L. Beneficial effects of probiotic and food borne yeasts on human health. Nutrients. 2010;**2**(4):449-473. DOI: 10.3390/nu2040449

[115] Butel MJ. Probiotics, gut microbiota and health. Médecine et Maladies Infectieuses. 2014;**44**(1):1-8. DOI: 10.1016/j.medmal.2013.10.002

[116] Buts JP. Twenty-five years of research on *saccharomyces boulardii* trophic effects: Updates and perspectives. Digestive Diseases and Sciences. 2009;**54**(1):15-18. DOI: 10.1007/s10620-008-0322-y

[117] Im E, Pothoulakis C. Progrès récents dans la recherche sur *Saccharomyces boulardii*. Gastroentérologie Clinique et Biologique. 2010;**34**(SUPPL. 1):S62-S70. DOI: 10.1016/S0399-8320(10)70023-3

[118] Mulero-Cerezo J, Briz-Redón Á, Serrano-Aroca Á. *Saccharomyces cerevisiae* Var. *Boulardii*: Valuable probiotic starter for craft beer production. Applied Sciences. 2019;**9**(16):3250. DOI: 10.3390/app9163250

[119] Canonico L, Zannini E, Ciani M, Comitini F. Assessment of non-conventional yeasts with potential probiotic for protein-fortified craft beer production. LWT. 2021;**145**(March):111361. DOI: 10.1016/j.lwt.2021.111361

[120] Hill C, Guarner F, Reid G, Gibson GR, Merenstein DJ, Pot B, et al. The international scientific association for probiotics and prebiotics consensus statement on the scope and appropriate use of the term probiotic. Nature reviews. Gastroenterology & Hepatology. 2014;**11**(8):506-514. DOI: 10.1038/nrgastro.2014.66

[121] Iorizzo M, Coppola F, Letizia F, Testa B, Sorrentino E. Role of yeasts in the brewing process: Tradition and innovation. PRO. 2021;**9**(5):839. DOI: 10.3390/pr9050839

[122] Garcia-Moreno H, Calvo JR, Maldonado MD. High levels of melatonin generated during the brewing process. Journal of Pineal Research. 2013;**55**(1):26-30. DOI: 10.1111/jpi.12005

[123] Maldonado MD, Moreno H, Calvo JR. Melatonin present in beer contributes to increase the levels of melatonin and antioxidant capacity of the human serum. Clinical Nutrition. 2009;**28**(2):188-191. DOI: 10.1016/j.clnu.2009.02.001

[124] Rodriguez-Naranjo MI, Gil-Izquierdo A, Troncoso AM, Cantos-Villar E, Garcia-Parrilla MC. Melatonin is synthesised by yeast during alcoholic fermentation in wines. Food Chemistry. 2011;**126**(4):1608-1613. DOI: 10.1016/j.foodchem.2010.12.038

[125] Benot S, Gobema R, Reiter RJ, Garcia-Mauriño S, Osuna C, Guerrero JM. Physiological levels of melatonin contribute to the antioxidant capacity of human serum. Journal of Pineal Research. 1999;**27**(1):59-64. DOI: 10.1111/j.1600-079X.1999.tb00597.x

[126] Reiter RJ, Tan DX, Maldonado MD. Melatonin as an antioxidant: Physiology versus pharmacology. Journal of Pineal Research. 2005;**39**(2):215-216. DOI: 10.1111/j.1600-079X.2005.00261.x

[127] Ramakrishna A, Giridhar P, Sankar KU, Ravishankar GA. Melatonin and serotonin profiles in beans of coffea species. Journal of Pineal Research. 2012;**52**(4):470-476. DOI: 10.1111/j.1600-079X.2011.00964.x

[128] Rodriguez-Naranjo MI, Gil-Izquierdo A, Troncoso AM, Cantos E, Garcia-Parrilla MC. Melatonin: A new bioactive compound in wine. Journal of Food Composition and Analysis. 2011;**24**(4-5):603-608. DOI: 10.1016/j.jfca.2010.12.009

[129] Venegas C, García JA, Escames G, Ortiz F, López A, Doerrier C, et al. Extrapineal melatonin: Analysis of its subcellular distribution and daily fluctuations. Journal of Pineal

Research. 2012;**52**(2):217-227. DOI: 10.1111/j.1600-079X.2011.00931.x

[130] Sanchez-Barcelo EJ, Mediavilla MD, Tan DX, Reiter RJ. Clinical uses of melatonin: Evaluation of human trials. Current Medicinal Chemistry. 2010;**17**(19):2070-2095. DOI: 10.2174/092986710791233689

[131] Juhnevica-Radenkova K, Moreno DA, Ikase L, Drudze I, Radenkovs V. Naturally occurring melatonin: Sources and possible ways of its biosynthesis. Comprehensive Reviews in Food Science and Food Safety. 2020;**19**(6):4008-4030. DOI: 10.1111/1541-4337.12639

[132] Postigo V, Sánchez A, Cabellos JM, Arroyo T. New approaches for the fermentation of beer: Non-*Saccharomyces* yeasts from wine. Fermentation. 2022;**8**(6):280. DOI: 10.3390/fermentation8060280

[133] Postigo V, Sanz P, García M, Arroyo T. Impact of non-*Saccharomyces* wine yeast strains on improving healthy characteristics and the sensory profile of beer in sequential fermentation. Food. 2022;**11**(14):2029. DOI: 10.3390/foods11142029

[134] Valera MJ, Morcillo-Parra MÁ, Zagórska I, Mas A, Beltran G, Torija MJ. Effects of melatonin and tryptophol addition on fermentations carried out by *Saccharomyces cerevisiae* and non-*Saccharomyces* yeast species under different nitrogen conditions. International Journal of Food Microbiology. 2019;**289**:174-181. DOI: 10.1016/j.ijfoodmicro.2018.09.013

[135] Fernández-Cruz E, Álvarez-Fernández MA, Valero E, Troncoso AM, García-Parrilla MC. Melatonin and derived l-tryptophan metabolites produced during alcoholic fermentation by different wine yeast strains. Food Chemistry. 2017;**217**:431-437. DOI: 10.1016/j.foodchem.2016.08.020

[136] Kocadağlı T, Yılmaz C, Gökmen V. Determination of melatonin and its isomer in foods by liquid chromatography tandem mass spectrometry. Food Chemistry. 2014;**153**:151-156. DOI: 10.1016/j.foodchem.2013.12.036

[137] Anisimov VN, Popovich IG, Zabezhinski MA, Anisimov SV, Vesnushkin GM, Vinogradova IA. Melatonin as antioxidant, geroprotector and anticarcinogen. Biochim Biophys Acta - Bioenerg. 2006;**1757**(5-6):573-589. DOI: 10.1016/j.bbabio.2006.03.012

[138] Koziróg M, Poliwczak AR, Duchnowicz P, Koter-Michalak M, Sikora J, Broncel M. Melatonin treatment improves blood pressure, lipid profile, and parameters of oxidative stress in patients with metabolic syndrome. Journal of Pineal Research. 2011;**50**(3):261-266. DOI: 10.1111/j.1600-079X.2010.00835.x

[139] Que Z, Ma T, Shang Y, Ge Q, Zhang Q, Xu P, et al. Microorganisms: Producers of melatonin in fermented foods and beverages. Journal of Agricultural and Food Chemistry. 2020;**68**(17):4799-4811. DOI: 10.1021/acs.jafc.0c01082

[140] Postigo V, García M, Arroyo T. Study of a first approach to the controlled fermentation for lambic beer production. Microorganisms. 2023;**11**(7):1681. DOI: 10.3390/microorganisms11071681

[141] Gerhäuser C. Beer constituents as potential cancer chemopreventive agents. European Journal of Cancer. 2005;**41**(13):1941-1954. DOI: 10.1016/j.ejca.2005.04.012

[142] Katalinić V, Milos M, Modun D, Musić I, Boban M. Antioxidant

effectiveness of selected wines in comparison with (+)-catechin. Food Chemistry. 2004;**86**(4):593-600. DOI: 10.1016/j.foodchem.2003.10.007

[143] Fegredo JA, Meynell R, Lai AKH, Wong MCY, Martin CR, Wiseman H, et al. The antioxidant capacity of beer: Relationships between assays of antioxidant capacity, color and other alcoholic and non-alcoholic beverages. In: Beer in Health and Disease Prevention. Amsterdam, The Netherlands: Elsevier; 2009. pp. 475-481. DOI: 10.1016/B978-0-12-373891-2.00046-8

[144] Viana AC, Pimentel TC, do Borges Vale R, Clementino LS, Januario Ferreira ET, Magnani M, et al. American pale ale craft beer: Influence of brewer's yeast strains on the chemical composition and antioxidant capacity. LWT. 2021;**152**(August):112317. DOI: 10.1016/j.lwt.2021.112317

[145] Pai TV, Sawant SY, Ghatak AA, Chaturvedi PA, Gupte AM, Desai NS. Characterization of Indian beers: Chemical composition and antioxidant potential. Journal of Food Science and Technology. 2015;**52**(3):1414-1423. DOI: 10.1007/s13197-013-1152-2

[146] Tafulo PAR, Queirós RB, Delerue-Matos CM, Sales MGF. Control and comparison of the antioxidant capacity of beers. Food Research International. 2010;**43**(6):1702-1709. DOI: 10.1016/j.foodres.2010.05.014

[147] Munteanu IG, Apetrei C. Analytical methods used in determining antioxidant activity: A review. International Journal of Molecular Sciences. 2021;**22**(7):3380. DOI: 10.3390/ijms22073380